New Insights into Earthen Site Conservation: Methods, Techniques, Management, and Key Case Studies

New Insights into Earthen Site Conservation: Methods, Techniques, Management, and Key Case Studies

Guest Editors

Stephen S.Y. Lau
Yile Chen
Liang Zheng
Jianyi Zheng

MDPI

Basel • Beijing • Wuhan • Barcelona • Belgrade • Novi Sad • Cluj • Manchester

Guest Editors
Stephen S.Y. Lau
Faculty of Humanities and Arts
Macau University of Science and Technology
Macau
China

Yile Chen
Faculty of Humanities and Arts
Macau University of Science and Technology
Macau
China

Liang Zheng
Faculty of Humanities and Arts
Macau University of Science and Technology
Macau
China

Jianyi Zheng
Faculty of Humanities and Arts
Macau University of Science and Technology
Macau
China

Editorial Office
MDPI AG
Grosspeteranlage 5
4052 Basel, Switzerland

This is a reprint of the Special Issue, published open access by the journal *Coatings* (ISSN 2079-6412), freely accessible at: https://www.mdpi.com/journal/coatings/special_issues/IK22749J11.

For citation purposes, cite each article independently as indicated on the article page online and as indicated below:

Lastname, A.A.; Lastname, B.B. Article Title. *Journal Name* **Year**, *Volume Number*, Page Range.

ISBN 978-3-7258-6472-0 (Hbk)
ISBN 978-3-7258-6473-7 (PDF)
https://doi.org/10.3390/books978-3-7258-6473-7

Contents

About the Editors

Stephen S.Y. Lau

Stephen S.Y. Lau is a doctoral supervisor in Architecture at the Faculty of Humanities and Arts, Macau University of Science and Technology. At the Faculty of Architecture, Professor Lau has served as Associate Dean for three terms; Chair of the PhD Committee at both Faculty and Department levels; and Admission Tutor for the BArch and MArch programmes. As Principal Investigator, he has completed 60 internally and externally funded research projects. He was the Primary Supervisor for twenty PhD students, two MPhil students, and seven Postdoctoral Fellows. In 2015-19, Professor Lau joined the Department of Architecture at the National University of Singapore as a Professor, Deputy Head of Research, Head of the Design Technology and Sustainability Research & Teaching Group, Chair of the Center of Advanced Study in Architecture (CASA), and Chair of the Architecture PhD programme. He had also been a Visiting Professor at the Department of Geo-sciences and Spatial Planning of Utrecht University; Honorary Professor of Fuzhou, Guangzhou, Soochow, Southwest, Tongji, and Wuhan Universities; and Visiting Senior Research Fellow at UNSW Australia. Professor Lau is a Registered Architect, Acoustician, and Lighting Designer. He is proficient in *LEED*, *BREEAM*, and *BEAM*. He is a Carbon Auditor, China GBL Assessor, and Vice Chair of the China GBC—Hong Kong Chapter. In recent years, he has contributed to the drafting of China National Standards, including the new standard for Architectural Acoustics (50118-2010), and two new standards for green building certification: Eco-district and Green Campus (2017, 2018). He was also a joint leader of a technical study on the Validation of National Standard GB by ISO Standard for Daylight Control at the Artificial Island Entrances/Exits of the Zhuhai–Guangzhou–Hong Kong–Macau Bridge.

Yile Chen

Yile Chen, Ph.D. in Architecture, is an Assistant Professor and Master's supervisor in Architecture at the Faculty of Humanities and Arts, Macau University of Science and Technology. She has published more than 50 papers as the first author or corresponding author in *SCIE*, *SSCI*, and *A&HCI*-indexed journals. In 2026, she served as editor for the Springer Geography book, *City-Sea Symbiosis: Settlement Morphology Types in Zhuhai and Macau*. She focuses on architectural technology in high-density urban form and water-faring community (Tanka), as well as the protection and renewal of urban and architectural heritage in Macau and Portuguese-speaking regions She is currently leading a Macau University of Science and Technology Research Fund (FRG) project focusing on the spatial morphological development of Macau and Timor-Leste. She has also participated in numerous projects, including the Science and Technology Development Fund (FDCT) Science and Technology Innovation Enhancement Program of the Macau Special Administrative Region, the Cultural Affairs Bureau of the Macau Special Administrative Region's Consulting Services, the Macau Special Administrative Region Government Higher Education Fund Research Project on Prevention and Response to Major Infectious Diseases, the China–Portugal Joint Laboratory for Cultural Heritage Conservation "Belt and Road" Open Project, the Zhejiang Provincial Philosophy and Social Sciences Planning Project, the Digital Rural Research Center of Hangzhou City University, and the Guangdong Provincial Department of Education's Key Scientific Research Platform Project for General Universities. At the same time, she is also an Editorial Board Member of *PLOS ONE* and *Humanities and Social Sciences Communications* and serves as a peer reviewer for 81 *SCIE/SSCI/A&HCI/ESCI*-indexed journals.

Liang Zheng

Liang Zheng, Ph.D. in Architecture, is an Assistant Professor at the Faculty of Humanities and Arts, Macau University of Science and Technology. His academic research centres on the integration of artificial intelligence and architectural design, with particular emphasis on the development of computer vision-based generative design systems, predictive models for urban environmental analysis, and technologies for the assessment of building material performance. His scholarly interests also extend to architectural history, heritage conservation, and the study of decorative arts. He has published more than 30 papers as the first author or corresponding author in SCIE, SSCI, and A&HCI-indexed journals, such as *International Journal of Architectural Heritage, Heritage Science, npj Heritage Science, Journal of Asian Architecture and Building Engineering, Island Studies Journal, International Journal of Low-Carbon Technologies, Buildings, Coatings, Applied Artificial Intelligence, Geocarto International, Frontiers in Earth Science, Frontiers in Microbiology, PLOS ONE, Atmosphere, Materials, Mathematics, AIMS Mathematics*, and *Sustainability*. He has also participated in numerous projects, including the Science and Technology Development Fund (FDCT) Science and Technology Innovation Enhancement Program of the Macau Special Administrative Region, the Cultural Affairs Bureau of the Macau Special Administrative Region's Consulting Services, the Macau Special Administrative Region Government Higher Education Fund Research Project on Prevention and Response to Major Infectious Diseases, the China–Portugal Joint Laboratory for Cultural Heritage Conservation "Belt and Road" Open Project, the Zhejiang Provincial Philosophy and Social Sciences Planning Project, the Digital Rural Research Center of Hangzhou City University, and the Guangdong Provincial Department of Education's Key Scientific Research Platform Project for General Universities.

Jianyi Zheng

Jianyi Zheng, PhD in urban planning and design, is an associate professor and the director of the Architecture Programme at the Faculty of Humanities and Arts (FA), Macau University of Science and Technology (MUST). He is also a Class-1 registered architect and registered urban and rural planner in mainland China and a registered architect in Macao SAR. He holds a Bachelor of Architecture and a master's degree in architectural history and theory from Huaqiao University, Quanzhou, China. Receiving his PhD in urban planning and design from the South China University of Technology, he has taught at FA MUST since 2018 and qualified as a Ph.D. supervisor of Architecture in 2020. His research interests focus on the field of urban morphology, including morphological analysis of HUL, smart heritage, and revitalization of historical buildings. Especially in the case of the Macao study, he is the PI of some related projects funded by the *NSFC, DSEDJ*, and *ICM*. His research explored the evolution of the urban core, morphological comparison with European cities, management unit divisions of HUL, Beco & Patio historical buildings, historical wharf buildings, grey bricks damage detection, and healthy high-density urban forms. Some AI and spatial information technologies have been widely employed in his smart heritage studies. In 2021, he worked hard to establish the Macao World Heritage Conservation and Development Research Centre at FA MUST, which further encourages more students and teachers in heritage conservation. He has had several papers published in prestigious Chinese journals and English journals indexed by *SCI, SSCI*, and *CSSCI*.

coatings

MDPI

Editorial

New Insights into Earthen Site Conservation: Methods, Techniques, Management, and Key Case Studies

Yile Chen * and Liang Zheng

Faculty of Humanities and Arts, Macau University of Science and Technology, Avenida Wai Long, Tapai, Macau 999078, China; zhengliang@must.edu.mo
* Correspondence: chenyile@must.edu.mo

1. Introduction

Earthen sites are ancient remains constructed using soil as the core building material. They carry profound historical, cultural, and scientific value and are important carriers of human civilization [1–3]. They not only occupy an important proportion of global cultural heritage resources but also contain rich historical information, providing key material evidence for exploring the origin and development of civilization. They possess irreplaceable historical, scientific, and artistic value. From a global perspective, earthen sites have formed a wide and regionally distinctive pattern due to the advantages of readily available raw materials and adaptability to diverse climates [4,5]. Their spatial distribution is highly consistent with the development trajectory of ancient civilizations and is also profoundly influenced by geographical environment, climate conditions, and cultural traditions. For example, in Africa, the Timbuktu and Djene-Djeno sites in West Africa and the Timgad (Algeria) and archeological site of Sabratha (Libya) in North Africa, etc. [6,7], adapted to arid and semi-arid climates with rammed earth and adobe techniques and witnessed the cultural exchanges during the trans-Saharan trade and the ancient Roman colonial period. Earthen sites in the Americas vividly demonstrate the wisdom of pre-Columbian indigenous peoples. Sites such as Mesa Verde National Park in North America [8], Paquimé in northern Mexico [9], and Sechin Bajo (Peru) in South America [10] showcase the innovative applications of earthen construction techniques by different civilizations. Eurasia, as the core cradle of ancient civilizations, boasts a greater diversity of earthen sites. East Asia, particularly China, has approximately 400 national-level earthen sites, covering 30 provinces (municipalities and autonomous regions), encompassing various types including ancient cities, tombs, and defensive structures [2]. The Mozu Kofun in Japan is a significant example [11]. Other sites include the Ashur (Qal'at Sherqat) in Iraq [12], the Persian Caravanserai in Iran in West Asia [13], the Boyne Valley site in Ireland [14], and the Çatalhöyük site in Turkey [15], recording the earthen construction traditions of different civilizations and the imprints of intercontinental exchange. While Oceania and Southeast Asia have relatively fewer sites, Aboriginal sites in Australia [16] and the ancillary buildings of Angkor Wat in Cambodia still retain traces of the application of earthen construction techniques [17]. Earthen heritage sites around the world are mostly concentrated in arid, semi-arid, and seasonally arid climate zones, while those in humid regions have been preserved through technological improvements [18]. Their functional diversity and wide distribution are not only important historical imprints but also provide important references for the protection and sustainable development of contemporary earthen architectural heritage.

Driven by this development, this Special Issue of *Coatings* brings together 10 original papers that collectively illustrate how earthen site conservation techniques and case studies are conducted. These articles cover fluidized catalytic cracking (FCC) technology, surface microchemical analysis, microbial diversity analysis, erosion resistance testing, lightweight crack detection and independent physical simulation, durability analysis, biocrust analysis, erosion resistance mechanisms, and composite biofilm ecological protection technology.

2. Overview of Contributions

This Special Issue primarily covers case studies from Spain, Malaysia, and China, as well as specific urban case studies from Zhuhai City, Lishui City, Quanzhou City, and Weiyuan City. This edition provides a reference for empirical analysis of conservation techniques for earthen sites across cities of different scales and types.

Frankly speaking, earthen site conservation techniques possess a rich and diverse technical framework, encompassing various experimental analytical methods. However, this Special Issue focuses on specific practical earthen site needs, emphasizing the following key models and technologies:

- X-ray diffraction (XRD);
- Scanning electron microscopy-energy dispersive spectrometer (SEM-EDS);
- High-throughput sequencing (based on the 16S rRNA gene V3–V4 region);
- QIIME2 (version 2019.4) for quality control, denoising, sequence merging, and chimera removal;
- Ion chromatography detection;
- Drone aerial photography;
- Microbial community analysis;
- Soil leaching simulations;
- Finite element mechanics analysis (FEM);
- X-ray fluorescence (XRF);
- XGRIDS Lixel L 3D laser scanner (Xgrids Limited, Shop 185 G/F, Hang Wai Ind Ctr, No 6 Kin Tai St Tuen Mun Nt, Hongkong).

By utilizing these technologies, they have provided practical methods and models for earthen site conservation and management.

2.1. New Technological Insights

María et al. [19] proposed a research scheme for characterizing earth mortars with experimentally added pozzolanic materials using fluidized bed catalytic cracking (FCC). Their ash exhibits excellent pozzolanic properties and can be used in protective processes. This paper provides a possibility for studying methods to enhance the material and structural stability of earth mortars.

Hu et al. [20] proposed a dual-path framework that combines lightweight crack detection with independent physical simulation. For detection, they developed an improved YOLOv12 model capable of lightweight and accurate identification of multiple crack types beneath complex wall textures. This framework provides an effective and interpretable solution for standardized recording and mechanistic analysis of cracks, offering a practical approach for the preventative protection and sustainable management of Fujian Tulou (earthen buildings).

Xiao et al. [21] also mentioned that future research on earthen site conservation is expected to integrate big data and cloud computing to achieve multidimensional prediction of surface degradation, while virtual reality (VR) will overcome the limitations of time and space and promote the development of conservation paradigms towards automation, intelligence, and sustainability.

2.2. Material Characterization Analysis

Zheng et al. [22] used microchemical analysis to determine that the rammed earth walls in Xiaochikan Village consist of three main minerals: Quartz (SiO_2, including alpha-type SiO_2), Calcite ($CaCO_3$, including synthetic calcite), and Gibbsite ($Al(OH)_3$). The core elemental composition of the four wall samples is consistent, mainly consisting of C, Si, Al, Ca, and Fe, with contents of over 93%. This reflects the matrix characteristics of the local alluvial soil, which consists of "silicate skeleton-carbonate cementation-organic matter residue", and also demonstrates the "local material" attribute of the rammed earth.

Wei et al. [23] combined high-throughput sequencing (based on the V3–V4 region of the 16S rRNA gene) and microbial community analysis in an experiment on Paishan Village, the largest existing rammed earth building complex in the Pearl River Delta. They discovered that different wall orientations resulted in different microbial community structures: the north wall was rich in hygrophilic microorganisms, while the south wall was dominated by thermophilic bacteria. The coexistence rate of core microorganisms across the four walls was only 0.29%. This study elucidates the distribution pattern of microorganisms in rammed earth walls in humid and hot regions, providing a scientific basis for the ecological restoration of rammed earth walls.

Wang et al. [24] continued the above research and added ion chromatography, finding that the accumulation of acid rain and sulfates caused cracks and spalling in the south wall of the rammed earth wall; salt spray and high humidity exacerbated the surface weathering of the west wall; active biological activity and high humidity made the minerals in the north wall easy to dissolve, resulting in a loose structure; and the east wall was more susceptible to weathering due to the dynamic changes in carbonate rocks.

2.3. Performance Analysis of Rammed Earth Walls

Zheng et al. [25] determined the local erosion resistance of rammed earth wall materials in Yangjiatang Village, Zhejiang Province, through sampling and testing. They concluded that the overall corrosion resistance order was W > S > N > E. The surface deterioration of the rammed earth walls in Yangjiatang Village was mainly caused by four factors: structural cracks, surface erosion, biological erosion, and roof damage. These factors collectively led to surface cracking and peeling (3–5 cm deep).

Yang et al. [26] used drone aerial photography, X-ray diffraction (XRD), field emission scanning electron microscopy (FE-SEM), and microbial diversity detection techniques to elucidate the erosion resistance mechanism of the gray green biocrust on the surface of traditional rammed earth houses in Chenjiapeng Village, Songyang County. They found that the algal crust forms a three-dimensional network through macroscopic coverage (reducing raindrop kinetic energy and moisture retention) and microscopic extracellular polysaccharide-cemented soil particles, thereby enhancing the compressive and shear strength of the rammed earth surface. Simultaneously, they proposed a "transparent membrane-biofiber-acid-suppressing layer" composite biofilm design. This design integrates a biodegradable polylactic acid (PLA) network structure, an algal attachment matrix, and calcium carbonate microparticles, enabling dynamic neutralization of acidic substances and achieving a synergistic effect between ecological protection and cultural heritage preservation. This provides an innovative solution for the erosion protection of traditional rammed earth structures, emphasizing environmental friendliness and sustainability.

Yang et al. [27] utilized optical microscopy, scanning electron microscopy, X-ray computed tomography, and image analysis of the Warring States Qin Great Wall site in Weiyuan City, Gansu Province, China. They quantified the three-dimensional pore structure characteristics of bare soil, cyanobacteria-lichen crusts, and moss crusts within a depth range of 0–50 mm, combining these with the precipitation infiltration process. They explored in depth how biocrusts affect the microstructure of the rammed earth wall soil, alter precipitation infiltration, promote vascular plant colonization, and thus change the erosion resistance of the rammed earth wall.

Ding et al. [28] explored the durability of brick-faced clay-core walls (BCWs) in traditional Quanzhou residential architecture. They found that the key to the durability of BCWs lies in the synergistic effect of brick and clay materials and the stability of silicate mineral materials, providing theoretical and methodological support for the sustainable development research of brick-and-clay buildings.

3. Outlook

As earthen sites remain irreplaceable carriers of human civilization, this Special Issue aims to lay a solid foundation for long-term, science-based protection practices, ultimately ensuring the enduring transmission of these precious cultural legacies for generations to come. This Special Issue highlights 10 successes in conservation methods for earthen sites, which are likely to encourage more progress in preserving built heritage and motivate future research. Because of the relatively short submission period, there were fewer results, but we have been greatly encouraged by these studies. We hope these rigorously validated case studies and innovative technical frameworks will serve as actionable references for heritage practitioners, researchers, and policymakers globally, fostering cross-regional knowledge exchange and collaborative efforts in addressing shared challenges such as climate-induced deterioration and sustainable conservation. We also look forward to more new insights into the protection of earthen sites in the future.

Author Contributions: Conceptualization, Y.C.; methodology, Y.C.; software, Y.C.; validation, Y.C.; formal analysis, Y.C.; investigation, Y.C.; resources, Y.C.; data curation, Y.C.; writing—original draft preparation, Y.C.; writing—review and editing, Y.C. and L.Z.; visualization, Y.C.; supervision, Y.C.; project administration, Y.C. and L.Z.; funding acquisition, Y.C. and L.Z. All authors have read and agreed to the published version of the manuscript.

Funding: This research was funded by the (1) Faculty Research Grants funded by Macau University of Science and Technology (FRG-MUST grant number: FRG-25-041-FA; FRG-25-067-FA); (2) Guangdong Provincial Department of Education's key scientific research platforms and projects for general universities in 2023: Guangdong, Hong Kong, and Macau Cultural Heritage Protection and Innovation Design Team (grant number: 2023WCXTD042); and (3) Guangdong Provincial Philosophy and Social Sciences Planning 2025 Lingnan Cultural Project (grant number: GD25LN30). The funders had no role in study conceptualization, data curation, formal analysis, methodology, software, decision to publish, or preparation of the manuscript. There was no additional external funding received for this study.

Institutional Review Board Statement: Not applicable.

Informed Consent Statement: Not applicable.

Data Availability Statement: Not applicable.

Acknowledgments: We sincerely thank the authors who contributed to this Special Issue and submitted high-quality manuscripts. We also thank the reviewers for their responsible and rigorous feedback, which has greatly improved each manuscript. Finally, we sincerely thank the editorial team of *Coatings* for their professional support, which ensured the success of this Special Issue.

Conflicts of Interest: The Guest Editors declare that they have no known competing financial interests or personal relationships that could have appeared to influence the work reported in this paper.

References

1. Oliver, A. Conservation of earthen archaeological sites. *Terra Lit. Rev.* **2008**, *80*, 80–95.
2. Guo, Z.; Qi, Q.; Zhang, S.; Chen, W.; Yang, L.; Zhang, Y. Characterization of spatial and chrono-cultural evolution of earthen sites construction technology in China based on GIS. *Archaeol. Anthropol. Sci.* **2025**, *17*, 142. [CrossRef]
3. Fabbri, A.; Morel, J.C. Earthen materials and constructions. In *Nonconventional and Vernacular Construction Materials*; Woodhead Publishing: Derbyshire, UK, 2016; pp. 273–299. [CrossRef]
4. Rael, R. *Earth Architecture*; Princeton Architectural Press: Princeton, NJ, USA, 2009.

5. Minke, G. *Building with Earth: Design and Technology of a Sustainable Architecture Fourth and Revised Edition*; Birkhäuser: Basel, Switzerland, 2021.
6. Siriman, D.; Wang, Y. The survival of earthen architecture in Malian Sahel, case study: The Historic City of Djenné. *Curr. Urban Stud.* **2021**, *9*, 83–106. [CrossRef]
7. Wilson, A. City sizes and urbanization in the Roman Empire. *Settl. Urban. Popul.* **2011**, *2*, 161.
8. Mesa Verde National Park. Available online: https://www.nationalparks.org/explore/parks/mesa-verde-national-park (accessed on 21 November 2025).
9. Minnis, P.E.; Whalen, M.E. (Eds.) *Discovering Paquimé*; University of Arizona Press: Tucson, AZ, USA, 2016.
10. Pozorski, T.; Pozorski, S. Architecture and Chronology at the Site of Sechín Alto, Casma Valley, Peru. *J. Field Archaeol.* **2005**, *30*, 143–161. [CrossRef]
11. Sasaki, K.I. Social stratification and the formation of mounded tombs in the kofun period of protohistoric Japan. In *Burial Mounds in Europe and Japan: Comparative and Contextual Perspectives*; Archaeopress: Bicester, UK, 2018; pp. 87–99.
12. Radner, K.; Squitieri, A. (Eds.) *Assur 2023: Excavations and Other Research in the New Town*; PeWe-Verlag: Gladbeck, Germany, 2024; Volume 1.
13. Mansouri, A.; Edgü, E.; Şalgamcıoğlu, M.E. Historic Persian Caravanserais. In Proceedings of the 10th International Space Syntax Symposium (SSS 10), London, UK, 13–17 July 2015.
14. Piera, M. Rathmullan excavation: A prehistoric landscape in the Boyne Valley. *J. Ir. Archaeol.* **2024**, *33*, 171–184.
15. Hodder, I.; Meskell, L. The symbolism of Çatalhöyük in its regional context. In *Religion in the Emergence of Civilization: Çatalhöyük as a Case Study*; Cambridge University Press: Cambridge, UK, 2010; pp. 32–72.
16. Corkill, T. Here and There: Links Between Stone Sources and Aboriginal Archaeological Sites in Sydney, Australia. Ph.D. Dissertation, The University of Sydney, Sydney, Australia, 1999.
17. Siedel, H.; Pfefferkorn, S.; von Plehwe-Leisen, E.; Leisen, H. Sandstone weathering in tropical climate: Results of low-destructive investigations at the temple of Angkor Wat, Cambodia. *Eng. Geol.* **2010**, *115*, 182–192. [CrossRef]
18. Nakhaei Ashtari, M.; Della Ventura, G.; Correia, M. From drought to deluge: The complex impacts of climate change on earthen heritage. *J. Cult. Herit. Manag. Sustain. Dev.* 2025; *ahead-of-print*. [CrossRef]
19. Magdalena, M.B.; García-Soriano, L.; Hueto-Escobar, A.; Mileto, C.; Vegas, F. Proposal for zeolite waste from fluid catalytic cracking as a pozzolanic addition for earth mortars: Initial characterization. *Coatings* **2025**, *15*, 1408. [CrossRef]
20. Hu, Y.; Chen, S.; Zhao, Z.; Cheng, S. Dual-Path Framework Analysis of Crack Detection Algorithm and Scenario Simulation on Fujian Tulou Surface. *Coatings* **2025**, *15*, 1156. [CrossRef]
21. Xiao, Y.; Chen, Y.; Huang, Y.; Yan, Y. Surface Protection Technologies for Earthen Sites in the 21st Century: Hotspots, Evolution, and Future Trends in Digitalization, Intelligence, and Sustainability. *Coatings* **2025**, *15*, 855. [CrossRef]
22. Zheng, L.; Deng, Q.; Liang, J.; Guo, Z.; Zhu, Y.; Liu, W.; Chen, Y. Microchemical Analysis of Rammed Earth Residential Walls Surface in Xiaochikan Village, Guangdong. *Coatings* **2025**, *15*, 1351. [CrossRef]
23. Wei, W.; Yang, S.; Song, J.; Ishak, M.S.B. Microbial Diversity Analysis on Rammed Earth Wall Surfaces in the Lingnan Region: A Case Study of Paishan Village, China. *Coatings* **2025**, *15*, 1236. [CrossRef]
24. Wang, Y.; Song, J.; Zhang, J.; Huang, Y.; Yang, S. Analysis of Rammed Earth Wall Erosion in Traditional Village Dwellings in Zhuhai City. *Coatings* **2025**, *15*, 526. [CrossRef]
25. Zheng, Y.; Song, J.; Zhang, X.; Hu, Y.; Chen, R.; Yang, S. Experimental Study on the Anti-Erosion of the Exterior Walls of Ancient Rammed-Earth Houses in Yangjiatang Village, Lishui. *Coatings* **2025**, *15*, 1173. [CrossRef]
26. Yang, J.; Wang, N.; Huang, Z.; Huang, Y.; Lv, W.; Yang, S. Anti-Erosion Mechanism of Biological Crusts and Eco-Protection Technology Using Composite Biofilms for Traditional Rammed Earth Dwellings in Songyang County. *Coatings* **2025**, *15*, 608. [CrossRef]
27. Yang, X.; Wu, F.; Li, L.; Shang, R.; Li, D.; Xu, L.; Cui, J.; Zhao, X. Biocrusts Alter the Pore Structure and Water Infiltration in the Top Layer of Rammed Soils at Weiyuan Section of the Great Wall in China. *Coatings* **2025**, *15*, 908. [CrossRef]
28. Ding, Y.; Guan, R.; Chen, L.; Wang, J.; Zhang, Y.; Fu, Y.; Zhang, C. Durability Analysis of Brick-Faced Clay-Core Walls in Traditional Residential Architecture in Quanzhou, China. *Coatings* **2025**, *15*, 909. [CrossRef]

coatings

MDPI

Article

Proposal for Zeolite Waste from Fluid Catalytic Cracking as a Pozzolanic Addition for Earth Mortars: Initial Characterisation

María Barros Magdalena [1], Lidia García-Soriano [2], Alicia Hueto-Escobar [2,*], Camilla Mileto [2] and Fernando Vegas [2]

1 Doctoral School, Universitat Politècnica de València, 46022 València, Spain; mabarmag@doctor.upv.es
2 Architecture, Heritage and Management for Sustainable Development Research Centre (PEGASO), Universitat Politècnica de València, 46022 València, Spain; ligarso@upv.es (L.G.-S.); cami2@cpa.upv.es (C.M.); fvegas@cpa.upv.es (F.V.)
* Correspondence: alhuees@upv.es

Abstract: This article presents the programme for the characterisation of earth mortars stabilised with experimental pozzolanic material from fluid catalytic cracking (FCC). This study aims to establish the optimal ratio for adding pozzolan to stabilise earth mortars. Ash may be used in conservation processes, as it presents suitable pozzolanic properties. Based on the starting premise that its application does not cause chromatic variations in the final mortar and displays resistance to damage from chlorides and extreme temperatures, it can be considered ideal for this purpose. The process of transformation into ash is linked to the production of naphthas and refined petroleum products, where the mineral is a catalyst for the reaction. With use, the mineral tends to shrink, losing the necessary properties for this process. Over the last decade, this process, which is widely used in the petrochemical industry, has generated a volume of waste of up to 3000 tons per day. The amount of waste generated is of interest for its reuse, and a rise is observed in preliminary studies, which confirm that this material is pozzolanic and non-toxic. This offers the possibility of studying this addition to stabilise materials and constructions manufactured with earth.

Keywords: FCC; characterisation; zeolite; pozzolan; material intervention

1. Introduction

Catalytic cracking is a crude oil-refining process that modifies and reduces the molecular structure of hydrocarbons to obtain naphthas. These are subsequently used to produce high-quality fuel such as gasolines, olefins and isobutanes. Given the constant high temperatures required for this process, a mineral catalyst such as zeolite is used during the separation stage. This process is known as fluid catalytic cracking (FCC). Each individual separation process reduces the average size of the catalyst particle, lessening its properties, so that it must be periodically replaced. The constant replacement results in large amounts of waste on a global scale, accounting for approximately 400,000 tons per year in 2000. The last twenty years have seen this waste production triple, going from 1000 tons per day to 3000 tons per day [1]. This type of waste, which has no second useful life cycle, is then stored.

These particles present a Y-type zeolite mineral structure, composed mainly of modified hydrated aluminosilicates [1], similar in molecular structure and mineralogical composition to metakaolin. These can be considered a highly pozzolanic material, as they actively react with lime hydroxide [2]. The calcium aluminate products obtained after the lime–pozzolan reaction are similar to those formed when Portland cement is hydrated [3]. In recent years, further research has highlighted properties such as fire resistance and the contribution to mechanical strength in lime and gypsum mortars. Pozzolanic reactivity with calcium hydroxide systems can be of interest in reducing setting time from 28 to 7 days so that strength can be achieved in shorter periods [4].

J. Payá et al. (2009) conducted a chemical characterisation of five fluid catalytic cracking (FCC) catalysts of different origins, concluding that no substantial differences were observed among them [5]. Pacewska et al. (2000) reported that the zeolite phase involved in FCC processes exhibits a chemical composition dominated by SiO_2 and Al_2O_3 [6] (75–90%) [7], thus indicating its close affinity to a Y-type zeolite [8], typically represented by the composition $Na_9(AlO_2)_9(SiO_2)_{15}{\cdot}27H_2O$, and is amorphous in nature. The overall chemical composition is consistent with a zeolitic material based on modified hydrated aluminosilicate frameworks [1]. This research project aims to assess the viability of proposing earth mortars stabilised with very low lime contents and incorporating zeolite derived from FCC processes. This study is framed within heritage conservation, where unstabilised earthen mortars are common despite their known physical limitations, particularly in the case of plasticity. The objective is to evaluate the final properties of earth lime zeolite mortars and ascertain whether adding zeolite enhances their resistance to mechanical and moisture-related degradation.

Elert's doctoral research (2014) [9], building on Doehne and Price (2010), emphasises strategies for consolidating clay that reduce water-induced deterioration by transforming expansive clays into non-expansive, cementitious phases such as calcium silicates or aluminosilicates, including amorphous phases and zeolites. This study examines property variations obtained through incorporating zeolite directly as a mortar additive, rather than through alkaline activation.

2. State of the Art

Particle size has a bearing on pozzolanic reactivity, as setting with a smaller particle size yields a greater reaction and in turn increases the percentage of lime set [10]. Several studies have sought to confirm this conclusion, experimenting with different prior treatments of ash, studying different crushing methods and studying the relation between time and the contribution or reduction of mechanical properties [11]. After establishing an optimal crushing time of 20 min, it was ascertained that these mechanical resistance values did not increase with longer time spans [11]. Other research focuses on the optimal percentage for the replacement of bonding agents in lime or gypsum systems, which it establishes as 15%–20% on dry weight replacement, obtaining high mechanical properties [11]. A subsequent study identified the optimal lime–FCC ratio as between 1:1 and 1:2 [12]. Durability was studied within the experimental field of lime-ash systems, establishing that adding these to lime mortar increased resistance to freeze–thaw cycles, as well as to $Na_2\ SO_4$ attacks, improving compression strength during these attacks [13]. Consequently, this was recommended for use in conservation mortars. As this is industrial waste, linked to the chemical separation processes of crude oil, tests have been carried out to establish possible toxicity. Nan et al. [14] and Al-Jabri et al. [15] conducted leaching studies on mortar, establishing that this industrial waste was neither dangerous nor polluting. It was also concluded that the concentration of potentially harmful heavy metals did not exceed the legal limits permitted by different EPA, WHO and EU legislations [16]. The Giquima research group characterised the chemical reactivity at curing ages below 28 days and determined that the material displays high reactivity. This reactivity is so pronounced that even low substitution ratios show a strong pozzolanic reaction [12]. The authors concluded that a 2% addition by weight is sufficient to achieve complete lime fixation. This characterisation shows a clear parallel with metakaolin, as evidenced by the high degree of free-lime fixation in the early curing stages [5].

3. Objectives

Over time the use of unsuitable materials has been observed in intervention processes on historic architecture, causing damage and alterations to the original material, ones that did not exist prior to the intervention. Therefore, modern materials should be included with full awareness of the original material and how it can chemically react to these new materials. In this regard, and as dictated by the conservation work, lime and lime–earth mortars are the most recommendable materials for conservation processes. As different properties will be required depending on the function of the mortar, there is a tendency in this process to use predosed natural hydraulic mortars, mixing them with earth as needed. The starting point for this study was to further research the improvement in earth–lime mortars through the addition of FCC waste zeolite. Piles et al. [13] concluded that adding FCC waste to lime mortars improves the mortar resistance and durability in relation to freeze–thaw cycles and aggressive agents. This study aims to ascertain the feasibility of adding this to earth mortars stabilised with low lime percentages, classifying their final properties. It was therefore necessary to develop a testing plan to identify the properties of the individual mortars. A secondary objective was to draw up a test programme that successfully identified the physical properties of the mortars and their behaviour in contact with water.

4. Methodology

4.1. Material and Test Times

The test programme proposed aimed to closely examine and quantify the contribution made to physical–mechanical properties and resistance to water action brought about by adding this FCC waste to earth mortars. Different samples were tested for this (Table 1), varying the lime–zeolite ratios to determine the extent to which this addition influences the final properties of the mortar. An earth mortar was prepared, and sand was added to adjust its plasticity, while its particle size distribution was used to develop the test material.

Table 1. Base weight for all percentage samples.

Reference Group	Reference Samples	
D1	1500 g soil + sand, 75 g lime, 316.5 mL water	
D2	1500 g soil + sand, 150 g lime, 7.5 g FCC, 316.5 mL water	
Test Group	**Test Samples**	**Reference**
D3	1500 g soil + sand, 75 g lime, 7.5 g FCC, 316.5 mL w	D1
D4	1500 g soil + sand, 75 g lime, 11.25 g FCC, 316.5 mL w	D1
D5	1500 g soil + sand, 150 g lime, 15 g FCC, 316.5 mL w	D2
D6	1500 g soil + sand, 150 g lime, 22.5 g FCC, 316.5 mL w	D2

In order to determine plasticity, the soil used for the development of the material was analysed according to the Unified Soil Classification System (USCS). The test evaluates the moisture range in which the material transitions from a brittle solid to a plastic and malleable state, indicating its ability to deform without fracturing. According to the results, the soil can be classified as a low-plasticity silty clay (CL–ML).

For the development of the test material, preliminary trials were conducted to characterise the properties of the soil and sand used to produce the test mixture. A granulometric analysis of both materials was carried out in order to adjust the particle-size distribution curve of the soil, adding a controlled amount of sand to make any necessary corrections.

The granulometric distribution of the soil and sand, as well as the resulting adjusted mixture, is presented in the corresponding Section 6. Following the recommendations of Houben and Guillaud for the ideal granulometric distribution for rammed earth [17], the sample was corrected by adding approximately 35% aggregate to the volume. This particle size study is included in Section 6.2.

An industrial mixer was used for 40 min to obtain a dry mix to guarantee a homogeneous material that could be used for the production of all earth mortars. A 3:1 sand-to-soil ratio (by volume) was mixed dry in an industrial mixer to prepare the material. The mixture was then stored in 15 L containers, to be used as needed. A particle size study was initially carried out on the earth to be used as a base material. This was chemically conducted through scanning electron microscopy and physically with a granulometric analysis according to specific European norms [18,19].

The percentage of lime to be added was calculated based on studies on tapia real, which recommend a maximum stabilisation of 10% [20]. In this case, the discrete or poor stabilisations chosen were 5%, while the maximum stabilisation was 10%, based on previous research [12]. The lime proportion was calculated based on the dry weight of the aggregate, and the optimum humidity context was set at 22% s.p.s. and applied through spraying [21]. After adding the proportion of lime and humidity, the mass was mixed again until it was homogeneous. Normalised moulds of 40 × 40 × 160 mm were used for the mortar tests. Lime selection was based on two key criteria: the availability of technical documentation enabling reliable assessment of material performance and its established use in restoration practice, ensuring proven durability in conservation works. The technical datasheet provides free-lime content, physical–mechanical parameters and a stable chemical composition. In turn, a natural hydraulic lime (NHL 3.5) was therefore selected to ensure compatibility with real architectural intervention conditions.

The FCC material was processed in accordance with the protocol described by J. Payá et al. [12]. The treatment consisted of a controlled grinding stage performed in a ball mill for 20 min, resulting in a substantial reduction in particle size, with an average diameter of approximately 6.37 μm.

4.2. Execution Methodology

To prepare the test specimens, the total capacity of the available moulds was first determined. Accordingly, 1.5 kg of a soil–sand mixture was established as the base mass, then the lime was incorporated at 5% and 10% by dry weight, corresponding to samples D1 and D2. The zeolite was then calculated relative to these lime contents, corresponding to 10% and 15% substitution levels based on the dry weight of lime. The dry constituents were blended using an industrial mixer to ensure complete homogenisation. This test established a behaviour pattern for lime–earth mortar for use when studying the variation in the properties of earth–lime–zeolite systems.

The mortar proposal for the four experimental samples with different lime–zeolite ratios was developed based on samples D1 and D2. This enabled further study of the behaviour of the basic and the modified or experimental materials, establishing the contribution of pozzolanic material to properties. Table 1 presents the test samples and the lime–zeolite ratios used.

Moisture was introduced by spraying the predetermined amount of water onto the mixture. While Schroeder (2011) [21] identifies an optimal moisture content of 15% for earthen construction, mortar formulation requires greater malleability and plasticity. Preliminary trials with varying moisture levels indicated that a 20% moisture content provided the most suitable consistency for mortar preparation, yielding a slightly plastic and cohesive mix.

Test material was produced manually, with a total of 50 specimens manufactured for each type of material, 150 in total. The test samples for earth mortars with 5% and 10% lime were tested with two percentages for addition, 10% and 15% zeolite (Table 2). Two groups of mortar, one with a high percentage and another with a low percentage of zeolite, were proposed. Thus, D3 and D4 displayed low values, 0.5 and 0.75%, respectively, while the percentages for D5 and D6 were higher, 1 and 1.5%, respectively.

Table 2. Reference and test samples.

Reference Group	Reference Samples			
D1	Earth–Lime 5%	9:0.5		
D2	Earth–Lime 10%	9:1		
Test Group	**Test Samples**		**% Zeolite**	**Reference Test Sample**
D3	Lime 5% + Zeolite 10%	9:0.5:1	0.5%	D1
D4	Lime 5% + Zeolite 15%	9:0.33:1	0.75%	D1
D5	Lime 10% + Zeolite 10%	9:1:1	1%	D2
D6	Lime 10% + Zeolite 15%	9:0.66:1	1.5%	D2

The test programme includes four testing stages with varying curing periods, aiming to characterise the evolution of the properties of the experimental mortar combining lime with spent catalyst. This approach was adopted due to the presence of calcium hydroxide in the formulation, whose carbonation process is slow, with significant mechanical strength typically developing after 28–30 days of curing.

In order to observe the evolution of material properties, the test was divided into four timeframes: early (7 days), intermediate (28 days) and mature (63 days) stages and two months after testing in the accelerated ageing chamber. These timespans are in keeping with the lime carbonation process. After 7 days, lime hydroxide is completely carbonated, and after 28 days, it begins to acquire strength. After 63 days, data were collected to study long-term behaviour.

The accelerated ageing procedure is designed to induce abrupt fluctuations in temperature and humidity. The climatic chamber conditions consist of two-week cycles in which temperature and humidity vary every 24 h. The conditions are described in the corresponding section and are shown in Table 3. Additionally, two specimens of each mortar type will be exposed to constant ultraviolet radiation.

Table 3. Environmental conditions during accelerated ageing.

	Duration	Temperature	Humidity
Cycle 1	24 h	20 °C	85%
	24 h	60 °C	85%
Cycle 2	24 h	5 °C	40%–45%
	24 h	−20 °C	55%

5. Test Plan

5.1. Test Programme

The test plan is composed of three groups of tests examining the types of properties analysed and the potential conclusions, considering relevant European regulations and other international recommendations for testing this type of element [18,19,22–31].

The first group focuses on prior characterisation, analysing visual appearance. A colorimetric test is proposed to study possible chromatic variations depending on the ratio used. The second group, focusing on the characterisation of mechanical properties, examines stress compression tests, Shore C hardness tests, and wetting–freezing and drying cycles. The third group of tests aims to understand behaviour upon contact with water, using Swinburne accelerated erosion tests (SAETs) to establish water absorption by capillarity and tests to establish water vapour permeability.

5.2. Prior Characterisation Tests

Colorimetry test. This test was made possible thanks to the help of Professor Ana Torres Barchino, from the research group on colour in architecture at the Instituto Universitario de Restauración del Patrimonio (UPV). The aim of this test was to establish whether the addition of zeolite modifies the colour compared to the original reference samples.

The test is also considered relevant to determine whether the ageing of the material leads to visual distortions from colour changes. Two colour ordering systems, NCS (Natural Colour System) and Munsell, were used in these tests.

5.3. Physical–Mechanical Tests

Stress compression tests. In order to ensure identical conditions for all tests [26], the test samples were previously exposed to a temperature of 100 °C for 24 h. Following this process, the test samples were stabilised under laboratory conditions prior to the actual tests, where continuous compression force was applied perpendicular to the transversal face. A loading rate of 0.03 MPa/s was applied, selected due to the brittleness of the test material (Ibertest compression machine, Madrid, Spain), obtaining maximum compression strength when the 0.03 Mpa/s test sample reaches failure.

Calculation of Shore C hardness. This test was proposed to establish an approximate surface hardness value [24]. In total, 10 measurements were carried out, designating one of the faces of the mould as a test face. The minimum separation between indentations was 2 cm, aiming to prevent pressure and the durometer indentation from altering the measurement. The test results correspond to the arithmetic mean of the measurements carried out, expressed in whole Shore C units.

5.4. Tests on Behaviour upon Contact with Water

Swinburne accelerated erosion test (SAET). This test classifies precipitation as moderate. The test executed established the level of erosion of test mortars when exposed to the erosive and kinetic action of water [27]. The test involves applying a continuous water stream for a duration of 10 min onto the surface of the specimen, at a 27° incline in relation to the horizontal. Water is supplied from a constant-level reservoir at a hydraulic head of 1 m and discharged through an outlet tube with a 5 mm internal diameter. Under these boundary conditions, the system delivers a flow rate of 5.217 L/min, corresponding to an exit velocity of 4.4287 m/s.

Cyclical wetting–freezing and drying test. The accelerated ageing test was carried out on the samples under extreme conditions, varying temperature and humidity parameters, seeking to create contrasting cycles. Despite the lack of relation between artificial cycles and real conditions, the test aims to subject the material to possible adverse conditions in order to subsequently establish a classification. The test studied partial exposure of the material to UV radiation. The experiments were conducted in a CCM 25/480 climatic chamber (Dycometal, Barcelona, Spain), fitted with an external ultraviolet irradiation module. The UV source operated within the UVA-340 spectral range, characterised by a peak emission at 340 nm. Cycles were programmed at two-week intervals for a two-month period (Table 3).

Test for establishing water absorption by capillarity. This test takes UNE-EN 15801 [29] as a reference norm, modifying exposure times and taking measurements in short exposure times. This variation is prompted by earth's sensitivity to contact with water, and the objective was to study behaviour during the initial phase of contact with water. Thus, weighing was carried out at 30 s; 1, 3, 5, 10, 20, 30 and 45 min; and 1, 2, 3 and 48 h, yielding more precise results.

Test for determining water vapour permeability. This test takes UNE-EN 15803 [30] as a reference norm, but with modifications due to the limitations in testing means. A saturation chamber (Kesternich corrosion chamber) was used to achieve constant temperature and humidity conditions in order to determine the degree of permeability and penetration of water vapour. This test aimed to determine the absorption capacity of water vapour for the test samples under saturation conditions. The results show the percentage of vapour absorption for each sample in different curing times in order to study possible behaviour variations. This is reflected in the mass water absorption and the absorption percentage of each case.

6. Results

6.1. SEM Analytical Results

The SEM analysis results of the corrected soil sample indicate a composition characteristic of a silty–clayey soil matrix (Figure 1), dominated by carbon (40–42 wt%) and oxygen (42–45 wt%), with moderate amounts of silicon and aluminium (Si = 5.8–6.2 wt%, Al = 2.2–2.3 wt%), consistent with aluminosilicate minerals such as phyllosilicates and feldspar-type phases.

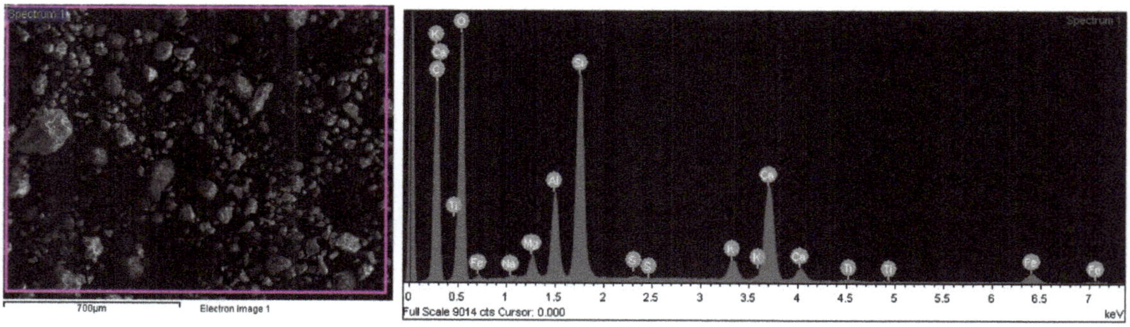

Figure 1. Macroscopic image and SEM micrograph of the sample.

The presence of calcium (=3.4–4.4 wt%) suggests minor contributions from carbonate phases or residual lime, while K, Mg, Fe, Ti and Mn appear only in trace concentrations. Overall, the elemental distribution shows no evidence of advanced pozzolanic reaction products, indicating that the material largely preserves the chemical signature of the base soil with limited interaction between the clay–silicate fraction and the calcium-bearing phases.

6.2. Particle Size Distribution

The initial earth sample is clayey, containing 50% clay and a low percentage of sand (Figure 2). The particle-size distribution exhibits a behaviour typical of a fine sand with a significant fraction of material passing through the N°200 sieve. After correcting the cumulative passing percentage, a P80 value of 0.204 mm was obtained, indicating that 80% of the particles are smaller than this size. The statistical mode corresponds to the 0.125 mm fraction, which represents the highest retained percentage after the fines (pan) fraction.

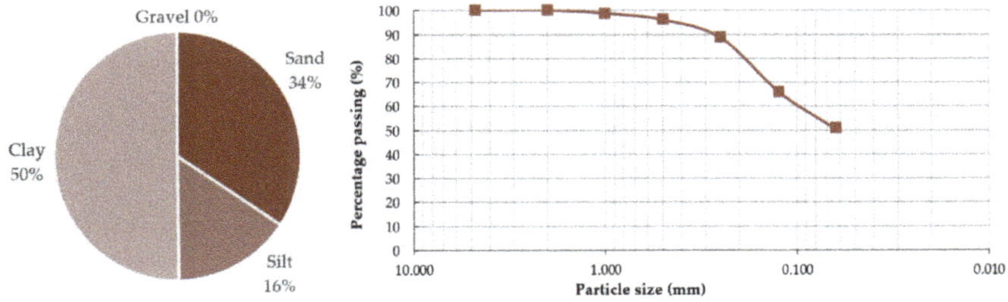

Figure 2. Particle size distribution of original earth.

The weighted standard deviation of particle size is 0.151 mm, indicating a moderate degree of dispersion within the sample. The granulometric classification parameters yield a uniformity coefficient Cu = 8.61 and a curvature coefficient Cc = 1.15, classifying the material as poorly graded, characteristic of sands dominated by fine fractions.

A suitable distribution with different aggregate sizes was obtained thanks to the corrected mix to which sand was added in fractions of between 2.00 mm and 0.125 mm. The particle size distribution of the second sample corresponds to a fine sand with a notable proportion of material retained on the N°200 sieve. The corrected cumulative passing curve yields a P80 of 0.353 mm. The characteristic diameters (D10 = 0.031 mm, D30 = 0.078 mm, D60 = 0.125 mm) result in Cu = 4.01 and Cc = 1.58, classifying the material as poorly graded sand (SP) under ASTM standards.

The modal fraction corresponds to the N°200 sieve (0.063 mm), confirming the predominance of very fine particles. The weighted mean grain size (0.160 mm) and high standard deviation (0.267 mm) indicate a dispersed distribution dominated by fine fractions that condition the material's physical behaviour. This is reflected in a particle size curve following Houben and Guillaud (1994) [17] (Figure 3).

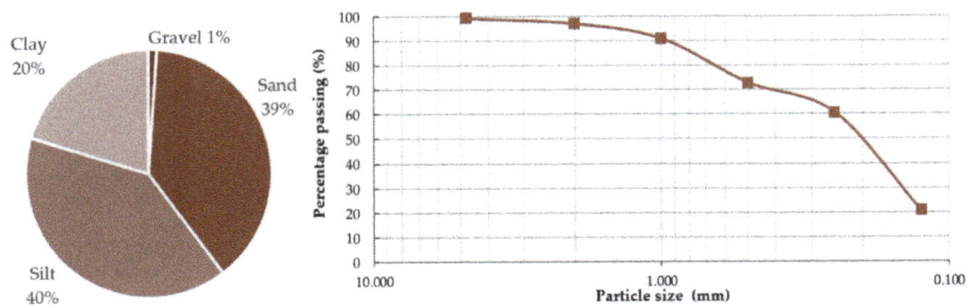

Figure 3. Particle size distribution of corrected earth.

6.3. Colorimetry Results

The table below shows the colour codes for each of the test samples at a natural curing age of 63 days (Figure 4), according to reference standard UNE-EN 15886 [28]. This representation presents the results obtained following analysis with two colour order systems, NCS and Munsell.

Treatment	Group	Specimens	Colour chart	Colour code	Colour
Stabilised earth mortar	Earth–lime mortar 5%	D1	NCS	NCS S 3020-Y40R	
			MUNSELL	7.5 YR 6/4	
		D3	NCS	NCS S 4020-Y40R	
			MUNSELL	7.5 YR 6/4	
		D4	NCS	NCS S 4020-Y40R	
			MUNSELL	7.5 YR 6/4	
	Earth–lime mortar 10%	D2	NCS	NCS S 4020-Y40R	
			MUNSELL	7.5 YR 6/4	
		D5	NCS	NCS S 4020-Y40R	
			MUNSELL	7.5 YR 6/4	
		D6	NCS	NCS S 4020-Y40R	
			MUNSELL	7.5 YR 6/4	

Figure 4. Colorimetric references with natural curing at 63 days.

This representation presents the results obtained following analysis with two colour order systems, NCS and Munsell.

Irrespective of the lime content, in the NCS analysis for the group of stabilised earth mortars, the samples with zeolite appeared more reddish than the reference samples with no addition. When analysed with the Munsell system, the same family presents barely any differences in tone. In terms of the evolution based on curing time, slight variations in colour are observed for the samples which include zeolite in the mix. The situation is stable after the 63 days of natural curing, although in the samples subjected to accelerated ageing, more notable variations are observed (Figure 5).

		D1		D3		D4	
		NCS	MUNSELL	NCS	MUNSELL	NCS	MUNSELL
Earth–lime mortar 5%	63 days	NCS S 4020-Y40R	7.5 YR 6/4	NCS S 4020-Y40R	7.5 YR 6/4	NCS S 4020-Y40R	7.5 YR 6/4
	A+UV	NCS S 4020-Y40R	7.5 YR 6/4	NCS S 3020-Y40R	7.5 YR 6/4	NCS S 3020-Y40R	7.5 YR 6/4

		D2		D5		D6	
		NCS	MUNSELL	NCS	MUNSELL	NCS	MUNSELL
Earth–lime mortar 10%	63 days	NCS S 4020-Y40R	7.5 YR 6/4	NCS S 4020-Y40R	7.5 YR 6/4	NCS S 4020-Y40R	7.5 YR 6/4
	A+UV	NCS S 4020-Y40R	7.5 YR 6/4	NCS S 3020-Y40R	7.5 YR 6/4	NCS S 3020-Y40R	7.5 YR 6/4

Figure 5. Comparative table. Colorimetric references with natural curing and UV ageing.

6.4. Results of Physical–Mechanical Tests

Mechanical resistance. The samples tested follow the samples of earth mortar and lime mortar. Furthermore, this was the only sample where tests were carried out at five curing ages: 7 days (1), 28 days (2), 63 days (3), accelerated ageing (4) and ageing with UV radiation (5) (Figures 6 and 7).

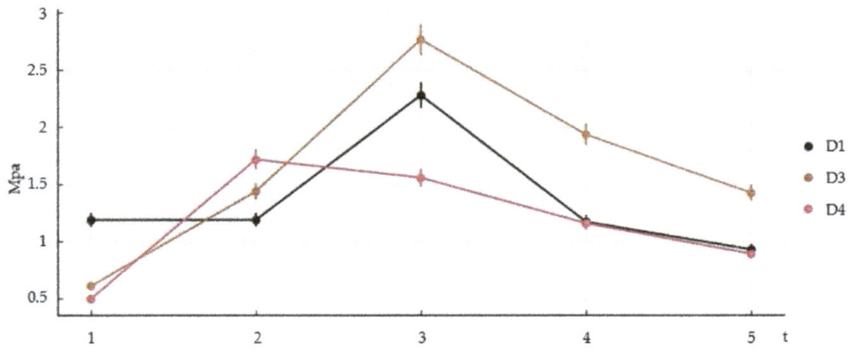

Figure 6. Compression strength at different curing ages. Samples with 0.5% zeolite (D3) and 0.75% zeolite (D4).

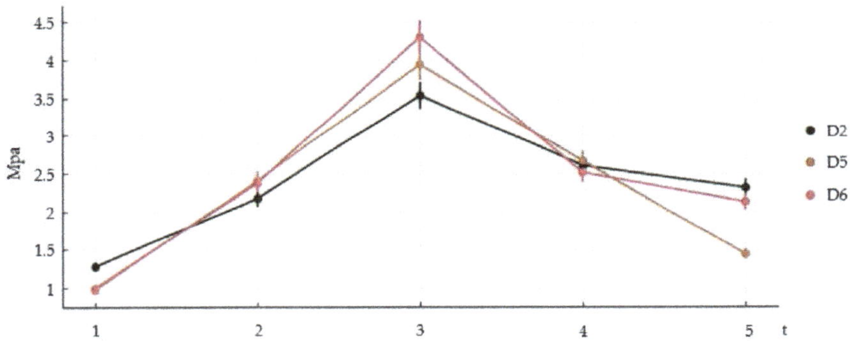

Figure 7. Compression strength at different curing ages. Samples with 1% zeolite (D5) and 1.5% zeolite (D6).

The samples with an addition of 0.5 and 0.75% zeolite began to outperform the properties of the reference samples after 28 days. The more stable properties were observed in D3, with the lowest proportion of zeolite, which at 63 days exceeded the reference values (D1) by 21.2%. After subjecting the samples to accelerated ageing conditions, strength dropped by 30% compared to natural ageing values.

After the 28 days of natural ageing, experimental samples exceed the reference values and display similar ageing values during accelerated ageing. Mortars with 1% and 1.5% zeolite display very similar properties to those of the reference values.

The increase in mechanical strength is directly related to the pozzolanic activity of the zeolite. Owing to its high SiO_2 and Al_2O_3 content, it is capable of reacting with the calcium hydroxide ($Ca(OH)_2$) released during hydration processes. Garcia de Lomas further explores the evolution of the chemical composition and proves that, through the pozzolanic reaction, the zeolite promotes the formation and growth of calcium silicate and calcium aluminate hydrate phases [3]. This leads to a denser hydration process, effectively filling pores and refining the microstructure, which ultimately results in a significant improvement in mechanical performance.

Establishing Shore C hardness. In earth mortars stabilised with 5% lime, it was observed that after the 63 days of natural curing, the surface resistance properties increased compared to the reference samples (Figure 8). In this case, the ageing test did not result in a reduction of the properties at surface level (Table 4). D3, with 0.5% zeolite, maintained stable properties following the accelerated ageing process with UV radiation.

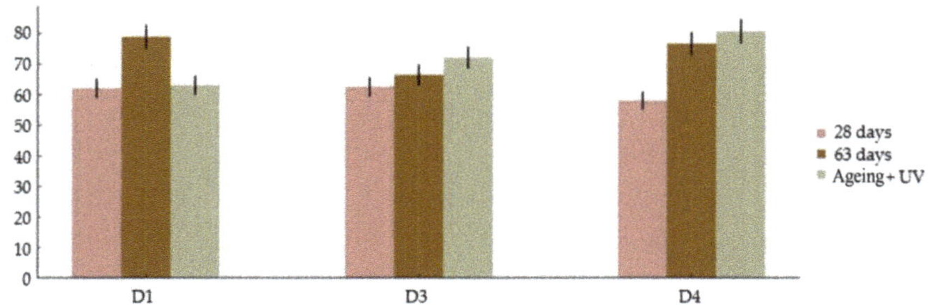

Figure 8. Surface hardness in earth mortars stabilised with 5% lime.

Table 4. Percentage of increase or reduction in surface resistance in relation to reference values.

	63 Days	Ageing
D3	−6.54%	14.46%
D4	8.07%	28.77%

Within the group of mortars stabilised with 10% lime, it was observed that at an early stage of curing, the properties of the experimental material outperformed the reference samples (Figure 9). The most stable formulation among these was D5 (1:1).

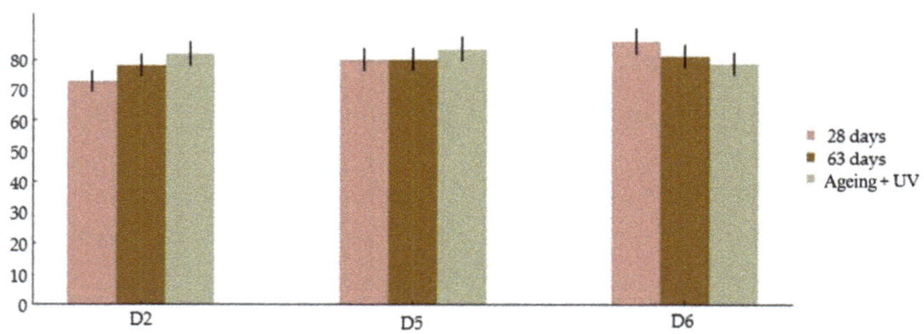

Figure 9. Surface hardness in earth mortars stabilised with 10% lime.

The results exhibit a trend consistent with the compression test, showing a progressive increase in strength as curing advances. It should be noted that mix D6—the formulation with the highest FCC-to-lime ratio—displays distinct behaviour: it presents markedly high early-age surface hardness that gradually diminishes with prolonged curing and exposure to radiation.

6.5. Results of Behaviour in Contact with Water

Swinburne accelerated erosion test (SAET): The results show that the higher the zeolite content, as in the cases of D5 with 1% or D6 with 0.66%, the greater the resistance to water erosion displayed by test samples (Figure 10). In these cases, no damage by erosion was observed for any of the test times (Table 5).

Among the test samples with a lower percentage of zeolite, D3 with 0.5% and D4 with 0.75% were more sensitive to water after natural curing than the samples subjected to the artificial ageing process. However, they still display greater resistance to erosive action than the reference samples with no addition. D4 showed increased resistance at 63 days and after accelerated ageing, reducing the erosion footprint in relation to initial values and obtaining the best results.

Figure 10. Erosion results for SAET.

Table 5. SAET results. Depth measurements and size of droplet footprint.

	28 Days			63 Days			UV Ageing		
	Dimensions		Footprint	Dimensions		Footprint	Dimensions		Footprint
D1	3.9 cm	1.9 cm	3 mm	1.6 cm	1.2 cm	2 mm	1.3 cm	1 cm	1.5 mm
D2	-	-	-	-	-	-	-	-	-
D3	-	-	-	7 mm	5 mm	<1 mm	0.9 mm	0.7 mm	1.5 mm
D4	-	-	-	-	-	-	-	-	-
D5	2.3 cm	1.2 cm	1 mm	-	-	-	1.6 cm	1.4 cm	2 mm
D6	-	-	-	-	-	-	-	-	-

Test for determining water absorption by capillarity. In the group with a low percentage of zeolite (Figure 11), it is worth noting that during the natural curing process absorption in the test samples was similar to that in reference values. After the ageing process, the entire group absorbs a greater volume of water, with the greatest difference in absorption found in aged samples exposed to radiation. The samples with 10% lime displayed similar capillary absorption characteristics after the natural curing period of 63 days (Figure 12). After being subjected to artificial ageing conditions, D5 (1:1) tended to display a greater absorption capacity for water volume.

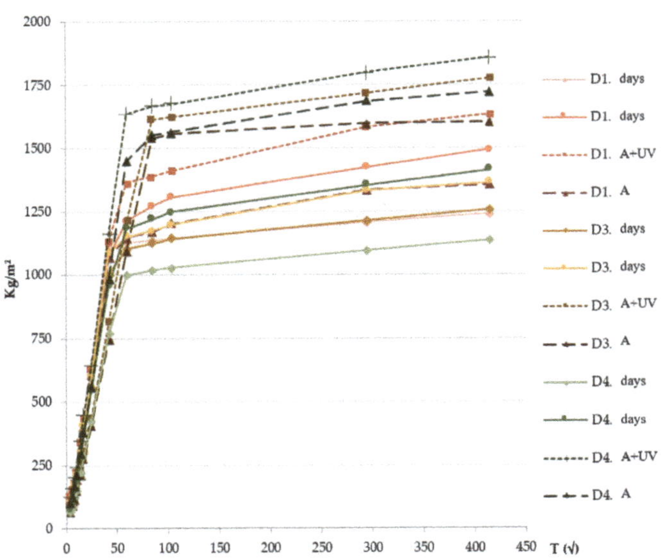

Figure 11. Water absorption by capillarity D3 and D4.

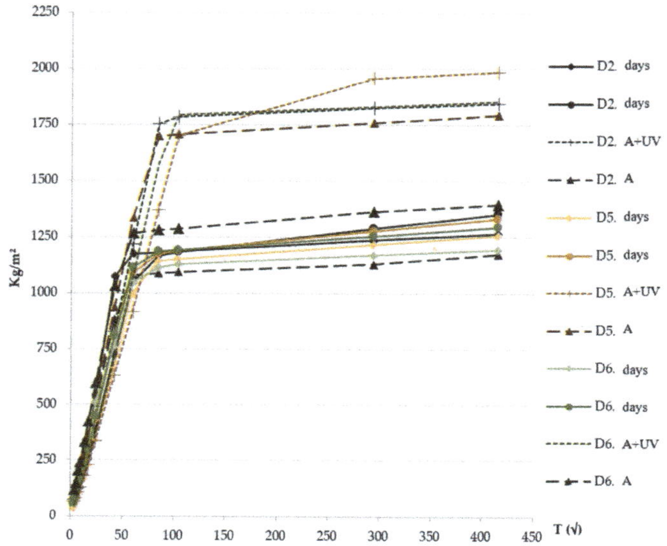

Figure 12. Water absorption by capillarity D5 and D6.

The incorporation of FCC significantly reduces the mortars' sorptivity and capillary absorption, reflecting a denser and less permeable microstructure. These improvements persist at 28 and 63 days and are retained following ageing and UV ageing. Although degradation cycles increase absorption across all formulations, mortars with higher FCC content display lower efficiency loss and exhibit greater stability under water ageing. Further research should deepen the analysis of capillary absorption to confirm FCC's contribution to enhanced water-suction resistance and overall mortar durability.

Water vapour permeability. The group of lime mortars stabilised with 5% lime showed a higher water absorption capacity during the early stages of curing (Figure 13). The more advanced the curing age, the lower the capacity for vapour capture. At 63 days of natural curing, the absorption percentage decreased. D4, with 0.75% zeolite, displayed less permeability.

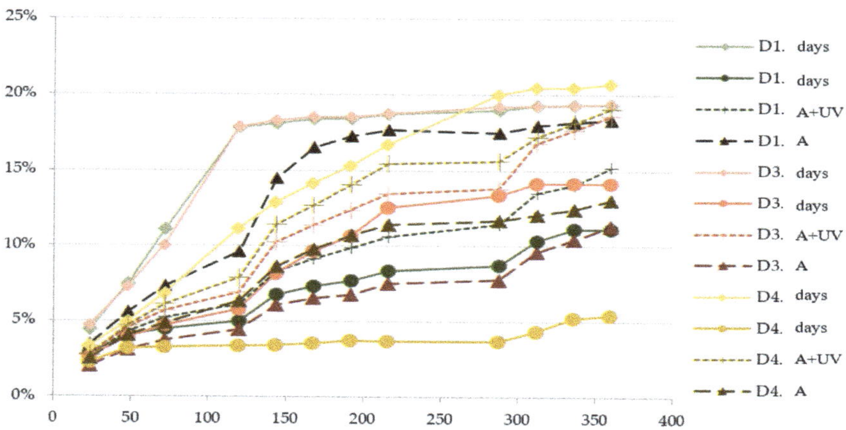

Figure 13. Water permeability by capillarity D3 and D4.

Following the ageing cycles, all the samples tended to recover vapour transmission properties, displaying similar absorption behaviour to reference samples.

Among the group stabilised at 10% (Figure 14), after the natural curing of 63 days, all the samples showed less permeability to water. D5 (1:1) showed more stable conditions throughout the evolution. All samples showed an increase in the percentage of water absorption after the ageing test.

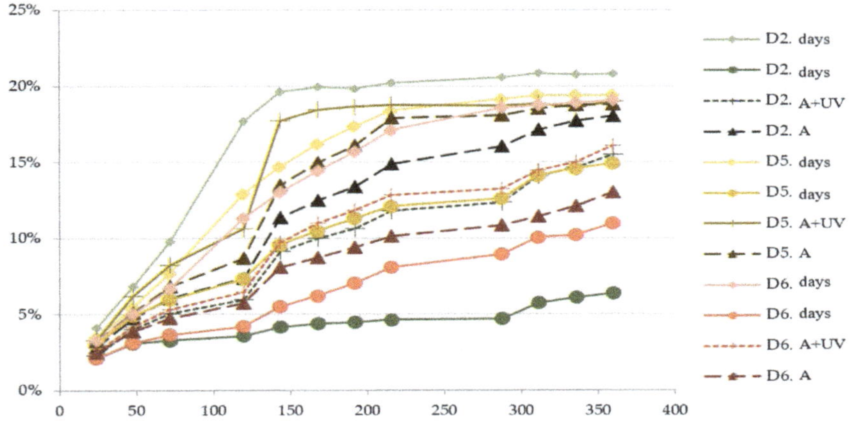

Figure 14. Water permeability by capillarity D5 and D6.

7. Conclusions

Based on the results obtained, the initial contributions regarding the properties of earth mortar with pozzolanic addition can be concluded, along with calculations for the optimal lime/zeolite ratio. D3 (0.33%) and D4 (0.5%) displayed the best properties and highest stability of the mortars tested.

Thanks to the test plan, it has been possible to quantify basic properties of the material. On a mechanical level higher compressive strength was observed, increasing by 21.2% compared to the reference values of the material, while surface hardness increased by 8% compared to initial values. As regards physical properties linked to porosity, the addition did not modify the permeability properties of the material, which meant it could be used without risk of damage due to incompatibility.

A difference in behaviour was also observed between artificial ageing with and without UV radiation. It was established that the material which undergoes radiation tends to show a slight deterioration of mechanical properties as well as chromatic variations with a tendency towards whiter tones, while capillary absorption is also increased, possibly due to modifications in porosity or loss of density in the material. It can thus be assumed that this material is sensitive to radiation exposure.

The material displays good properties in terms of mechanical resistance as well as in terms of permeability to water and presents stable characteristics except in the case of UV exposure. Accordingly, a second phase of this study is advised, continuing exposure to real conditions and enabling further research of the evolution with the simultaneous intervention of various factors.

Author Contributions: Conceptualisation, C.M. and M.B.M.; methodology, M.B.M. and L.G.-S.; analysis, M.B.M. and A.H.-E.; investigation, M.B.M., L.G.-S. and A.H.-E.; resources, M.B.M. and A.H.-E.; data curation, L.G.-S.; writing—original draft preparation, M.B.M.; writing—review and editing, L.G.-S., A.H.-E. and F.V.; visualisation, M.B.M. and A.H.-E.; supervision, F.V. and C.M.; project administration, F.V. and C.M.; funding acquisition, F.V. and C.M. All authors have read and agreed to the published version of the manuscript.

Funding: This research is included in the research project "Earth4Future: Sustainable reuse of earthen architecture and its lessons for contemporary architecture" (ref. PID2022-139154OB-I00; main researchers: Camilla Mileto and Fernando Vegas), funded by the Spanish Ministry of Science, Innovation and University. It is also included in the research project "Risk-Terra: Earthen architecture in the Iberian Peninsula: study of natural, social and anthropic risks and strategies to improve resilience" (ref. RTI2018-095302-B-I00; main researchers: Camilla Mileto and Fernando Vegas), funded by the Spanish Ministry of Science, Innovation and University.

Institutional Review Board Statement: Not applicable.

Informed Consent Statement: Not applicable.

Data Availability Statement: The original contributions presented in this study are included in the article. Further enquiries can be directed to the corresponding author.

Acknowledgments: This research would not have been possible without the guidance on pozzolanic materials and attention offered by José Mª Monzó Balbuena; the collaboration of D. Rafael Calabuig Pastor, a technician at the Materials Laboratory at the Higher Technical School of Building Engineering (ETSIE); D. Rafael Molina Nieto, a technician at the Architectural Constructions Laboratory at the Higher Technical School of Architecture (ETSA); and D. Joan Romero at the Geotechnics laboratory of the Ground Engineering department. The ash used in this project was provided by Lourdes Soriano Martínez from the Department of Building Engineering and Civil Engineering Projects.

Conflicts of Interest: The authors declare no conflicts of interest. The funders had no role in the design of the study; in the collection, analyses or interpretation of data; in the writing of the manuscript; or in the decision to publish the results.

References

1. Soriano Martínez, L. Nuevas Aportaciones en el Desarrollo de Materiales Cementantes con Residuo de Catalizador de Craqueo Catalítico (FCC). Ph.D. Thesis, Universitat Politècnica de València, Valencia, Spain, 2007. Available online: https://riunet.upv.es/handle/10251/2542 (accessed on 15 July 2024).
2. Escandino, A.; Amorós, J.L.; Moreno, A.; Sánchez, E. Utilizing the used catalyst from refinery FCC units as a substitute for kaolin in formulating ceramic frits. *Waste Manag. Res.* **1995**, *13*, 569–578. [CrossRef]
3. García de Lomas Gómez, M. Viabilidad Científica, Técnica y Medioambiental del Catalizador Gastado de Craqueo Catalítico (FCC) como Material Puzolánico. Ph.D. Thesis, Universidad Autónoma de Madrid, Madrid, Spain, 2016. Available online: http://hdl.handle.net/10486/671799 (accessed on 15 July 2024).
4. Pacewska, B.; Wilinska, I.; Bukowska, M. Use of spent catalyst from catalytic cracking in fluidized bed as a new concrete additive. *Thermochim. Acta* **1988**, *322*, 175–181. [CrossRef]
5. Payá, J.; Borrachero, M.V.; Monzó, J.; Soriano, L. Estudio del comportamiento de diversos residuos de catalizadores de craqueo catalítico (FCC) en cemento Portland. *Mater. Construcción* **2009**, *59*, 37–52. [CrossRef]
6. Pacewska, B.; Wilinska, I.; Bukowska, M. Hydration of cement slurry in the presence of spent cracking catalyst. *J. Therm. Anal. Calorim.* **2000**, *60*, 71–78. [CrossRef]
7. Su, N.; Chen, Z.H.; Fang, H.Y. Reuse of spent catalyst as fine aggregate in cement mortar. *Cem. Concr. Compos.* **2001**, *23*, 111–118. [CrossRef]
8. Omoniyi Odjo, A. Reciclado de Polímeros por Craqueo Catalítico; Estudio de la Viabilidad de Utilización de Reactores Convencionales de Craqueo Catalítico en Lecho Fluidificado. Ph.D. Thesis, Universidad de Alicante, San Vicente del Raspeig, Spain, 2016. Available online: http://hdl.handle.net/10045/68368 (accessed on 1 December 2024).
9. Elert, K. Alkaline Activation of Clays for the Consolidation of Earthen Architecture. Ph.D. Thesis, Universidad de Granada, Granada, Spain, 2014; p. 317. Available online: http://hdl.handle.net/10481/34114 (accessed on 1 March 2024).
10. Chen, H.L.; Tseng, Y.S.; Hsu, K.C. Spent FCC catalyst as a pozzolanic material for high-performance mortars. *Cem. Concr. Compos.* **2005**, *26*, 657–664. [CrossRef]
11. Payá, J.; Borrachero, M.V.; Monzó, J.; Gonzalez Lopez, E. Propiedades de morteros y hormigones fabricados con cementos tipo CEM II/A-Q a base de catalizador gastado de craqueo catalítico (FCC). In Proceedings of the III Congreso Nacional de Materiales Compuestos (MATCOMP99) (AEMAC), Gijón, Spain, 5–7 May 1999; pp. 493–500, ISBN 84-607-0078-X.
12. Payá, J.; Monzó, J.; Borrachero, M.V.; Velázquez, S. The chemical activation of pozzolanic reaction of fluid cracking catalyst residue (FC3R) in lime pastes. *Adv. Cem. Res.* **2007**, *19*, 9–16. [CrossRef]

13. Piles, V.; Borrachero, M.V.; Payá, J.; Monzó, J.; García-Codoñer, Á. Ensayos de envejecimiento acelerado sobre materiales compuestos con base de cal o con base mixta de cal y yeso. In Proceedings of the Actas del VI Congreso Nacional de Materiales Compuestos (MATCOMP05) (AEMAC), Zaragoza, Spain, 28–30 June 2005; pp. 947–954, ISBN 84-9705-821-6.
14. Nan, S.; Hung-Yuan, F.; Zong-Huei, C.; Fu-Sung, L. Reuse of waste catalysts from petrochemical industries for cement substitution. *Cem. Concr. Res.* **2000**, *30*, 1773–1783. [CrossRef]
15. Al-Jabri, K.; Baawain, M.; Taha, R.; Al-Kamyani, Z.S.; Al-Shamsi, K.; Ishtieh, A. Potential use of FCC spent catalyst as partial replacement of cement or sand in cement mortars. *Constr. Build. Mater.* **2013**, *39*, 77–81. [CrossRef]
16. Martínez-López, C.; Torres-Agredo, J.; Mejía- De Gutiéerrez, R.; Mellado-Romero, A.M.; PayáBernabeu, J.; Monzó-Balbuena, J.M. Uso de test de lixiviación para determinar la migración de contaminantes en morteros de sustitución con residuos de Catalizador de craqueo catalítico (FCC). *Rev. Fac. Minas Univ. Nac. Colomb.* **2013**, *80*, 163–170.
17. Houben, H.; Guillaud, H. *Earth Construction: A Comprehensive Guide*; Practical Action Publishing: London, UK, 1994.
18. *UNE 103101:1995*; Particle Size Analysis of a Soil by Screening. AENOR: Madrid, Spain, 1995.
19. *UNE-EN ISO 17892-4:2019*; Geotechnical Investigation and Testing—Laboratory Testing of Soil—Part 4: Determination of Particle Size Distribution. AENOR: Madrid, Spain, 2019.
20. Mileto, C.; Vegas, F.; López, J.M. Criterios y técnicas de intervención en tapia. La restauración de la torre Bofilla de Bétera (Valencia). *Inf. Construcción* **2011**, *63*, 81–96. [CrossRef]
21. Schroeder, H. Transferencia de humedad y el cambio en la resistencia durante la construcción de edificios de tierra. *Inf. Construcción* **2011**, *63*, 107–116. [CrossRef]
22. *UNE-EN 772-1:2011*; Methods of Test for Masonry Units—Part 1: Determination of Compressive Strength. AENOR: Madrid, Spain, 2011.
23. *UNE-EN 1097-3:1999*; Tests for Mechanical and Physical Properties of Aggregates—Part 3: Determination of Loose Bulk Density and Voids. AENOR: Madrid, Spain, 1999.
24. *UNE 102042:2023*; Gypsum Plasters. Other Test Methods. AENOR: Madrid, Spain, 2023.
25. *UNE-EN 13286-2:2010*; Unbound and Hydraulically Bound Mixtures—Part 2: Test Methods for Laboratory Reference Density and Water Content—Proctor Compaction. AENOR: Madrid, Spain, 2010.
26. *UNE-EN 13279-2:2014*; Gypsum Binders and Gypsum Plasters—Part 2: Test Methods. AENOR: Madrid, Spain, 2014.
27. *UNE-EN 41410:2023*; Compressed Earth Blocs for Walls and Partitions. Definitions, Specifications and Test Methods. AENOR: Madrid, Spain, 2023.
28. *UNE-EN 15886:2011*; Conservation of Cultural Property—Test Methods—Colour Measurement of Surfaces. AENOR: Madrid, Spain, 2011.
29. *UNE-EN 15801:2010*; Conservation of Cultural Property—Test Methods—Determination of Water Absorption by Capillarity. AENOR: Madrid, Spain, 2010.
30. *UNE-EN 15803:2010*; Conservation of Cultural Property—Test Methods—Determination of Water Vapour Permeability (dp). AENOR: Madrid, Spain, 2010.
31. *UNE-EN 17036:2019*; Conservation of Cultural Heritage—Artificial Ageing by Simulated Solar Radiation of the Surface of Untreated or Treated Porous Inorganic Materials. AENOR: Madrid, Spain, 2019.

coatings

MDPI

Article

Dual-Path Framework Analysis of Crack Detection Algorithm and Scenario Simulation on Fujian Tulou Surface

Yanfeng Hu [1,†], Shaokang Chen [2,†], Zhuang Zhao [3,*] and Si Cheng [1]

1 Faculty of Design and Architecture, Universiti Putra Malaysia, Serdang 43400, Malaysia; gs62156@student.upm.edu.my (S.C.)
2 School of the Arts, Universiti Sains Malaysia, Gelugor 11800, Malaysia
3 Sydney Water Corporation, 1 Smith Street, Parramatta, NSW 2150, Australia
* Correspondence: brad.zhao@sydneywater.com.au
† These authors contribute equally to this work.

Abstract: Fujian Tulou, a UNESCO World Heritage Site, is highly vulnerable to environmental and anthropogenic stresses, with its earthen walls prone to surface cracking that threatens both structural stability and cultural value. Traditional manual inspection is inefficient, subjective, and may disturb fragile surfaces, highlighting the need for non-destructive and automated solutions. This study proposes a dual-path framework that integrates lightweight crack detection with independent physical simulation. On the detection side, an improved YOLOv12 model is developed to achieve lightweight and accurate recognition of multiple crack types under complex wall textures. On the simulation side, a two-layer RFPA^{3D} model was employed to parameterize loading conditions and material thickness, reproducing the four-stage crack evolution process, and aligning well with field observations. Quantitative validation across paired samples demonstrates improved consistency in morphology, geometry, and topology compared with baseline models. Overall, the framework offers an effective and interpretable solution for standardized crack documentation and mechanistic interpretation, providing practical benefits for the preventive conservation and sustainable management of Fujian Tulou.

Keywords: crack detection; cultural heritage protection; Fujian Tulou; YOLOv12; RFPA^{3D}; surface inspection

1. Introduction

Fujian Tulou, as a representative form of traditional rammed-earth architecture in China, has been inscribed on the UNESCO World Heritage List due to its distinctive defensive structure, ecological adaptability, and socio-cultural significance [1]. However, prolonged exposure to natural weathering and anthropogenic disturbances inevitably leads to surface deterioration of the earthen walls, manifesting as strip cracks, intersecting fissures, polygonal crack networks, exfoliation, and discoloration [2]. These forms of damage not only compromise structural stability but also diminish cultural and historical value, and in severe cases may trigger partial or overall collapse, posing safety risks [3,4]. Thus, the accurate detection and reliable prediction of crack initiation and evolution are crucial for the long-term preservation and scientific conservation of earthen heritage sites.

Structural health monitoring (SHM) has long considered cracking as a core research topic [5]. Traditional inspection methods largely rely on visual surveys or manual interpretation of surface morphology, sometimes assisted by basic image processing operations such as grayscale enhancement, histogram equalization, edge detection (e.g., Canny), or threshold segmentation [6,7]. For example, grayscale enhancement and histogram equalization can highlight surface crack features [8,9], the Canny algorithm can assist in edge recognition [10], and threshold segmentation can be used for crack area extraction [11]. While effective in controlled laboratory conditions, these methods are highly sensitive to lighting, shadows, and material heterogeneity, and require extensive manual parameter tuning, limiting their applicability in complex heritage environments.

The rapid development of deep learning has provided new opportunities for non-contact, automated detection of surface damage in cultural heritage [5]. Non-destructive techniques such as drones, video monitoring, and smartphone imaging, when combined with convolutional neural networks, allow efficient remote crack detection. Two-stage detectors (e.g., Fast R-CNN) can achieve high accuracy in complex scenarios but struggle to meet real-time monitoring needs [12]. In contrast, single-stage detectors such as SSD, RetinaNet, and the YOLO family offer a better trade-off between speed, accuracy, and cross-platform deployment [13,14]. With successive iterations, YOLO models have achieved competitive performance in small-scale crack detection under challenging textures [15,16]. For example, studies on materials such as steel and concrete have shown that the improved YOLO model can significantly improve the detection capabilities of small-scale cracks and complex texture backgrounds [17–19]. Nevertheless, due to the heterogeneity of earthen materials and the complexity of weathered surfaces, current YOLO-based methods remain limited in capturing early crack initiation and reproducing full-scale crack evolution.

Beyond detection accuracy, recent advances emphasize that AI technologies can also play a broader role in the real-time monitoring and digital management of heritage structures [20,21]. In this regard, the integration of computer vision outputs into Heritage Building Information Modeling (HBIM) frameworks has gained increasing attention [22]. HBIM provides a comprehensive and user-friendly platform that federates geometric data, environmental conditions, structural monitoring records, and conservation histories, thereby offering interpretable tools for both researchers and practitioners [23–25]. For example, HBIM can incorporate parametric modeling scripts to represent complex architectural geometries [26,27], and Revit plug-ins now enable integration with real-time data collected from IoT devices and AI models, improving both workflow efficiency and monitoring accuracy [28]. Such environments have already been coupled with sensor networks and analysis modules to visualize real-time conditions, generate automatic reports, and support preventive decision-making [29]. Furthermore, AI-assisted model updating has demonstrated the potential to turn monitoring data into reliable numerical surrogates for decision support, for instance by calibrating finite element models with modal parameters optimized through swarm intelligence algorithms [30,31]. These studies highlight that the value of AI lies not only in enhancing detection but also in linking field evidence to digital twins that integrate monitoring, simulation, and conservation planning. However, despite these advances, existing studies remain largely fragmented. Most research efforts focus either on improving the accuracy of AI-based crack detection or on integrating monitoring data into HBIM and digital twin platforms, but rarely address both dimensions in a unified framework. Moreover, current approaches often treat cracks as static surface features, lacking mechanisms to explain or predict their progressive evolution. This study differs from previous work by introducing a dual-path framework. By linking detection outputs with mechanistic parameters, the framework not only standardizes multi-class crack documentation but also reproduces the four-stage evolution process observed in situ. To the best of our knowledge, this is the first attempt to establish such a detection–simulation loop tailored for earthen heritage architecture, thereby extending AI applications from isolated recognition tasks to interpretable, evolution-aware, and conservation-oriented decision support.

Building on these advances, this study proposes a dual-path parallel framework. The first path employs an improved YOLOv12 detection model (YOLOv12-MLE) for multi-class identification and standardized documentation of cracks on Tulou facades, covering strip, intersecting, and polygonal patterns. The second path introduces an independent physical simulation using RFPA^{3D}, parameterizing the loading ratio ($\lambda = \Delta y / \Delta x$) and overlay thickness (t) to mechanistically reproduce the four crack evolution stages—initiation, propagation, coalescence, and saturation—and validate against field observations. Both outputs can be integrated into HBIM-centered workflows, thereby extending the applicability of AI beyond isolated detection tasks.

The framework offers three main innovations: (i) a lightweight YOLOv12-MLE model enabling accurate, edge-deployable detection of Tulou cracks, benchmarked against mainstream detectors using F1, precision, *recall*, and *mAP@50*; (ii) a parameterized RFPA^{3D} simulation protocol systematically examining the effects of λ and t on crack morphology, density, and stage transitions, yielding qualitative consistency with on-site observations; and (iii) a cross-validation strategy linking detection results with physical mechanisms, producing reusable datasets and visual reports to support targeted conservation strategies such as thickness control and localized reinforcement.

In summary, the proposed dual-path framework not only extends the application of deep learning in cultural heritage conservation but also provides an efficient, interpretable, and scientifically grounded approach for the standardized recording and mechanistic study of crack evolution in Fujian Tulou. The remainder of this paper is organized as follows: Section 2 presents the framework architecture, experimental design, dataset construction, and evaluation metrics; Section 3 discusses the results and analysis; Section 4 concludes the contributions and outlines future research directions.

2. Materials and Methods

2.1. Case Study Description

The samples for this study were collected from the Gaopi and Nanxi tulou clusters in southern Fujian, including Chaoyang, Chengqi, Fuxing, Kuiju, Yanxiang, and Zhencheng, as shown in Figure 1. These Fujian tulou were officially inscribed on the UNESCO World Heritage List in 2008. Chengqi and Zhencheng are among Fujian's most representative and renowned tulou [32]. These large, multi-story dwellings, characterized by circular or rectangular layouts, combined defensive functions with communal living spaces, reflecting their ecological adaptability and sociocultural traditions [33]. Despite their historical and architectural significance, tulou are highly susceptible to environmental and human influences. Long-term exposure to rainfall, temperature fluctuations, and human intervention leads to the gradual deterioration of earthen walls, often manifesting as linear cracks, intersecting fissures, and polygonal crack networks [34]. These defects not only compromise the structural stability of the walls but also threaten the cultural value and authenticity of these heritage sites.

Field investigations revealed a wide range of crack patterns, from thin surface fissures in the plaster overlay to large-scale vertical cracks propagating through the rammed-earth substrate, as shown in Figure 2. In panel (a), thin polygonal cracks are visible in the plaster overlay, with widths generally below 2 mm. These micro-cracks often result from surface shrinkage and represent the initiation stage of deterioration. In panel (b), localized surface loss and cavity formation are observed, due to long-term moisture penetration and material erosion. Such defects can accelerate crack propagation and undermine the protective surface layer. Panel (c) illustrates a large vertical crack penetrating the rammed-earth substrate, with a width exceeding 10 mm. This type of through-crack is strongly correlated with differential settlement or thermal stresses, posing significant risks to structural stability.

These examples highlight the heterogeneity of Tulou wall degradation, where both fine-scale surface cracks and deep structural fissures coexist. Documenting such variations provides essential ground-truth evidence for validating the YOLOv12-MLE crack detection model and supports the parameterized RFPA^{3D} simulations in replicating progressive crack evolution.

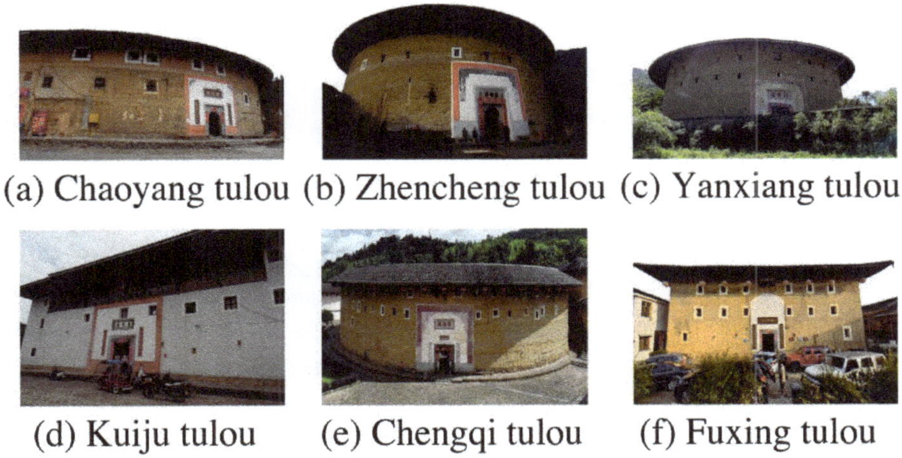

Figure 1. Case studies from the Gaopi and Nanxi Tulou. (**a**) Chaoyang tulou; (**b**) Zhencheng tulou; (**c**) Yanxiang tulou; (**d**) Kuiju tulou; (**e**) Chengqi tulou; (**f**) Fuxing tulou.

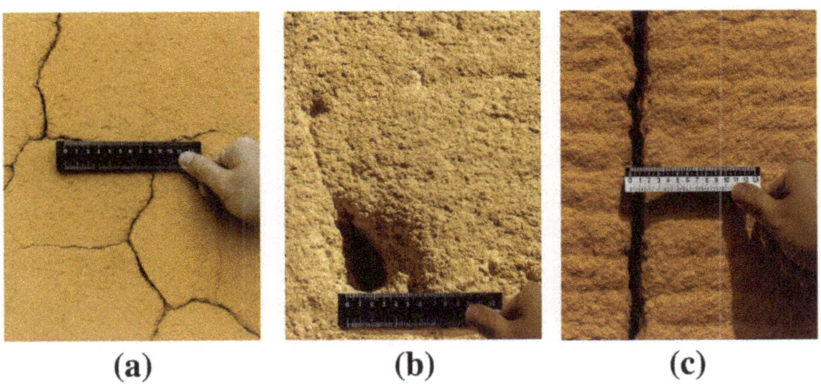

Figure 2. Representative crack types observed in Fujian Tulou. (**a**) Polygonal cracks forming irregular networks on the wall surface; (**b**) Surface erosion and cavity-like defects caused by material loss; (**c**) Vertical linear crack extending along the earthen wall.

To support standardized recording and physics-aware interpretation, we classify cracks by width and report representative cases observed in Yongding and Nanjing clusters together with indicative environmental and structural parameters, as shown in Table 1. (i) Micro cracks (<1 mm). Frequently observed as fine polygonal patterns on well-maintained facades (e.g., Zhencheng Tulou). Dominant drivers are seasonal humidity swings (annual RH $\approx$ 70%–85% with $\pm$10%–20% excursions) and daily solar heating ($\Delta T \approx$ 10–15 °C), causing reversible surface shrink-swell. Typical seasonal widening is 0.1–0.3 mm in sun-exposed sectors; wall-thickness transitions around openings act as stress concentrators. (ii) Small cracks (1–5 mm). Common on inner walls of square Tulou (e.g., Wenchang building at Tianluokeng). Moisture variability (valley RH $\approx$ 80%,

±20% seasonally) and minor differential settlement (≈ 0.5–1 mm/yr) jointly drive 2–4 mm openings; legacy cement patches (1950s–1980s) can introduce interface incompatibility, adding ≈ 0.5–1 mm. (iii) Moderate cracks (5–20 mm). Observed on tall or slightly leaning walls (e.g., Yuchang). Thick-to-thin wall gradients (base 2.0 m → top 0.8 m) amplify tensile demand; high moisture near rivers (RH swing up to 30%) plus settlement (≈ 5 mm/yr) and historic cement jackets (5–10 cm) lead to 10–15 mm openings and debonding along repair interfaces. (iv) Large cracks (>20 mm). Found on neglected/abandoned sections (e.g., remnants in the Gaobei area). Lack of maintenance, intense wetting–drying cycles (±25% RH), and extreme temperatures (−2 to 38 °C) accelerate fabric degradation and foundation softening; annual widening can exceed 10 mm, with local collapse risk.

Table 1. Crack width classes, typical causes, and indicative parameters in Fujian Tulou.

Width Class	Typical Locations/Cases	Dominant Drivers	Indicative Parameters
Micro <1 mm	Overlay surface	RH swings; daily solar ΔT; thickness transitions	RH 70%–85%, ±10–20%; ΔT 10–15 °C; seasonal widening 0.1–0.3 mm
Small 1–5 mm	Inner walls, near foundations	Moisture + minor settlement; legacy cement patches	RH ≈ 80%, ±20%; settlement 0.5–1 mm/yr; +0.5–1 mm at cement interfaces
Moderate 5–20 mm	Tall/leaning walls	Thickness gradient; high moisture; settlement; rigid repairs	Base 2.0 m → top 0.8 m; RH swing up to 30%; settlement ≈ 5 mm/yr; repair jacket 5–10 cm
Large >20 mm	Neglected sections/partial ruins	Wetting–drying cycles; extreme ΔT; no maintenance	RH ±25%; −2 to 38 °C; widening > 10 mm/yr; local collapse risk

The width-based typology provides standardized labels for the detection pipeline (training/validation strata) and priors for the IEIoU-enhanced localization of long, high-aspect-ratio cracks. The parameter ranges (RH, ΔT, thickness) give physically plausible bounds for RFPA^{3D}, easing calibration of λ and t. The representative cases create repeatable test patches for cross-checking prediction consistency (skeleton IoU, Chamfer, Hausdorff, spacing MAE, orientation EMD) against real façades.

2.2. Data Prepossessing and Dataset Construction

To support crack detection and evolution prediction in Fujian Tulou, a high-quality image dataset of earthen-site cracks was constructed and annotated. Given the variability of rammed-earth materials across climates, construction techniques, and historical periods, crack patterns exhibit significant complexity and diversity. Based on conservation practice and fracture mechanics theory, a multi-class semantic labeling scheme was established, as shown in Figure 3, covering strip cracks, intersecting cracks, and polygonal crack networks to provide standardized samples for deep learning models.

The final dataset comprised 2800 images, annotated into four evolutionary stages: initiation (700), propagation (700), coalescence (700), and saturation (700). Images were sourced from three complementary channels: (i) Digital archives from heritage agencies (38%, 1064 images), primarily from the National Cultural Heritage Administration platform and the Fujian Tulou Conservation Center; (ii) High-resolution field surveys (27%, 756 images) conducted by conservation technicians at representative Tulou sites; (iii) Open-access resources (35%, 980 images), including Wikimedia Commons, academic databases, and archaeological repositories, restricted to Fujian Tulou or structurally equivalent rammed-earth heritage buildings. This combination ensured that the dataset not only reflected authentic case-study conditions but also captured broader variations in crack morphology, surface texture, and environmental contexts. Such integration of multiple sources was essential to improve model robustness and generalization, reducing overfitting to a single site and enabling applicability to a wider range of Tulou structures.

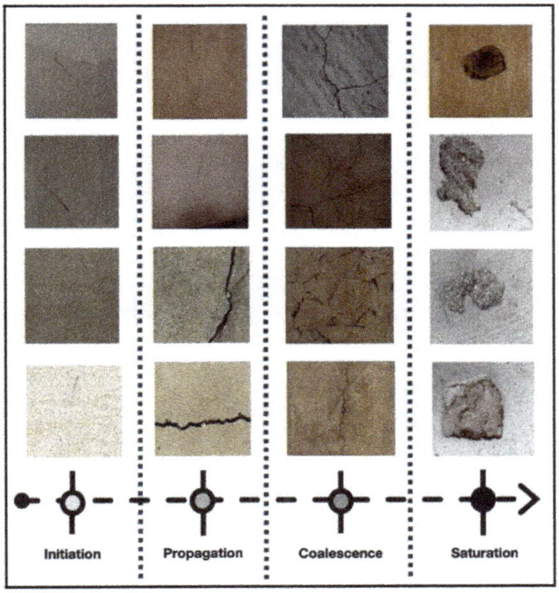

Figure 3. Crack semantic annotation system.

To improve data quality and adaptability, the GrabCut algorithm was applied to remove complex backgrounds, and all images were resized to 512 $\times$ 512 pixels for compatibility with YOLO models and mainstream detectors. Data augmentation strategies including rotation, mirroring, color perturbation, and perspective transformation were employed to expand sample distributions across crack stages and alleviate imbalance in saturation samples.

Annotation was conducted collaboratively by heritage conservation experts, structural engineers, and computer vision specialists. Each image was independently labeled by two experts, cross-validated, and refined through three rounds of review. The inter-annotator agreement reached a Cohen's Kappa of 0.88, indicating high consistency. All bounding boxes and labels were converted into a unified JSON format compatible with YOLOv12 and RFPA^{3D} inputs. To further optimize anchor box matching, K-means clustering was applied to the distribution of bounding box dimensions, yielding improved detection accuracy and convergence. Consistency evaluation was performed using IoU $\geq$ 0.5 as the matching criterion, with each region annotated by three experts; the mean κ value was 0.89, confirming robust agreement. To prevent data leakage, time-series images of the same crack object were assigned to the same subset. The dataset was finally split into training (80%) and validation (20%) sets, with 20% of the training set reserved as a tuning subset. Early stopping was adopted to terminate training if validation loss failed to improve for five consecutive epochs, effectively reducing overfitting and enhancing generalization.

2.3. Model Comparison and Improvement

To validate the effectiveness of the proposed improvements, we conducted both ablation studies and comparative experiments with mainstream object detection models, including YOLOv5, YOLOv6, YOLOv7, YOLOv9, YOLOv10, YOLOv11, YOLOv12, DEtection TRansformer (DETR), Faster R-CNN, and SSD. These algorithms vary in precision, speed, generalization capacity, and computational cost, and were selected to provide a comprehensive benchmark on the Fujian Tulou crack dataset. The evaluation considered: (i) capability to identify multiple crack types (e.g., vertical, horizontal, step-shaped, ring, and radial cracks); (ii) robustness to complex wall backgrounds, as Tulou facades consist of rammed earth, timber beams, and cobblestone with highly heterogeneous textures; and (iii) adaptability to cultural heritage scenarios, which was evaluated through dataset diversity and robustness testing. Specifically, the crack dataset incorporated images from multiple sources (digital archives, field surveys, and open-access repositories), covering a wide range of wall conditions such as variations in thickness, moisture stains, and historical repair traces. To enhance adaptability, data augmentation techniques (e.g., rotation, mirroring, color perturbation, perspective transformation) were applied to simulate heterogeneous surfaces and environmental interferences. Furthermore, stratified cross-validation ensured that each subset of wall conditions was represented in both training and validation sets, allowing performance indicators (F1, precision, *recall*, *mAP@50*) to be systematically compared across scenarios.

Based on overall performance, YOLOv12 was chosen as the baseline model, upon which we developed an improved framework, YOLO-MLE (YOLO for Multi-scale Long-crack Extraction), specifically tailored for Tulou crack detection, as shown in Figure 4. YOLO-MLE employs MobileNetV4 as a lightweight yet effective backbone, structured into five hierarchical blocks (Block1–Block5) to extract multi-scale features from $512 \times 512 \times 3$ input images. Shallow-to-deep features capture low-level textures through to high-level semantics, while skip connections preserve spatial detail and facilitate cross-scale information flow. To enhance spatial representation and semantic refinement, A2C2f modules are embedded after feature cascades at levels P3, P4, and P5, integrating channel and spatial attention for improved context fusion.

In the intermediate layer (P4), a SaturatedLSTMCA module is introduced to perform sequential modeling on flattened spatial vectors, enabling the network to capture long-range spatial dependencies. This design is particularly suited for thin, elongated, or curved cracks exhibiting sequential patterns on Tulou facades. The detection head operates at three scales: P3 (8×8 grid) for small targets, P4 (16×16) for medium targets, and P5 (32×32) for large targets. Each branch outputs classification and localization predictions supervised by an enhanced EIoU loss function (EnEIoU). Unlike standard IoU-based losses, EnEIoU incorporates geometric priors, including aspect ratio sensitivity, directional consistency, and adaptive weight adjustment, making it highly effective for crack-like targets with high aspect ratios and directional variance. Detailed implementation steps are provided in the Supplementary Materials.

Specifically, lightweight, and efficient MobileNetV4 was selected as the backbone to support high-precision feature extraction of Fujian Tulou cracks across multiple structural scales, as shown in Figure 5. As a UNESCO World Heritage Site, Tulou walls are composed of rammed earth, timber beams, and cobblestones, resulting in highly complex textures, strong noise, and pronounced material heterogeneity. Crack patterns are diverse, ranging from micro-surface fissures to vertical, horizontal, radial, and penetrating structural cracks. Therefore, the YOLOv12-MLE framework must strike a balance between lightweight design and robust performance to ensure accurate crack recognition under mobile deployment conditions.

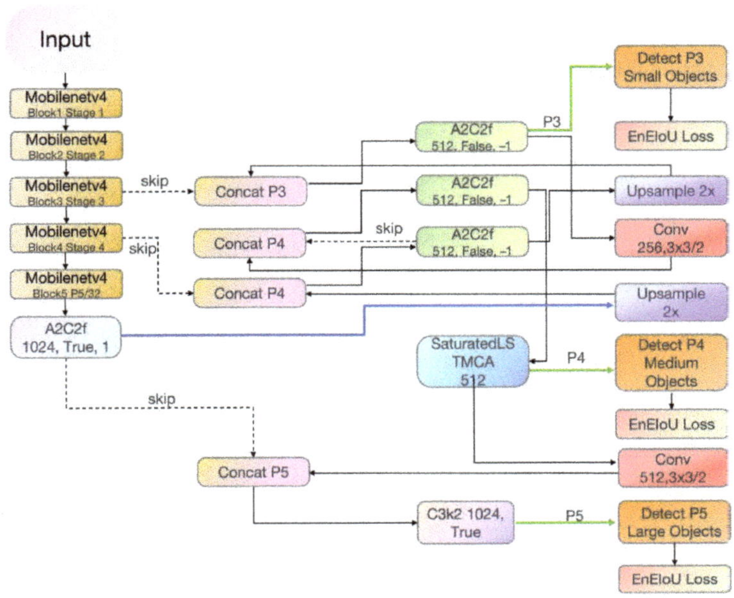

Figure 4. YOLO-MLE framework diagram. The architecture integrates a MobileNetV4 backbone with an Improved Multi-channel Attention (IMCA) module and an Enhanced EIoU loss function. The diagram shows five hierarchical feature extraction blocks (Block1–Block5) with skip connections, three-scale detection heads (P3, P4, P5), and sequential modeling via a SaturatedLSTMCA module. This design enables multi-scale feature extraction, enhances sensitivity to elongated cracks, and improves localization under heterogeneous Tulou wall textures.

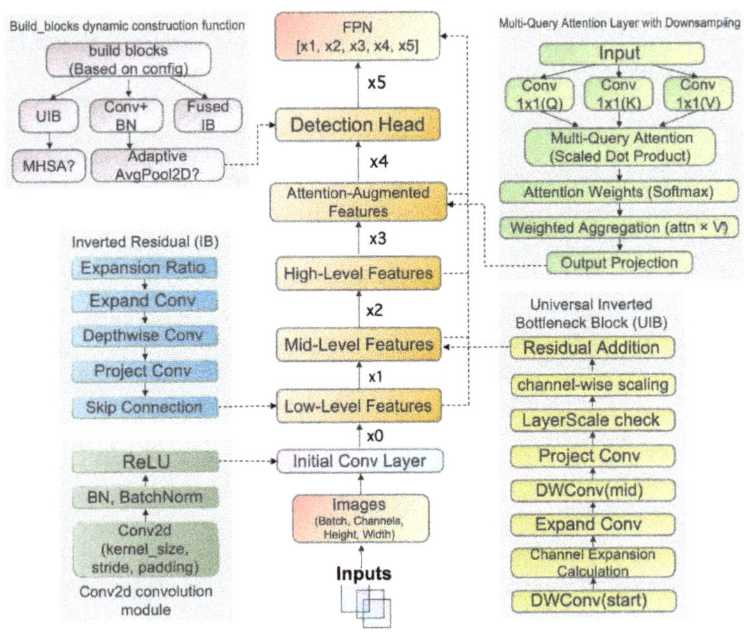

Figure 5. MobileNetV4 backbone network framework diagram. The backbone employs Universal Inverted Bottleneck (UIB) blocks across four convolutional stages, balancing computational efficiency

and feature richness. Depthwise and expansion convolutions capture surface textures and crack patterns, while projection layers compress features for lightweight deployment. The figure also highlights the integration of Multi-Query Attention modules and a dual-path down-sampling mechanism, which strengthen spatial context modeling under noisy wall conditions.

The architecture employs Universal Inverted Bottleneck (UIB) blocks with variable depth configurations, organized into a four-stage convolutional structure to balance computational efficiency and representational power. First, initial depth wise convolutions capture low-level surface textures such as rammed-earth imprints and timber grain. Second, expansion convolutions increase channel capacity to extract multi-scale crack details. Third, intermediate depth wise convolutions model higher-order interactions between cracks and earthen textures. Finally, projection convolutions compress high-dimensional features, reducing computational cost while preserving critical crack semantics.

At key stages, Multi-Query Attention modules are embedded and combined with a dual-path down-sampling mechanism, significantly enhancing spatial context modeling under noisy wall backgrounds. This design is particularly effective for low-contrast cracks and highly textured surfaces. Additionally, a Layer Scale module introduces a learnable parameter (γ) to fine-tune channel-level feature responses, allowing the model to adaptively adjust crack detection sensitivity across varying Tulou materials and illumination conditions.

The backbone adopts a hierarchical feature extraction strategy, ultimately producing a five-level feature pyramid that provides rich multi-scale semantics for downstream crack detection and evolution modeling. To ensure real-time inference on edge devices such as Jetson AGX Orin, convolution channel numbers are optimized with the make divisible function to multiples of eight, maintaining high performance while preserving lightweight efficiency.

The improved multi-channel attention mechanism (IMCA) was designed to enhance the representational power of conventional attention modules by incorporating a complete LSTM sequence encoder, a dual-path adaptive fusion strategy, and dimensional alignment, as illustrated in Figure 6. This module was specifically optimized for crack detection on Fujian Tulou walls, where surfaces are composed of rammed earth, timber beams, and cobblestones, characterized by complex textures, high noise levels, and strong heterogeneity. Crack patterns are diverse and exhibit long-range dependencies and staged variations, including initiation, propagation, coalescence, and saturation.

IMCA employs a dual-path feature modeling strategy to balance global context and local crack detail. In the first path, multiple statistical pooling operations (average, max, and standard deviation pooling) are used to extract complementary global descriptors, which are then adaptively fused through a learnable weighting scheme. This design significantly improves discrimination of crack features under noisy Tulou wall backgrounds. In the second path, a lightweight LSTM network models spatial sequences along rows or columns of feature maps, capturing crack directionality and continuity. To ensure compatibility across feature layers, an adaptive sequence processing strategy is introduced, automatically padding or truncating inputs for stable multi-level modeling.

After dual-path modeling, outputs are integrated via learnable fusion coefficients and a Sigmoid activation function to generate spatial attention maps. These maps emphasize crack-related regions while suppressing high-frequency noise caused by rammed-earth imprints, timber grains, and repair marks. To further enhance modeling of crack anisotropy, IMCA computes and fuses attention responses along both height and width axes, thereby improving sensitivity to vertical, radial, and curved cracks typical of Tulou walls.

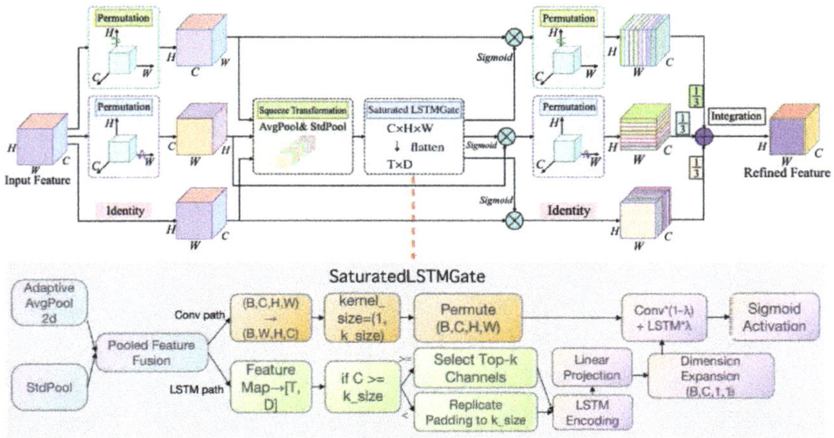

Figure 6. Improved Channel Attention Mechanism (MCA) framework diagram. The figure illustrates the feature modeling strategy: (i) statistical pooling operations (average, max, and standard deviation) fused adaptively to enhance global descriptors; and (ii) a lightweight LSTM path for spatial sequence modeling to capture crack directionality. Outputs are combined through learnable fusion coefficients and generate spatial attention maps, shown as highlighted regions emphasizing cracks while suppressing background noise; Conv*$(1-\lambda)$ + LSTM*λ represents a weighted sum of convolutional and LSTM paths, where * denotes element-wise scaling.

In the Fujian Tulou crack detection task, IMCA demonstrated clear advantages: it effectively suppressed false detections under noisy rammed-earth backgrounds, improved modeling of crack directionality and morphology, and supported differentiation of crack evolution stages. In the Fujian Tulou crack detection task, the improved multi-head cross-attention (IMCA) module demonstrated clear advantages. As shown in the ablation experiments (Table 2), introducing IMCA into the baseline model increased the F1 score from 85.5% to 87.4% and improved precision from 88.7% to 90.9%, while also enhancing *recall* to 84.1%. These gains indicate that IMCA effectively suppresses false detections under noisy rammed-earth backgrounds and improves the modeling of crack directionality and morphology. Moreover, when IMCA was combined with MobileNetV4 and EnEIoU, the full YOLOv12-MLE framework achieved the best overall performance (F1 = 91.8%, *mAP@50* = 95.5%), supporting the claim that IMCA contributes not only to geometric feature extraction but also to distinguishing different crack evolution stages.

Table 2. Ablation experiment results.

Model	Mv4	IMCA	EnEIoU	F1 Score	P (%)	*R* (%)	*mAP@50* (%)	GFLOPs	Params (MB)
Baseline				85.5	88.7	82.6	91.5	5.99	2.52
Mobilenetv4	√			89.1	93.7	85.0	93.7	4.68	2.21
IMCA		√		87.4	90.9	84.1	92.8	5.99	2.52
EnEIoU			√	89.0	92.2	86.0	94.0	5.99	2.52
Mobilenetv4 + IMCA	√	√		90.9	94.2	87.8	95.3	4.68	2.21
Mobilenetv4 + EnEIoU	√		√	90.7	92.9	88.6	95.1	4.68	2.21
IMCA + EnEIoU		√	√	88.2	87.5	88.9	93.9	5.99	2.52
Mv4 + IMCA + EnEIoU	√	√	√	91.8	91.5	92.0	95.5	4.68	2.21

Furthermore, its lightweight design enables real-time performance on mobile and edge devices, offering reliable technical support for non-destructive crack detection and structural health assessment in earthen heritage conservation.

The enhanced EIoU loss function introduces an aspect-ratio-sensitive mechanism and a directional consistency penalty, as shown in Figure 7. This design is particularly suited for the linear crack patterns commonly observed on Tulou walls, strengthening the model's ability to capture complex geometric morphologies. Specifically, the squared differences in width (p_w^2) and height (p_h^2) between the predicted and ground-truth boxes are first computed and normalized by the diagonal length of the minimum enclosing box, ensuring scale invariance. Next, an aspect-ratio discrepancy metric is introduced as a criterion for identifying elongated structures. When the detected target exceeds a predefined aspect-ratio threshold, a linear mask is activated to further determine whether the object corresponds to a slender crack.

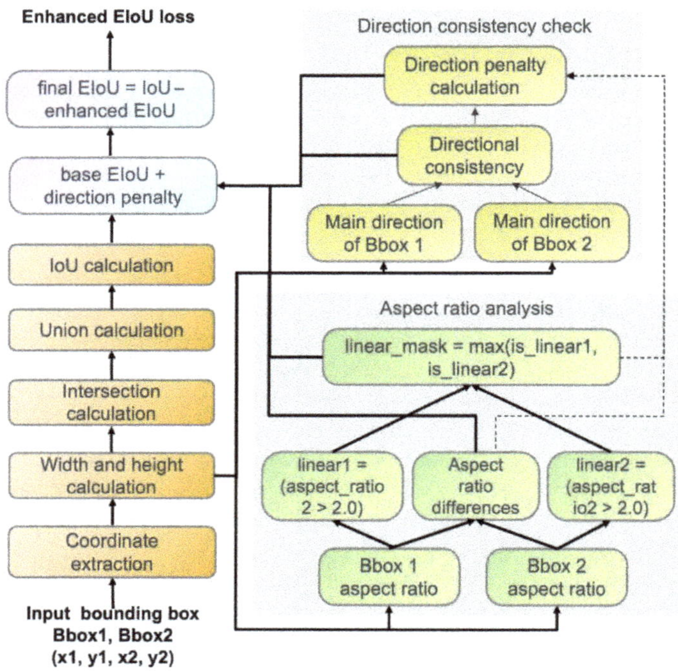

Figure 7. Enhanced EIoU loss function framework.

On this basis, a directional consistency penalty is constructed to evaluate the angular deviation between the principal axes of predicted and ground-truth boxes. Even when IoU values are high, large orientation discrepancies are penalized, preventing misclassification of elongated cracks due to incorrect alignment. Finally, the overall EIoU loss is formulated as a weighted combination of the baseline EIoU term and the proposed orientation penalty, as expressed in Equation (1):

$$\text{Enhanced EIoU} = \text{EIoU} + \lambda \cdot \text{DirectionPenalty} \quad (1)$$

where λ is a weight coefficient adaptively adjusted based on the target's geometric characteristics, enabling the model to dynamically adjust the optimized center of gravity for targets of varying morphology. This improvement is particularly critical in crack detection tasks, especially in linear crack structures caused by stress evolution, which often exhibit distinct directionality and high aspect ratios. By incorporating these geometric prior constraints into the loss function, not only does it improve detection accuracy, but it also enhances the model's structural interpretation capabilities, aligning with the actual evolution of cracks in cultural heritage sites.

2.4. Evaluation Metrics and Statistical Formulations

To ensure the robustness and statistical validity of our results, we conducted ablation and comparison experiments for the YOLOv12-MLE configuration. In each run, we trained and evaluated the model using a fixed train-test split and different random seeds. For each performance metric (F1 score, precision, *recall*, and *mAP@50*), the formula is as follows:

Precision measures the proportion of correctly classified "crack" features among all predicted instances, as shown in Equation (2):

$$Precision = \frac{TP}{TP + FP} \tag{2}$$

The *recall* rate evaluates the model's ability to correctly retrieve all relevant crack attributes, as shown in Equation (3):

$$Recall = \frac{TP}{TP + FN} \tag{3}$$

The F1-Score is a balanced metric that combines precision and *recall*. It comprehensively evaluates a model's ability to accurately identify small cracks and fully cover all types of cracks in the Fujian Tulou crack detection task. It is particularly suitable for evaluating the model's robustness and generalization performance in high-noise and complex texture backgrounds, as shown in Equation (4):

$$\text{F1 Score} = 2 \cdot \frac{Precision \times Recall}{Precision + Recall} \tag{4}$$

Object Detection Metrics (Bounding Box Evaluation). Because YOLOv12 performs both classification and object localization, bounding box accuracy must be evaluated using the following metric: mean average precision at an intersection-over-union (IoU) of 0.5 (*mAP@50*). This metric evaluates the model's object detection performance by measuring the average precision at which the predicted bounding box achieves at least 50% accuracy. The intersection-over-union (IoU) with the ground truth reflects the model's ability to correctly detect and classify cracks at moderate localization thresholds. A higher *mAP@50* score indicates greater accuracy in identifying and localizing crack classes, as shown in Equation (5):

$$mAP@50 = \frac{1}{N}\sum_{i=1}^{N} AP_i \tag{5}$$

2.5. Crack Prediction Method and Model Description

After incorporating the proposed improvements, the YOLOv12-MLE model achieved finer-grained feature parsing in the Fujian Tulou crack detection task. Unlike conventional detectors that only output bounding boxes, the improved framework additionally classifies crack evolution stages and estimates the orientation angle (θ), thereby providing richer inputs for subsequent physical modeling and three-dimensional crack evolution simulations. Based on this output, the detected crack coordinates and orientation information are translated into the initial defect distribution of $RFPA^{3D}$, combined with boundary loading conditions to conduct quasi-static evolution simulations. Specifically, the orientation angle θ is used to initialize the crack growth direction, while the stage labels are mapped to the damage variable D, controlling the rate of crack evolution across different loading steps.

Within the modeling framework, RFPA^{3D} [35] identifies tensile or shear damage of material elements under three-dimensional stress states and quantifies crack development through the damage variable D, as shown in Figure 8. By parameterizing the loading ratio ($\lambda = \Delta y/\Delta x$) and overlay thickness (t), the model extrapolates the transition from micro-scale damage propagation to macro-scale crack patterns—ranging from parallel cracks to ladder-like networks and polygonal crack meshes. This achieves a unified modeling objective of "physics-driven mechanisms coupled with structural response regulation."

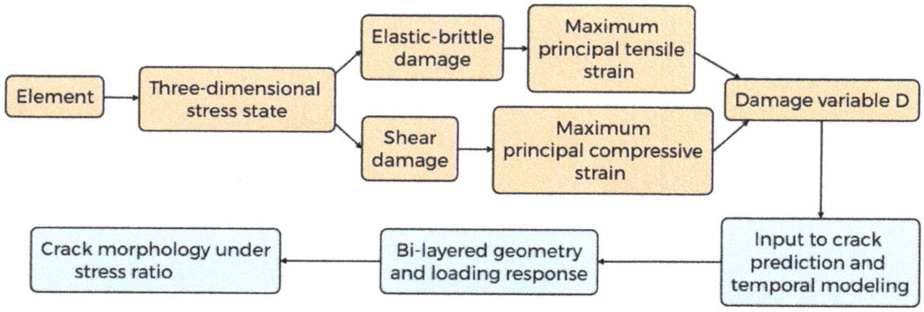

Figure 8. Crack evolution modeling process based on three-dimensional damage mechanism (RFPA^{3D}) and double-layer structure model. The figure depicts the simulation workflow used to reproduce four stages of crack evolution. The bi-layer slab model includes a brittle overlay and a ductile rammed-earth substrate, with loading ratio (λ) and thickness (t) as key parameters. Stress distribution and crack propagation pathways are shown step-by-step, illustrating how RFPA3D captures micro-to-macro transitions consistent with field observations.

RFPA^{3D} adopts a finite element method based on elastic damage theory to simulate the progressive failure of quasi-brittle materials such as rock, rammed-earth walls, and ceramic fragments under external disturbances. The essence of the crack evolution process lies in the heterogeneous development of element-level damage. To reflect the inherent heterogeneity of Tulou materials, RFPA^{3D} assigns a strength threshold to each mesoscopic element according to a Weibull probability distribution function, as shown in Equation (6):

$$\varphi(u) = \frac{m}{u_0}\left(\frac{u}{u_0}\right)^{m-1} exp\left[-\left(\frac{u}{u_0}\right)^{m}\right] \tag{6}$$

u_0 is the average mechanical parameter of the element (such as strength, Young's modulus, etc.). m is the material homogeneity index, where $m \rightarrow \infty$ indicates complete homogeneity and m $\rightarrow$ 0 indicates high heterogeneity. $\varphi(u)$ is the failure probability density function of the element. RFPA^{3D}, based on the elastic-brittle damage theory, defines the following elastic modulus degradation model for each element, as shown in Equation (7):

$$E = (1 - D)E_0 \tag{7}$$

E_0 is the original elastic modulus. D is the damage variable (D = 0 indicates no damage, D = 1 indicates complete failure). When $D \rightarrow 1$, to avoid solution interruption, the system sets a minimum value of E $\geq$ 5–10.

When the tensile strain reaches the threshold ε_{t0}, the element begins to damage. The damage evolution expression is shown in Equation (8):

$$D = \begin{cases} 0,\varepsilon > \varepsilon_{tu} \\ 1 - \frac{\sigma_{tr}}{E_{0^{\varepsilon}}}\varepsilon_{tu} < \varepsilon < \varepsilon_{t0} \\ 1,\varepsilon \leq \varepsilon_{tu} \end{cases} \quad (8)$$

σ_{tr} is the residual tensile stress. ε_{t0} and ε_{tu} are the damage initiation strain and ultimate strain, respectively. For failure in shear and compression modes, the Mohr-Coulomb criterion is used, with a threshold of the maximum compressive strain ε_c set to control compressive-shear failure.

To simulate the crack evolution mechanism of the surface layer of Tulou under various loading conditions, this paper comprehensively considers the two dominant failure mechanisms, tensile failure, and compressive-shear failure, in the RFPA^{3D} three-dimensional finite element system based on the elastic-brittle damage constitutive theory. Under one-dimensional uniaxial tension, the evolution of the damage variable D is defined as above. When the crack propagates into a three-dimensional stress state, the constitutive relationship must introduce an equivalent principal tensile strain to replace the uniaxial tensile strain, as shown in Equation (9):

$$\varepsilon_{eq} = \sqrt{\langle\varepsilon_1\rangle^2 + \langle\varepsilon_2\rangle^2 + \langle\varepsilon_3\rangle^2}\ with\ \langle\varepsilon_i\rangle = \max(0, -\varepsilon_i) \quad (9)$$

ε_1, ε_2, and ε_3 are the principal strains. This approach allows damage evolution to preserve the tensile-dominated failure characteristics of heterogeneous materials.

To reflect the crack initiation and propagation behavior of an element under compression and shear stress, this paper introduces the Mohr-Coulomb criterion as a second damage threshold. If the element is not damaged in tension, the following expression is used to determine whether shear damage has occurred, as shown in Equation (10):

$$D = \begin{cases} 0,\varepsilon < \varepsilon_{c0} \\ 1 - \frac{\sigma_{cr}}{E_{0^{\varepsilon}}}\varepsilon \geq \varepsilon_{c0} \end{cases} \quad (10)$$

where ε is the three-dimensional principal compressive strain. ε_{c0} is the ultimate compressive principal strain of the material under uniaxial compressive stress. σ_{tr} is the residual strength; and E_0 is the initial Young's modulus.

For the Tulou surface material, considering that the surface plaster and rammed earth matrix are typical quasi-brittle materials, their compression and shear behavior will affect the crack morphology and propagation direction. Therefore, ε_{c0} is calculated based on the following Equation (11) [36]:

$$\varepsilon_{c0} = \frac{\sigma_c + \left(\frac{1+sin\varnothing}{1-sin\varnothing}\right)(\sigma_3 - \mu(\sigma_2 + \sigma_3))}{E_0} \quad (11)$$

σ_c is uniaxial compressive strength of the material. σ_2, σ_3 are the secondary principal stress and minimum principal stress. $\varnothing$ is Angle of internal friction. μ is Poisson's ratio.

The proposed model not only reproduces the four-stage crack evolution process (initiation–propagation–coalescence–saturation) under various stress states but also allows physical labels to be embedded into the crack prediction model (e.g., YOLOv12), thereby enhancing structural interpretability and trend-fitting accuracy. To further simulate the cracking behavior of Tulou walls, a flat double-layer model was adopted following

Helgeson and Aydin [37], representing in-plane tensile stresses on the surface layer of the structure. In this model, the upper layer corresponds to plaster or decorative coatings, which are more brittle and prone to cracking, while the lower layer represents the rammed-earth substrate that provides structural support and constraint.

To examine the influence of overlay thickness (t) on crack spacing and propagation, four thickness settings were used: t = {5, 10, 15, 30 mm}, with substrate thicknesses T = {15, 20, 25, 30 mm}, following standard ratios in layered material fracture studies. Each model was discretized into mesoscopic elements of $1 \times 1 \times 1$ mm^3. For example, the case of t = 5 mm and T = 15 mm consisted of a $160 \times 160 \times 20$ mm^3 slab subdivided into eight-node hexahedral elements, as shown in Figure 9.

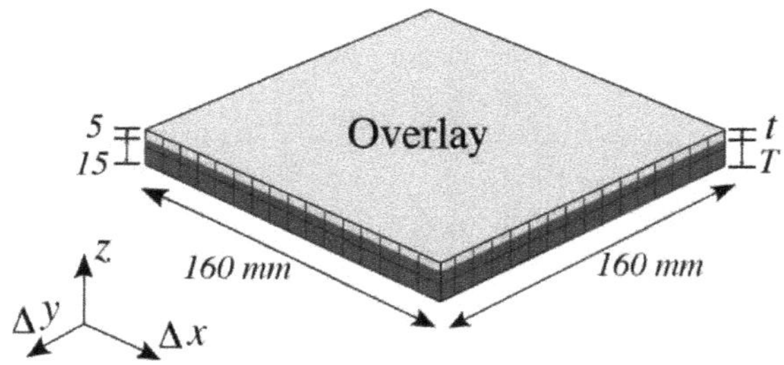

Figure 9. Dimensions and boundary conditions for the double-layer plate model.

To simulate environmental stresses such as thermal expansion–contraction, drying shrinkage, and structural displacement, quasi-static biaxial tensile loads were applied. Displacement increments were imposed in both x and y directions, generating in-plane tensile stresses. The loading ratio λ was defined as the ratio of orthogonal displacement increments, as shown in Equation (12):

$$\lambda = \frac{\Delta y}{\Delta x} \tag{12}$$

with Δx fixed at 0.004% and Δy set to 0, 0.0008%, 0.002%, and 0.004%, corresponding to λ = 0, 0.2, 0.5, and 1.0. λ = 0 represents uniaxial tension, while λ = 1 approximates isotropic biaxial tension, which reproduces the polygonal crack patterns commonly observed on Tulou surfaces under natural weathering [38].

To further examine the influence of geometric proportions, the ratio $\eta = t/T$ was introduced, where t is overlay thickness and T is substrate thickness. Four sets of slab models were constructed with t = {5, 10, 15, 30 mm} and T = {15, 30, 45, 90 mm}, maintaining $\eta \approx 0.33$ as a standard layered ratio. The η parameter reflects the mechanical dominance of the cracking region; higher η values imply stronger deformation capacity of the overlay and greater potential for pattern transitions and saturation. This model thus captures the transition of crack networks from parallel to ladder-like and eventually polygonal configurations under asymmetric thickness and biaxial loading, providing a meso-scale physical foundation for coupling with deep learning-based crack detection and evolution prediction.

The material properties of the double-layer Tulou wall model are summarized in Table 3. The overlay represents plaster or coating layers with high material heterogeneity, lower compressive strength, and brittle characteristics, making them more susceptible to cracking under minor perturbations. The substrate simulates the rammed-earth base with higher strength, better uniformity, and greater ductility, providing effective constraint and support. During loading, cracks predominantly initiate and develop within the overlay. The homogeneity index m characterizes material variability, with smaller values indicating greater heterogeneity. To better approximate real Tulou conditions, the overlay Young's modulus was set higher than that of the substrate, ensuring that cracks preferentially nucleate in the overlay under the same applied stress, while the substrate—possessing significantly higher compressive strength—remains intact to provide structural resistance.

Table 3. Tulou crack model material parameters.

Parameter	Overlay	Substrate (Rammed-Earth Base)
Material homogeneity index (m)	3–8	10–30
Young's modulus (E, MPa)	1000–3000	300–1200
Compressive strength (MPa)	1–5	2–4
Poisson's ratio	0.2–0.25	0.25–0.30
Compression-to-tension ratio	8–12	10–12
Friction angle (°)	28–32	30–36

2.6. Quantitative Validation and Cross-Check

To overcome the subjectivity of relying solely on visual similarity and expert interpretation, a three-tier quantitative evaluation system was established, comprising: (i) morphological consistency (pixel/skeleton level), (ii) geometric statistical consistency (spacing/orientation), and (iii) topological consistency (loops/connectivity). The evaluation was conducted on binary crack maps derived from both simulated predictions and field observations (or subsequent-stage images).

Specifically, for the same Tulou wall surface or regions under comparable conditions, pairs of "early-stage" images (I_t) and "late-stage" images ($I_{t+\Delta}$) were selected as the validation set. The detection features (spacing, density, and orientation) extracted from early-stage cracks were used to drive the RFPA^{3D} simulation, generating predicted binary crack maps $\hat{B}_{t+\Delta}$. The crack binary image $\hat{B}_{t+\Delta}$ obtained by $I_{t+\Delta}$ detection is used as the "quasi-true value". To reduce the influence of segmentation thickness, the binary image is morphologically skeletonized S(·).

Tier 1: Morphological consistency (pixel/skeleton level): The skeleton Intersection-over-Union (IoU) was employed, as defined in Equation (13):

$$IoU_{skeleton} = \frac{|S(\hat{B}) \cap S(B)|}{|S(\hat{B}) \cup S(B)|} \tag{13}$$

$S(\hat{B})$ is the set of pixels in the predicted crack skeleton. $S(B)$ is the set of pixels in the actual crack skeleton. This is an "overlap" metric, like the intersection-over-union ratio of two shapes, but only for the crack skeleton. The value range is from 0 to 1, with values closer to 1 indicating a closer match between the predicted and actual skeletons.

Tier 2: Bidirectional Chamfer distance, as shown in Equation (14):

$$d_{ch}(P,Q) = \frac{1}{|P|}\sum_{p\in P} \min_{q \in Q} \|p-q\|_2 + \frac{1}{|Q|}\sum_{q\in Q} \min_{p \in P} \|p-q\|_2 \tag{14}$$

$P = S(\hat{B})$ is the predicted skeleton point set, and $Q = S(B)$ is the predicted true skeleton point set. For each predicted point p, find the nearest true point q, calculate the distance, and then average the distance across all points. This bidirectional calculation ensures that both the predicted and true points can find each other.

Tier 3: Hausdorff Distance, as shown in Equation (15):

$$d_H(P,Q) = max\left\{ \sup_{p \in P} \inf_{q \in Q} \|p-q\|, \sup_{q \in Q} \inf_{p \in P} \|p-q\| \right. \tag{15}$$

The Hausdorff distance captures the worst-case error, defined as the maximum distance from any point on the predicted skeleton to the closest point on the ground-truth skeleton, and vice versa. Full implementation details and parameter settings are provided in the Appendix A.

In summary, the three morphological metrics serve complementary purposes: (i) IoU evaluates global overlap, answering whether "the predicted cracks align with the ground truth". (ii) Chamfer distance measures average error, assessing "how precisely the aligned cracks match". (iii) Hausdorff distance quantifies extreme deviations, indicating "how large the worst-case error is." Together, these measures provide a comprehensive evaluation of crack evolution predictions, capturing global consistency, average accuracy, and worst-case performance.

Next, this study designed two validation protocols. Protocol P1 (real-time post-test control): The metrics were run for each plot (I_t, $I_{t+\Delta}$), reporting the mean $\pm$ standard deviation with a 95% CI. These metrics were compared to a baseline of uncalibrated physical parameters (e.g., fixed $\lambda = 1$), with paired t-tests and significance stars. Protocol P2 (synthetic-measured hybrid): A batch of synthetic "post-tests" was generated in RFPA^{3D} and parameters (λ, t, m) were perturbed to test the metrics' sensitivity to parameter deviations and generate robustness curves. Parameter inversion and calibration: Using $EMD_\theta + a{\cdot}MAE_{spacing}$ as the objective function, grid search and Bayesian optimization of λ and t were performed to minimize the difference. The improvement and cost of the metrics before and after calibration were reported.

All results are reported as mean $\pm$ standard deviation with 95% CI. The Shapiro–Wilk test was used for normality assessment. Depending on distribution characteristics, paired *t*-tests or Wilcoxon signed-rank tests were applied. Effect sizes (Cohen's d) were further reported. All evaluation steps—including image binarization, skeletonization, orientation estimation, distance transforms, and topological analysis—were implemented in Python (v3.10.12) using OpenCV, scikit-image, and SciPy. Thresholds and algorithmic parameters are detailed in the Table A1 to ensure reproducibility.

3. Results

3.1. Ablation Experiments

An ablation study was conducted to systematically evaluate the contributions of the MobileNetV4 backbone (Mv4), the improved multi-head cross-attention module (IMCA), and the EnEIoU to overall detection performance. The baseline model achieved an F1 score of 83.2%, precision (P) of 85.6%, *recall* (R) of 80.1%, and *mAP@50* of 84.7%, serving as a performance reference for subsequent improvements, as shown in Table 2.

The best performance was obtained when all three modules (Mv4 + IMCA + EnEIoU) were combined, yielding an F1 score of 91.8%, precision of 91.5%, *recall* of 92.0%, and *mAP@50* of 95.5%. In addition to significantly surpassing the baseline in detection accuracy, this configuration accurately captured the full crack evolution pathway (initiation–propagation–coalescence–saturation). Importantly, these gains were achieved while maintaining computational efficiency, requiring only 4.68 GFLOPs and 2.21 MB of parameters, highlighting the lightweight and deployable nature of the framework. This makes it particularly suitable for real-time crack detection on edge devices deployed in heritage conservation sites.

For individual modules, introducing only the MobileNetV4 backbone improved performance to an F1 score of 89.1% and *mAP@50* of 93.7%, while reducing computational cost from 5.99 GFLOPs to 4.68 GFLOPs, confirming the backbone's role in balancing efficiency and accuracy. The EnEIoU loss alone also led to a notable improvement (F1 = 89.0%, *mAP@50* = 94.0%), whereas IMCA contributed more modestly (F1 = 87.4%). When IMCA and EnEIoU were combined without Mv4, performance was relatively lower (F1 = 88.2%, *mAP@50* = 93.9%) and computational complexity was the highest (5.99 GFLOPs, 2.52 MB). These findings demonstrate that while each module enhances detection accuracy to some extent, their synergistic integration is most effective, and that the lightweight MobileNetV4 backbone plays a critical role in strengthening feature representation while optimizing computational efficiency.

3.2. Model Comparison Results

As shown in Table 4, the models exhibited substantial differences in performance on the tasks of Tulou surface defect detection and crack prediction. Overall, YOLOv12-MLE (YOLO for Multi-scale Long-crack Extraction) achieved the best results across all metrics, with an F1 score of 91.8% and *mAP@50* of 95.5%. It maintained an excellent balance between precision and *recall*, while requiring only 4.68 GFLOPs and 2.21 MB of parameters, demonstrating outstanding lightweight efficiency. These characteristics make it particularly well-suited for deployment on edge devices or in real-time heritage monitoring scenarios.

Table 4. Model comparison results.

Model	F1 Score	P (%)	*R* (%)	*mAP@50* (%)	GFLOPs	Params (MB)
DETR	85.5	88.7	82.6	91.5	5.99	2.52
YOLOv5n	89.1	93.7	85.0	93.7	4.68	2.21
YOLOv6n	87.4	90.9	84.1	92.8	5.99	2.52
YOLOv9n	89.0	92.2	86.0	94.0	5.99	2.52
YOLOv10n	90.9	94.2	87.8	95.3	4.68	2.21
YOLOv11n	90.7	92.9	88.6	95.1	4.68	2.21
YOLOv12n	88.2	87.5	88.9	93.9	5.99	2.52
YOLO-MLE	91.8	91.5	92.0	95.5	4.68	2.21

By contrast, DETR achieved competitive accuracy (F1 = 85.2%, *mAP@50* = 89.9%) but demanded 54.12 GFLOPs and 19.01 MB, limiting its applicability in resource-constrained or low-latency conservation contexts. Within the YOLO family, YOLOv12n also performed strongly, offering a favorable trade-off between accuracy and efficiency. YOLOv11n and YOLOv5n obtained F1 scores of 83.3% and 84.0%, with *mAP@50* values of 89.5% and 90.0%, respectively, making them viable for real-time crack detection. In contrast, YOLOv10n and YOLOv9n showed weaker performance (F1 = 74.5% and 78.0%; *mAP@50* = 81.0% and 84.4%), and despite lower computational costs (8.41 and 7.86 GFLOPs), they lack sufficient accuracy for demanding Tulou crack prediction tasks. YOLOv6n delivered moderate results (F1 = 80.5%, *mAP@50* = 87.2%) but required more computational resources than YOLOv5n and YOLOv12n, reducing its efficiency advantage.

In summary, while some YOLO models outperform larger detectors such as DETR in terms of lightweight efficiency and detection accuracy, the improved YOLOv12-MLE demonstrated the most balanced performance. Its high accuracy, low computational cost, and compact size make it particularly suitable for on-site real-time protection and long-term monitoring of Fujian Tulou.

3.3. Performance Evaluation of CAM, Grad-CAM, XGrad-CAM, SSCAM for Deep Learning Model

In practical deployment, the proposed system operates by capturing façade images of Tulou walls through fixed cameras which are then processed in real time by the YOLOv12-MLE detection model to identify cracks and by the RFPA^{3D} module to simulate their likely evolution. This workflow provides conservation teams with rapid diagnostic outputs directly on site. As illustrated in Figure 10, four mainstream class activation mapping techniques (CAM, Grad-CAM, SSCAM, and XGrad-CAM) were employed to enhance model interpretability in the Fujian Tulou crack detection task. These methods were used to compare the spatial focus and discriminative regions of the baseline YOLOv12n with the improved YOLOv12-MLE. Results show that YOLOv12-MLE consistently exhibited stronger structural focus and boundary sensitivity across all visualization methods, enabling more accurate localization of crack-related regions. Compared with YOLOv12n, YOLOv12-MLE produced more concentrated attention responses and significantly suppressed background texture interference, particularly in regions with wall cavities and surface roughness. This demonstrates the model's ability to effectively distinguish cracks from non-crack features, thereby improving robustness under the highly heterogeneous conditions of earthen wall surfaces.

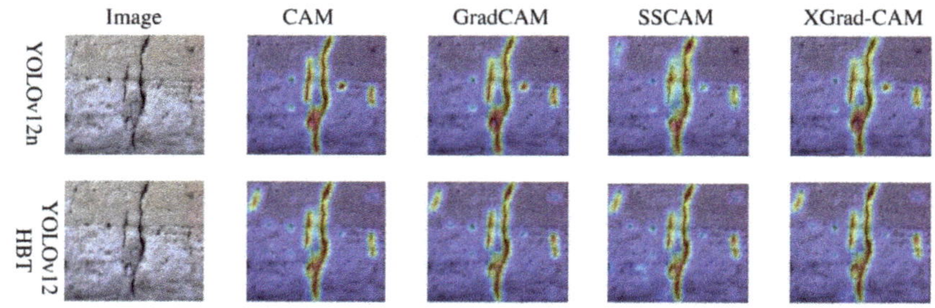

Figure 10. Visualization of attention maps for YOLOv12n and YOLOv12-MLE using CAM, GradCAM, SSCAM, and XGrad-CAM. In the attention maps, red and yellow regions indicate areas of high model attention corresponding to crack features, while blue regions represent low attention or background areas.

Closer examination of Grad-CAM and SSCAM results further indicated that the high-response areas of YOLOv12-MLE aligned closely with main crack trunks and branching fissures, whereas YOLOv12n displayed dispersed attention and false hotspots. These findings confirm that the improved model not only enhances semantic focus on crack structures but also achieves greater sensitivity in capturing boundary features, providing transparent and traceable visual evidence for crack detection and evolution prediction.

Despite these advantages, certain limitations were observed in complex scenarios. As shown in Figure 11, the input image depicts an outer Tulou wall with multiple intersecting cracks accompanied by windows, material heterogeneity, and illumination reflections. Under such conditions, YOLOv12-MLE attention maps generated by CAM, XGrad-CAM, Grad-CAM, and SSCAM were primarily concentrated on large, salient cracks and structural openings (e.g., vertical crack trunks and window edges), while failing to adequately activate fine branching cracks or subtle micro-fissures at higher wall elevations. Particularly in Grad-CAM, attention hotspots were overly concentrated and neglected peripheral micro-cracks that often represent early indicators of material deterioration.

Notably, strong activations were also observed around window frames and small openings, suggesting a tendency toward misclassification. Since Tulou walls frequently contain small windows, ventilation holes, or repair marks, which share visual similarities with cracks in terms of vertical, high-contrast edges, the model occasionally misidentified these features as defects. This highlights a limitation in distinguishing high-contrast non-crack features from true crack signals.

Overall, these observations indicate that the model's semantic focus mechanism tends to prioritize high-contrast, large-scale, or opening-adjacent regions, while devoting less attention to low-contrast, dispersed, or peripheral crack signals. In other words, the spatial sensitivity and feature separation capacity of the current model diminish under noisy or occluded conditions. This suggests a potential blind spot in practical Tulou conservation: although YOLOv12-MLE performs reliably on prominent cracks, it remains vulnerable to incomplete focus and interpretability gaps when dealing with complex wall environments or early-stage micro-cracks.

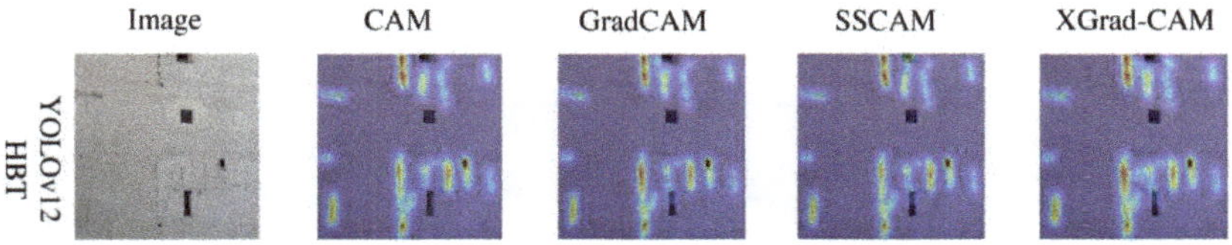

Figure 11. Visualization of YOLOv12-MLE revealing attention limitations under occlusion and ambiguous postures using CAM, GradCAM, SSCAM, and XGrad-CAM. In the attention maps, red and yellow regions indicate areas of high model attention, while blue regions represent low attention or background areas.

In addition to these interpretability findings, it is important to consider how the system would operate under real field conditions. Data acquisition in Fujian Tulou environments is often affected by variable illumination, surface moisture, vegetation occlusion, and restricted accessibility of higher wall sections. While YOLOv12-MLE showed strong robustness for large and salient cracks, its performance decreased for microcracks and shaded or partially occluded regions, indicating the necessity of standardized acquisition protocols and potentially multi-modal sensing (e.g., UAV-based photogrammetry, multi-spectral or thermal imaging) to ensure reliable input quality. Furthermore, maintaining model performance in practice is challenging because crack morphology evolves with seasonal weathering and conservation interventions. This requires periodic dataset updates and retraining to prevent performance degradation, as well as lightweight strategies such as model compression and distillation to support long-term operation on edge devices. Addressing these practical challenges will be essential for ensuring sustainable monitoring and decision support in heritage conservation workflows.

3.4. Crack Simulation Results

In RFPA^{3D}, quasi-static biaxial tensile loading was applied to the double-layer slab model, with Δx fixed at 0.004% and Δy set to 0, 0.0008%, 0.002%, and 0.004%, corresponding to four typical loading ratios: λ = 0, 0.2, 0.5, and 1.0. To investigate the influence of overlay thickness on crack density, the overlay thickness was varied as t = {5,10,15,30} mm, while the substrate thickness was fixed as T = {15,20,25,30} mm. As summarized in Table 5, crack morphology was strongly controlled by the loading ratio λ. At λ = 0, uniaxial strain generated parallel cracks, consistent with the strip-shaped cracking patterns often observed in Tulou walls under unidirectional stress or temperature gradients. When λ increased to 0.2–0.5, cracks evolved into ladder-like or intersecting networks, closely resembling field observations of intersecting cracks caused by uneven settlement or biaxial loading of outer walls. At λ = 1.0, representing isotropic biaxial tension, the morphology further developed into closed polygonal crack networks, consistent with the surface cracking observed on Tulou facades under long-term thermal expansion–contraction or multi-axial stress constraints.

Table 5. Different fracture patterns of models with varying overlay thicknesses (5, 10, 15, and 20 mm) under different loading ratios (λ = 0, 0.2, 0.5, 1). The 3D schematic blocks illustrate simulated crack morphologies: darker/red regions correspond to more severe fracture development, while lighter and yellow regions indicate relatively intact areas.

Overlay Thickness t/mm	0	0.2	0.5	1
5				
10				
15				
20				

The simulations also revealed the effect of overlay thickness on crack density. In thinner overlays (e.g., t = 5 mm, T = 15 mm), cracks were significantly more numerous, forming dense polygonal networks. In contrast, in thicker overlays (e.g., t = 30 mm, T = 30 mm), the number of cracks was greatly reduced, although the critical λ-dependent transitions in crack morphology remained unchanged. This finding aligns with in situ observations of Fujian Tulou: plaster layers on facades (typically 5–10 mm thick) tend to generate dense crack networks, whereas thicker rammed-earth substrates (exceeding 30 mm) show fewer cracks, though their morphology is still primarily governed by the loading ratio λ. The strong consistency between simulations and field observations reinforces the reliability of the proposed dual-path framework combining YOLOv12-MLE detection and RFPA^{3D} simulation for predicting crack evolution in earthen heritage structures.

On this basis, to further elucidate the evolution of cracks from microscopic defects to macroscopic patterns, Figure 12 presents the progressive four-stage failure process and maximum principal stress (σ_1) distribution for a representative case with λ = 1, overlay thickness t = 5 mm, and substrate thickness T = 15 mm.

In the initial stage (Step 3), the inherent heterogeneity of Tulou wall materials produced uneven stress distributions, with local weak zones experiencing elevated tensile stresses, serving as potential crack nucleation sites. By Step 8–12, micro-cracks gradually emerged and extended along the direction of maximum principal tensile stress, forming skeleton-like crack patterns like the orthogonal fissures observed on Tulou facades. During the propagation stage (Step 16–22), crack density increased, and coalescence became evident: on the one hand, secondary cracks merged into primary ones at near-right angles, leading to localized block segmentation; on the other hand, sequential connections between parallel cracks promoted the development of honeycomb-like or polygonal crack networks, resembling typical field patterns on Tulou walls. Finally, in the saturation stage (Step 30), cracks penetrated the interface between layers, inducing overlay detachment and forming isolated "island" units. Stress was progressively released, marking the transition into the saturation phase.

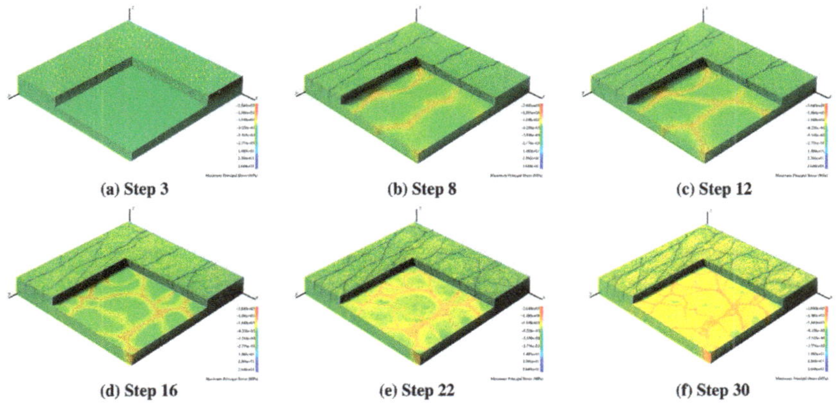

Figure 12. Stress distribution and crack evolution at different simulation steps in the bi-layered slab model under biaxial loading (λ = 1, t = 5 mm, T = 15 mm). Subfigures (**a**–**f**) correspond to Step 3, 8, 12, 16, 22, and 30, respectively, illustrating the progressive formation and propagation of cracks. The color bar represents the maximum principal stress (MPa), where values are shown in scientific notation (e.g., −2.640e+03). Red indicates regions of high tensile stress, while blue indicates compressive or low-stress areas.

This simulation not only reveals the progressive evolution of cracks, from stress concentration to interfacial delamination, but also provides a mechanistic explanation of defect formation in Tulou walls. The strong agreement between simulated results and observed crack morphologies validates the physical fidelity of the proposed framework.

3.5. Quantitative Results of Evolutionary Predictions

Quantitative validation of the proposed dual-path framework (YOLOv12-MLE + RFPA^{3D}) was conducted on 20 paired "early–late" Tulou wall crack samples. Results demonstrated stable performance across three evaluation levels: global morphological consistency, geometric statistical consistency, and topological agreement.

As shown in Table 6, calibrated predictions achieved a skeleton IoU of 0.80 $\pm$ 0.05 (baseline is 0.72 $\pm$ 0.06), significantly improving overall crack skeleton overlap. The Chamfer distance decreased from 5.9 px to 3.7 px, while the Hausdorff distance was reduced from 15.2 px to 9.8 px, indicating improvements in both mean and extreme deviations. Geometric metrics also showed consistent gains: spacing MAE was controlled at 0.95 mm, orientation EMD decreased to 0.09, and density error was reduced to 3.2%. These results confirm that the framework effectively reproduces the spatial distribution and geometric features of Tulou cracks.

In the parameter sensitivity analysis, predictive performance declined sharply when overlay thickness t and loading ratio λ deviated from true conditions, but calibration effectively restored accuracy (Table 7). For example, at λ = 0.5, the skeleton IoU dropped to 0.69; after calibration ($\lambda \approx 1.1$, $t \approx 1.0$ cm), it recovered to 0.79.

Overall, these findings demonstrate that the proposed framework not only reproduces overall crack morphology but also quantitatively aligns spatial distribution, orientation features, and topological structures. Parameter calibration further confirmed the intrinsic coupling between the model and physical constraints, highlighting the framework's interpretability and potential for generalization.

Table 6. Performance metrics for paired Tulou crack samples (mean ± SD).

Metric	Baseline Model	Calibrated Model	Improvement
Skeleton IoU	0.72 ± 0.06	0.80 ± 0.05	+11.1%
Chamfer distance (px)	5.9 ± 1.1	3.7 ± 0.8	−37.3%
Hausdorff distance (px)	15.2 ± 3.6	9.8 ± 2.4	−35.5%
Spacing MAE (mm)	1.6 ± 0.5	0.95 ± 0.3	−40.6%
Orientation EMD	0.15 ± 0.04	0.09 ± 0.02	−40.0%
Density error (%)	7.5 ± 2.8%	3.2 ± 1.4%	−57.3%

Table 7. Parameter sensitivity and calibration effects.

Parameter Setting	Skeleton IoU	Chamfer Distance (px)	Hausdorff Distance (px)	Orientation EMD
$\lambda = 0.5$	0.69	7.0	16.1	0.16
$\lambda = 1.0$ (baseline)	0.72	5.9	15.2	0.15
$\lambda = 1.5$	0.74	5.5	13.9	0.14
Calibrated ($\lambda \approx 1.1$, $t \approx 1.0$ cm)	0.79	3.8	9.5	0.09

4. Discussion

This study proposes a dual-path framework tailored for Fujian Tulou conservation: (i) lightweight crack detection based on YOLOv12-MLE and (ii) independent physical simulation using RFPA^{3D}. The two paths are designed to complement each other, with the shared goal of providing reliable detection outcomes and mechanistic insights for heritage protection.

In the detection path, ablation and comparative experiments demonstrated that YOLOv12-MLE, through the integration of the MobileNetV4 backbone, the improved multi-head cross-attention module (IMCA), and the EnEIoU loss, achieved superior detection accuracy and robustness under heterogeneous wall textures while maintaining a compact and edge-deployable architecture. Class activation mapping further confirmed the model's spatial focus, showing enhanced crack localization and boundary sensitivity under complex backgrounds. Nevertheless, limitations remain: the model exhibited insufficient activation for fine-scale cracks and occasional misclassification of small openings or windows as cracks. Thus, interpretability analysis both validated the effectiveness of the improvements and revealed directions for future optimization.

In the simulation path, RFPA^{3D} reproduced crack evolution patterns strongly governed by the loading ratio (λ), while crack density was primarily influenced by overlay thickness (t). These findings closely match field observations of strip-shaped, ladder-like, and polygonal cracks on Tulou facades, indicating that the combined detection–simulation strategy provides a mechanistic explanatory framework and physically validates observed crack generation and propagation.

From a conservation perspective, the proposed framework offers three major advantages: (i) structured crack archives that facilitate cross-temporal comparison and rapid reinspection; (ii) mechanistic evidence for differentiated effects of λ and t, informing thickness control, localized reinforcement, and boundary-zone interventions; and (iii) a lightweight and interpretable detection model deployable on mobile or unattended devices, meeting real-time and sustainable monitoring requirements in heritage sites. Compared with previous studies that focused primarily on either detection accuracy or HBIM integration, our approach establishes a detection–simulation loop that not only identifies cracks but also explains their physical evolution. This novelty extends AI applications from isolated recognition tasks to interpretable, mechanism-aware, and decision-oriented conservation tools.

Quantitative evaluation further confirmed consistency at multiple levels. Skeleton IoU, Chamfer, and Hausdorff distances demonstrated strong agreement with observed crack morphologies; orientation EMD and spacing MAE validated alignment of directional and spacing statistics; and topological measures (e.g., Betti number $\beta 1$ and loop rate) revealed consistent trends in connectivity. It is emphasized that these assessments represent cross-sectional consistency diagnostics rather than strict temporal prediction. Since current paired samples approximate temporal evolution rather than deriving from long-term monitoring, future work should integrate multi-epoch fixed-point imaging and thermo–hydro–mechanical coupled field measurements, as well as explicit parameter-mapping protocols, to test parameter identifiability and move toward quantifiable prediction.

5. Conclusions

This study introduced a dual-path framework for the conservation of Fujian Tulou, integrating lightweight crack detection with YOLOv12-MLE and physics-based simulation with RFPA^{3D}. Compared with previous studies that have focused mainly on either detection performance or digital model integration, our framework establishes a detection–simulation loop that both identifies cracks and explains their underlying physical mechanisms. This novelty extends the role of AI in heritage protection from isolated recognition tasks to an integrated, mechanism-aware, and decision-support system.

Despite promising results, limitations remain. RFPA^{3D} simulations rely on an idealized double-layer slab assumption and neglect the multilayered composite structure, pore heterogeneity, and environmental degradation processes of Tulou walls, potentially leading to bias under extreme climate conditions. Future studies should incorporate high-resolution 3D modeling, microstructural characterization, and multi-physics coupling to improve physical accuracy. Moreover, this study focused mainly on geometric crack features (direction, spacing, density) while omitting factors such as moisture, salt migration, and repair traces. Multi-modal data sources (e.g., hyperspectral imaging, thermal infrared, and 3D scanning) could be integrated with deep learning to construct a more comprehensive predictive framework. Finally, although YOLOv12-MLE achieved lightweight performance, further optimization is required for ultra-low power and extreme environments. Techniques such as model distillation, quantization, and neural architecture search (NAS) may facilitate efficient deployment under highly constrained computational resources.

Supplementary Materials: The following supporting information can be downloaded at website: https://www.mdpi.com/article/10.3390/coatings15101156/s1.

Author Contributions: Conceptualization, Y.H. and S.C. (Shaokang Chen); methodology, Y.H.; software, S.C. (Shaokang Chen); validation, Y.H., S.C. (Si Cheng) and S.C. (Shaokang Chen); formal analysis, S.C. (Si Cheng); investigation, Y.H.; resources, Z.Z.; data curation, S.C. (Si Cheng); writing—original draft preparation, Y.H.; writing—review and editing, S.C. (Shaokang Chen) and Z.Z.; visualization, S.C. (Shaokang Chen); supervision, Z.Z.; project administration, Z.Z. All authors have read and agreed to the published version of the manuscript.

Funding: This research received no external funding.

Institutional Review Board Statement: Not applicable.

Informed Consent Statement: Not applicable.

Data Availability Statement: The data supporting the findings of this study are openly available in Figshare at https://doi.org/10.6084/m9.figshare.30104506 (accessed on 20 September 2025).

Conflicts of Interest: Authors Zhuang Zhao is employed by the company Sydney Water Corporation. The remaining authors declare that the research was conducted in the absence of any commercial or financial relationships that could be construed as potential conflicts of interest.

Abbreviations

The following abbreviations are used in this manuscript:

YOLO	You Only Look Once
SSD	Single Shot MultiBox Detector
DETR	DEtection TRansformer
$RFPA^{3D}$	Realistic Failure Process Analysis in 3D
CAM	Class Activation Mapping
Grad-CAM	Gradient-weighted Class Activation Mapping
SSCAM	Score-Weighted Class Activation Mapping
XGrad-CAM	Extended Gradient-weighted Class Activation Mapping
IoU	Intersection over Union
mAP@50	Mean Average Precision at 50% IoU threshold
GFLOPs	Giga Floating Point Operations per Second
UIB	Universal Inverted Bottleneck
IMCA	Improved Multi-head Cross Attention
EnEIoU	Enhanced Efficient Intersection over Union
LSTM	Long Short-Term Memory
SaturatedLSTMCA	Saturated Long Short-Term Memory Convolutional Attention
MAE	Mean Absolute Error
EMD	Earth Mover's Distance

Appendix A

Table A1 summarizes the parameter settings used in the quantitative evaluation workflow for Fujian Tulou crack detection and evolution prediction. The workflow integrates image preprocessing, skeletonization, feature extraction, and quantitative metrics, ensuring reproducibility of results and consistency across validation protocols. Each step is listed with the corresponding algorithm, key parameters, and implementation notes, covering operations from binary conversion and skeleton IoU to Chamfer/Hausdorff distances and topological analysis (e.g., Betti numbers, loop ratio).

Table A1. Parameter settings for quantitative evaluation workflow.

Step	Operation	Algorithm/Function	Parameters	Notes
Binary conversion	Otsu thresholding	cv2.threshold (. . . , cv2.THRESH_OTSU)	N/A	Remove small regions <20 px
Skeletonization	Thinning	Zhang-Suen/cv2.ximgproc.thinning	mode = THINNING_ZHANGSUEN	Produces 1-pixel skeleton
Orientation estimation	Structure tensor	3×3 window	Histogram bins = 18 (10° per bin)	Orientation in $[0^\circ, 180^\circ)$
Spacing extraction	Euclidean distance transform	cv2.distanceTransform	Pixel to mm scaling factor = s	Ridge detection = local maxima
Density calculation	Skeleton length/area	np.count_nonzero (skeleton)	Normalize by area	Gives cracks per cm^2
Skeleton IoU	Intersection over union	Binary maps	N/A	Evaluates overlap
Chamfer distance	Symmetric Chamfer	cv2.distanceTransform + NN search	N/A	Average distance both ways
Hausdorff distance	Max distance	scipy.spatial.cKDTree	N/A	Worst-case mismatch
Betti numbers	Connected components and loops	scipy.ndimage.label + contour hierarchy	N/A	Gives β0\beta_0, β1\beta_1
Graph abstraction	Skeleton to graph	Nodes = end-points/branchpoints	N/A	Compute avg. degree, loop ratio

Table A1. *Cont.*

Step	Operation	Algorithm/Function	Parameters	Notes
Statistics	Significance tests	Paired *t*-test/Wilcoxon	$\alpha = 0.05$	With Shapiro–Wilk normality check
Binary conversion	Otsu thresholding	cv2.threshold (. . ., cv2.THRESH_OTSU)	N/A	Remove small regions <20 px
Skeletonization	Thinning	Zhang-Suen/cv2.ximgproc.thinning	mode = THIN-NING_ZHANGSUEN	Produces 1-pixel skeleton
Orientation estimation	Structure tensor	3×3 window	Histogram bins = 18 (10° per bin)	Orientation in [0°, 180°)
Spacing extraction	Euclidean distance transform	cv2.distanceTransform	Pixel to mm scaling factor = s	Ridge detection = local maxima
Density calculation	Skeleton length / area	np.count_nonzero (skeleton)	Normalize by area	Gives cracks per cm^2
Skeleton IoU	Intersection over union	Binary maps	N/A	Evaluates overlap
Chamfer distance	Symmetric Chamfer	cv2.distanceTransform + NN search	N/A	Average distance both ways
Hausdorff distance	Max distance	scipy.spatial.cKDTree	N/A	Worst-case mismatch
Betti numbers	Connected components and loops	scipy.ndimage.label + contour hierarchy	N/A	Gives β0\beta_0, β1\beta_1
Graph abstraction	Skeleton to graph	Nodes = end-points/branchpoints	N/A	Compute avg. degree, loop ratio
Statistics	Significance tests	Paired *t*-test/Wilcoxon	$\alpha = 0.05$	With Shapiro–Wilk normality check

References

1. Porretta, P.; Pallottino, E.; Colafranceschi, E. Minnan and Hakka Tulou: Functional, typological and construction features of the rammed earth dwellings of Fujian (China). *Int. J. Archit. Herit.* **2022**, *16*, 899–922. [CrossRef]
2. Chen, W.; Yan, B.; Guo, S.; Liu, Y.; Yang, F.; Zhang, K.; Mao, W. An in-situ conservation method of the rammed earth sites using a new silica protective agent. *Constr. Build. Mater.* **2024**, *452*, 138960. [CrossRef]
3. Zhang, Y.; Jiang, S.; Quan, D.; Fang, K.; Wang, B.; Ma, Z. Properties of sustainable earth construction materials: A state-of-the-art review. *Sustainability* **2024**, *16*, 670. [CrossRef]
4. Golewski, G.L. The phenomenon of cracking in cement concretes and reinforced concrete structures: Mechanisms, causes, types, and detection methods—A review. *Buildings* **2023**, *13*, 765. [CrossRef]
5. Azouz, Z.; Honarvar Shakibaei Asli, B.; Khan, M. Evolution of crack analysis in structures using image processing technique: A review. *Electronics* **2023**, *12*, 3862. [CrossRef]
6. Pala, G.K.; Kesana, N.S.; Gopalapurapu, K.S. Advancements in optimizing the quality assurance and crack detection by leveraging the application of emerging artificial intelligence trends for enrichment of civil infrastructure—A recapitulation. In *Recent Developments and Innovations in the Sustainable Production of Concrete*; Woodhead Publishing: Cambridge, UK, 2025; pp. 635–655.
7. Miao, P.; Srimahachota, T. Cost-effective system for detection and quantification of concrete surface cracks by combination of convolutional neural network and image processing techniques. *Constr. Build. Mater.* **2021**, *293*, 123549. [CrossRef]
8. Vijayan, V.; Joy, C.M.; Shailesh, S. A survey on surface crack detection in concretes using traditional, image processing, machine learning, and deep learning techniques. In Proceedings of the 2021 International Conference on Communication, Control and Information Sciences (ICCISc), Idukki, India, 16–18 June 2021; Volume 1, pp. 1–6.
9. Sun, Z.; Caetano, E.; Pereira, S.; Moutinho, C. Employing histogram of oriented gradient to enhance concrete crack detection performance with classification algorithm and Bayesian optimization. *Eng. Fail. Anal.* **2023**, *150*, 107351. [CrossRef]
10. Han, H.; Deng, H.; Dong, Q.; Gu, X.; Zhang, T.; Wang, Y. An advanced Otsu method integrated with edge detection and decision tree for crack detection in highway transportation infrastructure. *Adv. Mater. Sci. Eng.* **2021**, *2021*, 9205509. [CrossRef]
11. Wang, Y.; Zhang, J.Y.; Liu, J.X.; Zhang, Y.; Chen, Z.P.; Li, C.G.; Yan, R.B. Research on crack detection algorithm of the concrete bridge based on image processing. *Procedia Comput. Sci.* **2019**, *154*, 610–616. [CrossRef]
12. Xu, X.; Zhao, M.; Shi, P.; Ren, R.; He, X.; Wei, X.; Yang, H. Crack detection and comparison study based on faster R-CNN and mask R-CNN. *Sensors* **2022**, *22*, 1215. [CrossRef]
13. Yu, F.; Du, C.; Hua, A.; Jiang, M.; Wei, X.; Peng, T.; Hu, X. EnCaps: Clothing image classification based on enhanced capsule network. *Appl. Sci.* **2021**, *11*, 11024. [CrossRef]
14. Kim, H.J.; Lee, D.H.; Niaz, A.; Kim, C.Y.; Memon, A.A.; Choi, K.N. Multiple-clothing detection and fashion landmark estimation using a single-stage detector. *IEEE Access* **2021**, *9*, 11694–11704. [CrossRef]
15. Sohaib, M.; Arif, M.; Kim, J.M. Evaluating YOLO models for efficient crack detection in concrete structures using transfer learning. *Buildings* **2024**, *14*, 3928. [CrossRef]

16. Liu, Y.; Zhou, T.; Xu, J.; Hong, Y.; Pu, Q.; Wen, X. Rotating target detection method of concrete bridge crack based on YOLOv5. *Appl. Sci.* **2023**, *13*, 11118. [CrossRef]
17. Yin, Z.; Li, H.; Qi, B.; Shan, G. BBW YOLO: Intelligent detection algorithms for aluminium profile material surface defects. *Coatings* **2025**, *15*, 684. [CrossRef]
18. Karimi, N.; Mishra, M.; Lourenço, P.B. Automated surface crack detection in historical constructions with various materials using deep learning-based YOLO network. *Int. J. Archit. Herit.* **2025**, *19*, 581–597. [CrossRef]
19. Ren, W.; Zhong, Z. LBA-YOLO: A novel lightweight approach for detecting micro-cracks in building structures. *PLoS ONE* **2025**, *20*, e0321640. [CrossRef]
20. Bruno, S.; Galantucci, R.A.; Musicco, A. Decay detection in historic buildings through image-based deep learning. *Vitruvio* **2023**, *8*, 6–17. [CrossRef]
21. Croce, V.; Caroti, G.; Piemonte, A.; De Luca, L.; Véron, P. H-BIM and Artificial Intelligence: Classification of Architectural Heritage for Semi-Automatic Scan-to-BIM Reconstruction. *Sensors* **2023**, *23*, 2497. [CrossRef] [PubMed]
22. Laohaviraphap, N.; Waroonkun, T. Integrating Artificial Intelligence and the Internet of Things in Cultural Heritage Preservation: A Systematic Review of Risk Management and Environmental Monitoring Strategies. *Buildings* **2024**, *14*, 3979. [CrossRef]
23. Odgers, D.; Henry, A. *Practical Building Conservation: Stone*; Ashgate: Farnham, UK, 2012; p. 338.
24. D'Agostino, D.; Congedo, P.M.; Cataldo, R. Computational Fluid Dynamics (CFD) Modeling of Microclimate for Salts Crystallization Control and Artworks Conservation. *J. Cult. Herit.* **2014**, *15*, 448–457. [CrossRef]
25. Pocobelli, D.P.; Boehm, J.; Bryan, P.; Still, J.; Grau-Bové, J. BIM for Heritage Science: A Review. *Herit. Sci.* **2018**, *6*, 30. [CrossRef]
26. Ceccarelli, L.; Bevilacqua, M.G.; Caroti, G.; Castiglia, R.B.F.; Croce, V. Semantic Segmentation through Artificial Intelligence from Raw Point Clouds to H-BIM Representation. *Disegnarecon* **2023**, *16*, 171–178. [CrossRef]
27. Yiğit, A.Y.; Uysal, M. Automatic Crack Detection and Structural Inspection of Cultural Heritage Buildings Using UAV Photogrammetry and Digital Twin Technology. *J. Build. Eng.* **2024**, *94*, 109952. [CrossRef]
28. Rodrigues, F.; Cotella, V.; Rodrigues, H.; Rocha, E.; Freitas, F.; Matos, R. Application of Deep Learning Approach for the Classification of Buildings' Degradation State in a BIM Methodology. *Appl. Sci.* **2022**, *12*, 7403. [CrossRef]
29. Standoli, G.; Salachoris, G.P.; Masciotta, M.G.; Clementi, F. Modal-Based FE Model Updating via Genetic Algorithms: Exploiting Artificial Intelligence to Build Realistic Numerical Models of Historical Structures. *Constr. Build. Mater.* **2021**, *303*, 124393. [CrossRef]
30. Gara, F.; Nicoletti, V.; Arezzo, D.; Cipriani, L.; Leoni, G. Model Updating of Cultural Heritage Buildings through Swarm Intelligence Algorithms. *Int. J. Archit. Herit.* **2025**, *19*, 259–275. [CrossRef]
31. Salehi, H.; Burgueño, R. Emerging Artificial Intelligence Methods in Structural Engineering. *Eng. Struct.* **2018**, *171*, 170–189. [CrossRef]
32. Fan, J.; Chen, Y.; Zheng, L. Artificial Intelligence for Routine Heritage Monitoring and Sustainable Planning of the Conservation of Historic Districts: A Case Study on Fujian Earthen Houses (Tulou). *Buildings* **2024**, *14*, 1915. [CrossRef]
33. Luo, Y.; Yin, B.; Peng, X.; Xu, Y.; Zhang, L. Wind-Rain Erosion of Fujian Tulou Hakka Earth Buildings. *Sustain. Cities Soc.* **2019**, *50*, 101666. [CrossRef]
34. Zhou, Q. Research on Traditional Reinforcement Techniques for Rammed Earth Walls in China. *Int. J. Archit. Herit.* **2025**, *19*, 496–514. [CrossRef]
35. Tang, C.A. Numerical simulation of progressive rock failure and associated seismicity. *Int. J. Rock Mech. Min. Sci.* **1997**, *34*, 249–261. [CrossRef]
36. Tang, C.A.; Liang, Z.Z.; Zhang, Y.B.; Xu, T. Three-dimensional material failure process analysis. *Key Eng. Mater.* **2005**, *297–300*, 1196–1201. [CrossRef]
37. Helgeson, D.E.; Aydin, A. Characteristics of joint propagation across layer interfaces in sedimentary rocks. *J. Struct. Geol.* **1991**, *13*, 897–911. [CrossRef]
38. Tang, C.A.; Zhang, Y.B.; Liang, Z.Z.; Xu, T.; Tham, L.G.; Lindqvist, P.-A.; Kou, S.Q.; Liu, H.Y. Fracture spacing in layered materials and pattern transition from parallel to polygonal fractures. *Phys. Rev. E* **2006**, *73*, 056120. [CrossRef]

coatings

Review

Surface Protection Technologies for Earthen Sites in the 21st Century: Hotspots, Evolution, and Future Trends in Digitalization, Intelligence, and Sustainability

Yingzhi Xiao [1], Yi Chen [1], Yuhao Huang [2,*] and Yu Yan [2,*]

1 Faculty of Art, Xi'an University of Architecture and Technology, No. 13, Middle Section of Yanta Road, Beilin District, Xi'an 710055, China; lennon@xauat.edu.cn (Y.X.); chenney61@xauat.edu.cn (Y.C.)
2 Institute of Urban and Sustainable Development, City University of Macau, Avenida Padre Tomás Pereira, Taipa, Macau 999078, China
* Correspondence: u24092120176@cityu.edu.mo (Y.H.); yuyan@cityu.edu.mo (Y.Y.)

Abstract: As vital material carriers of human civilization, earthen sites are experiencing continuous surface deterioration under the combined effects of weathering and anthropogenic damage. Traditional surface conservation techniques, due to their poor compatibility and limited reversibility, struggle to address the compound challenges of micro-scale degradation and macro-scale deformation. With the deep integration of digital twin technology, spatial information technologies, intelligent systems, and sustainable concepts, earthen site surface conservation technologies are transitioning from single-point applications to multidimensional integration. However, challenges remain in terms of the insufficient systematization of technology integration and the absence of a comprehensive interdisciplinary theoretical framework. Based on the dual-core databases of Web of Science and Scopus, this study systematically reviews the technological evolution of surface conservation for earthen sites between 2000 and 2025. CiteSpace 6.2 R4 and VOSviewer 1.6 were used for bibliometric visualization analysis, which was innovatively combined with manual close reading of the key literature and GPT-assisted semantic mining (error rate < 5%) to efficiently identify core research themes and infer deeper trends. The results reveal the following: (1) technological evolution follows a three-stage trajectory—from early point-based monitoring technologies, such as remote sensing (RS) and the Global Positioning System (GPS), to spatial modeling technologies, such as light detection and ranging (LiDAR) and geographic information systems (GIS), and, finally, to today's integrated intelligent monitoring systems based on multi-source fusion; (2) the key surface technology system comprises GIS-based spatial data management, high-precision modeling via LiDAR, 3D reconstruction using oblique photogrammetry, and building information modeling (BIM) for structural protection, while cutting-edge areas focus on digital twin (DT) and the Internet of Things (IoT) for intelligent monitoring, augmented reality (AR) for immersive visualization, and blockchain technologies for digital authentication; (3) future research is expected to integrate big data and cloud computing to enable multidimensional prediction of surface deterioration, while virtual reality (VR) will overcome spatial–temporal limitations and push conservation paradigms toward automation, intelligence, and sustainability. This study, grounded in the technological evolution of surface protection for earthen sites, constructs a triadic framework of "intelligent monitoring–technological integration–collaborative application," revealing the integration needs between DT and VR for surface technologies. It provides methodological support for addressing current technical bottlenecks and lays the foundation for dynamic surface protection, solution optimization, and interdisciplinary collaboration.

Keywords: earthen sites; surface conservation of earthen sites; cutting-edge technologies; sustainable development; intelligent systems; digitization

1. Introduction

As vital material carriers of human civilization, earthen sites have become a central topic in the implementation of the United Nations Sustainable Development Goals (SDGs) and the Convention Concerning the Protection of the World Cultural and Natural Heritage under the global sustainable development agenda [1]. Their conservation must strictly adhere to the principles of "authenticity, integrity, and minimal intervention" as established in the Venice Charter [2], not only because the survival status of earthen sites is directly linked to the preservation of global cultural diversity but also because it directly affects the implementation effectiveness of SDG Target 11.4: "Strengthen efforts to protect and safeguard the world's cultural and natural heritage" [3]. At the 2025 international conference, the International Council on Monuments and Sites (ICOMOS) noted that approximately 17% of world heritage sites contain earthen components, and the concept of surface conservation for earthen sites has now fully transitioned into a digital, intelligent, and sustainable system [4]. This transformation is reflected in the use of RS for regional environmental baseline monitoring and ecological change detection [5], millimeter-level spatial data archiving through LiDAR [6], and the construction of real-time environmental monitoring networks via IoT technology [7]. These technological systems have become tools for climate change risk management (SDG 13) and, through mechanisms of data sharing, have also promoted community participation (SDG 16), creating a new global paradigm for protection that integrates digital technology with humanistic care. Under the pressures of global warming and extreme weather events, earthen sites are increasingly challenged by natural weathering and anthropogenic threats, with worsening surface deterioration and deformation. Traditional conservation approaches, due to their poor compatibility and limited reversibility, are unable to address such complex degradation processes. For instance, the earthen walls of the ancient city site of Jiaohe in Xinjiang, China, have experienced intensified cracking and exfoliation due to significant diurnal temperature variations [8]. Such ecological damage not only threatens the structural integrity of the sites but also leads to the loss of historical and cultural information, highlighting the urgency of advancing digital and intelligent technologies for surface conservation of earthen sites [9]. It is necessary, on the one hand, to respond to the convention's demand for "authenticity" in conservation and, on the other, to fulfill SDG mandates for climate action and sustainable development by implementing interdisciplinary technologies to establish a protection paradigm that fosters both cultural resilience and ecological adaptability [10]. However, current research lacks an integrated "digital–intelligent–sustainable" development pathway. Challenges include the absence of standards for data interoperability, the lack of intelligent closed-loop decision making systems, and the deficiency of frameworks that coordinate ecological and cultural evaluations. To address these gaps, it is essential to accelerate the deep integration of digital, intelligent, and sustainable technologies for surface conservation across different types of earthen sites, thereby forming a comprehensive protection system characterized by data sharing, intelligent decision making, and ecological–cultural synergy, ultimately achieving the dual goals of heritage conservation and ecological restoration.

Currently, under the dual pressures of structural deterioration and deformation, earthen site surfaces reveal a lack of deep integration among digital, intelligent, and sustainable conservation technologies. The technical perspective of earthen site surface conservation unfolds in three dimensions: first, digital methods such as RS and LiDAR are employed to authentically archive the spatial characteristics of site surfaces [11,12]; second, intelligent monitoring and early warning systems are built using IoT technologies and machine learning [4]; third, from the standpoint of sustainable development, a multidisciplinary approach combining archaeology, ecology, and mechanics is adopted to promote joint progress in both surface protection and ecological management [13]. The value dimension is constructed hierarchically from local to international levels; at the level of local cultural roots, earthen sites serve as spatial narratives of regional civilizations—for example, in the loess cave settlements of the Loess Plateau, the construction techniques using raw earth materials embody the ecological wisdom of agrarian culture [14]; at the national strategic level, earthen sites represent a critical component of land-use planning strategies adopted by various countries, such as the "International Campaign to Save the Nubian Monuments" initiated in the 1960s by UNESCO and the Egyptian government to rescue the Abu Simbel temples threatened by flooding due to the Aswan Dam; at the level of global governance, mechanisms such as UNESCO's "Transnational Joint Application for the Silk Road to be Listed as a World Heritage" [15] correspond with the Convention Concerning the Protection of the World Cultural and Natural Heritage, which considers earthen sites as indicators for measuring global sustainable development. Therefore, earthen sites are not only material foundations for preserving local cultural identities and essential elements of national strategic planning but also key objects of UNESCO-led international collaboration and benchmarks within the World Heritage Convention for evaluating cultural heritage worldwide [16]. Empowered by digital technology and supported by interdisciplinary collaboration, earthen sites have become a civilizational paradigm for addressing ecological crises and fulfilling the shared responsibilities of humanity.

From isolated preservation efforts to comprehensive governance, a trend of increasingly mature theoretical systems and continuous innovation in technical applications has been shown. In terms of theoretical development, heritage protection guidelines such as the Venice Charter and The Burra Charter, issued in the mid-to-late 20th century, established the three fundamental principles of authenticity, integrity, and minimal intervention [17], providing foundational value orientations for the surface protection of earthen sites [18,19]. Entering the 21st century, research on the surface protection of earthen sites has made notable breakthroughs and advancements [20], with conservation concepts gradually evolving from "static protection" to "dynamic sustainable development" [21]. Scholars have also begun to explore how to coordinate the relationship between preservation and development in the context of globalization, informatization, and the era of intelligent technologies [22–24].

At the methodological and technical level, despite continuous technological iterations, the surfaces of earthen sites still reveal key technical limitations under multi-source damage, such as insufficient accuracy in digital data acquisition and lagging responses in intelligent sensing. In the early stages, surface protection of heritage sites mainly relied on physical reinforcement, characterized by single, unsystematic approaches. Since the beginning of the 21st century, digital and intelligent technologies have become major driving forces, forming a mutually reinforcing innovation framework. The application level of advanced digital technologies in the spatial information acquisition and recording of earthen site surfaces is relatively high [25]. Among them, RS is primarily used to obtain the location and surrounding environmental information of earthen sites [26,27], serving as a foundational tool for preliminary investigations and subsequent monitoring. Technologies such as LiDAR and photogrammetry (PG) can partially compensate for the lack of detail in visual imagery, enabling more accurate scanning of earthen site surfaces, extracting fine surface features, and constructing high-precision 3D models [28–30], thereby realizing the digital archiving of spatial surface characteristics. At the level of intelligent applications, smart methods are being used to guide decision making and management of earthen site surface protection [4]. For example, by combining big data analysis and machine learning, researchers can perform intelligent diagnostics and predictions regarding disease types and their development trends on site surfaces [31]. The application of IoT technology facilitates the establishment of real-time monitoring systems, enabling continuous feedback of internal and external environmental data (e.g., temperature, humidity, light, wind speed) and structural conditions [32], which serve as a scientific basis for rational conservation strategies. The core breakthrough of this stage lies in the organic integration of traditional conservation knowledge with modern technologies, promoting a paradigm shift from reactive emergency measures to proactive prevention in the surface protection of earthen heritage sites [33].

At present, the existing literature on the surface protection of earthen sites primarily focuses on conservation concepts and management strategies [34,35], the development of protective materials and restoration techniques [36,37], and the mechanisms of surface deterioration and preventive conservation [33]. Although significant breakthroughs have been achieved in both theoretical research and technical approaches, and a multidimensional technical integration system has been initially established [38], there remains a substantial gap in the integration of digitalization and intelligent technologies for earthen site surface protection from a sustainable development perspective. The current research lacks a systematic framework for technical integration, making it difficult to support the dynamic needs of spatial conservation.

To gain a deeper understanding of the literature and recent developments in the field of earthen site surface protection, identify existing achievements and unresolved issues, and explore future research trends, this study aims to systematically trace the evolution of digital, intelligent, and sustainable technologies in surface protection of earthen sites from 2000 to 2025 through bibliometric and multidimensional analysis. Drawing on data from both the Web of Science and Scopus core databases, this study employs tools such as CiteSpace 6.2 R4 and VOSviewer 1.6 to conduct a bibliometric analysis of the research progress on earthen site protection from multiple perspectives—including publication volume, keyword co-occurrence, and geographic distribution of contributing countries or regions. These tools facilitate the efficient processing of large volumes of data from the literature and allow for the generation of visual maps to clearly reflect research hotspots and trends [39]. In addition, this study combines a close reading of the frontier literature with GPT-assisted extraction and analysis of key points to ensure that no critical content is overlooked. A dynamic interpretation of the hotspot literature is conducted to identify core research themes and trending topics, clarify the evolution of research priorities across different stages, and summarize the development of research methodologies. Furthermore, by leveraging both the Web of Science and Scopus databases, this study identifies the key triggering factors behind highly cited literature and analyzes the driving mechanisms—such as policy guidance, technological breakthroughs, and major archaeological discoveries—that have contributed to paradigm shifts in the field. It also establishes causal links between research contexts and technological developments across different periods and traces the origins of key variables, thereby overcoming the limitations of traditional literature review approaches. Ultimately, this study provides valuable references for scholars in the field of earthen site surface protection, helping them to understand the research trajectory from various dimensions and better anticipate future trends. Therefore, the study focuses on the following three aspects.

(1) Systematic Literature Review and Knowledge Mapping: This section provides a comprehensive analysis of the research progress in the field of earthen site surface protection in multiple dimensions, including publication trends, national/regional institutional outputs, co-citation networks, and keyword co-occurrence analysis. The aim is to offer scholars a systematic literature review, intuitive visualized knowledge maps, and insights into the evolution of research trends.

(2) In-Depth Analysis of Research Hotspots: This study focuses on identifying key research hotspots from a technical perspective. It aims to offer more methodological options for the application of surface protection technologies in earthen sites, as well as valuable references for understanding the current application status and developmental trajectory of relevant technologies.
(3) Insights into Future Research Directions: This study also seeks to summarize potential future research directions and emerging hotspots, providing scholars with effective guidance for further studies in the field of earthen site surface protection.

2. Materials and Methods

2.1. Data Source

This study conducts a multifaceted analysis of academic research related to the earthen site surface conservation using the core collections of the Web of Science and Scopus databases as the primary data sources [40,41]. These two databases were selected because they offer broad disciplinary coverage, large-scale data repositories, high-quality literature, and comprehensive citation information. Additionally, both platforms provide robust technical support and user-friendly retrieval experiences, coupled with powerful analytical functions. Their authoritative status in the field and high level of academic recognition make them particularly well suited for bibliometric research [42–44]. These characteristics ensure that the data extracted for this study are both comprehensive and reliable, thereby laying a solid foundation for the analysis. To more accurately capture cutting-edge developments in the field of earthen site surface conservation, this study focuses on the literature published between 2000 and 2025. This period coincides with the widespread application of modern GIS and RS technologies and represents a critical phase of rapid advancement in digitization and intelligent systems. Moreover, it marks the emergence of technologies such as VR and AR. Therefore, the selected timeframe effectively reflects the academic evolution and emerging trajectories of earthen site surface conservation research on earthen sites during this key era of technological transformation [45].

During the literature retrieval process, Boolean operators were used to construct the search strategy. The search query applied was the following: TS = ("Earth heritage" AND "Spatial Protection" OR "Earthen Heritage" OR "Protective Shelter" OR "Earthen Ruins" AND "Shelter Design" AND "Earth Site" AND "Structural Reinforcement" OR "Surface Technology") AND PY = ("2000–2025") AND DT = ("Article"). Here, TS refers to the search topic, PY denotes the publication year, and DT indicates the document type. The language was limited to English. Using this Boolean combination, a total of 1544 publications were initially retrieved. After applying a time filter covering the period from 1 January 2000, to 1 April 2025, the dataset was narrowed down to 1216 publications. When restricting the document type to journal articles, 1019 records remained. Finally, after limiting the language to English, 967 eligible articles were identified. These 967 articles were included in the subsequent bibliometric analysis based on both quantitative and visual methods. The analysis focused on various dimensions, including publication volume, countries/regions, institutions, journals, authors, and keywords. A PRISMA 2020 flow diagram for the identification and selection of studies is illustrated in Figure 1. These 967 articles constitute the core dataset for this study.

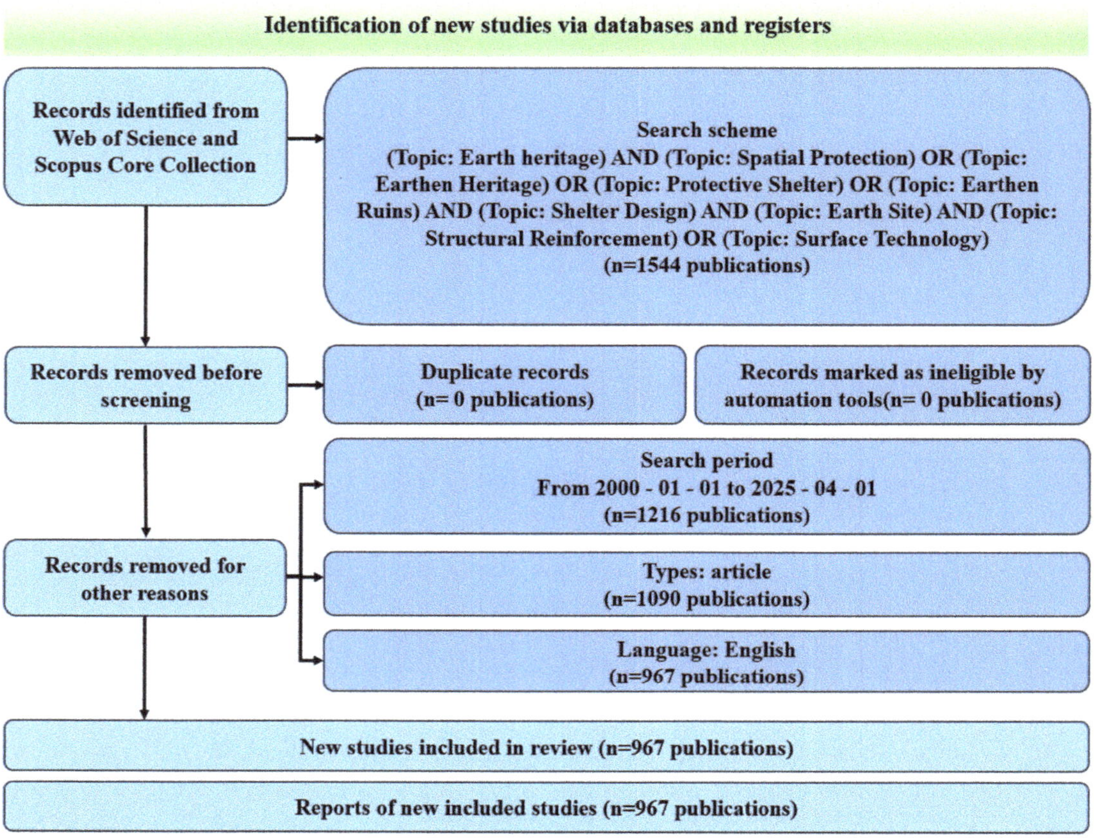

Figure 1. PRISMA 2020 flow diagram for the identification and selection of studies (source: created by the authors).

2.2. Research Method

This systematic review was conducted in accordance with the PRISMA 2020 guidelines. The search strategy, study selection, data extraction, and synthesis processes followed the PRISMA 2020 flow diagram (Figure 1).

Bibliometrics refers to the quantitative analysis of literature, aiming to reveal patterns and developmental processes within a statistical framework [46]. Based on this foundation, mathematical modeling, statistical analysis, and computer technologies are applied to examine the distribution, disciplinary linkages, and temporal evolution of academic publications [47], thereby uncovering the underlying mechanisms of scholarly communication. The core of bibliometric analysis lies in the quantitative evaluation of publication volume, citation frequency, and related indicators. Techniques such as co-occurrence analysis, clustering, and network analysis are integrated to construct knowledge maps, which visualize research hotspots, collaboration networks, and interdisciplinary trends. Bibliometrics provides robust data support for scientific research and contributes to the advancement of research toward greater standardization, visualization, and sustainability [48].

This study adopts a dual-tool bibliometric analysis framework using CiteSpace 6.2 R4 (https://citespace.podia.com/ (accessed on 17 July 2025)) and VOSviewer 1.6 (https://www.vosviewer.com/ (accessed on 17 July 2025)). VOSviewer facilitates interactive identification of research hotspots through visual network maps, allowing users to intuitively observe associations among keyword nodes. CiteSpace, on the other hand, is well suited for exploring scientometric relationships among publications, enabling the tracking of research hotspots and trend shifts through co-citation networks, burst keywords, and other metrics. The combined use of these two tools offers a complementary, multidimensional analytical perspective; VOSviewer provides a macroscopic visualization of the research landscape, while CiteSpace uncovers deeper logical connections between documents from a scientometric standpoint. However, due to differences in their underlying algorithms and visualization methods, the simultaneous use of both tools may yield differing analytical outcomes. Therefore, interpretation and cross-validation of results are essential to ensure analytical accuracy. In the visualization outputs, both software platforms represent keywords as circular nodes, with the node diameter reflecting the centrality or weight of the keyword [49]. Links between nodes indicate co-occurrence relationships, with the link thickness being positively correlated with co-occurrence strength [50]. This study conducts statistical analyses of multidimensional data, including keyword co-occurrence, geographic distribution by country/region, author collaboration networks, and publication clustering. The bibliometric findings are then visually presented through the knowledge maps generated by VOSviewer and CiteSpace [51].

In addition, this study incorporates a manual review of the high-impact literature combined with GPT-assisted analysis (https://chatgpt.com/ (accessed on 17 July 2025)). In this process, GPT is positioned as a semantic analysis assistant to support researchers in interpreting and analyzing the literature content. The workflow begins with the input of a substantial body of literature on the spatial conservation of earthen site surface conservation published between 2000 and 2025, drawn from the Web of Science and Scopus core databases, along with relevant supplementary data, such as policy documents and technical standards. Subsequently, the entire literature corpus is semantically segmented into several layers, with core keywords extracted from each. Context-sensitive prompts are then designed to indicate the degree of semantic correlation among concepts, and this framework is used to infer future research pathways. The initial and secondary outputs generated by GPT are manually reviewed by researchers to remove irrelevant content and verify the accuracy of the extracted information. Cross-validation is performed to ensure an error rate of less than 5%. If the depth of analysis is found to be insufficient, the input is refined, and the prompts are optimized for iterative deep learning inference. Through this combined process of quantitative data analysis and manually curated theoretical interpretation, the study identifies the developmental trajectories of research hotspots and extracts valuable insights into future research directions. In the process of GPT-assisted semantic mining, this study adopts a three-level verification mechanism to ensure analytical accuracy. 1. Clear evaluation criteria are established, with the domain-specific hotspot literature serving as the benchmark dataset. The core themes extracted by GPT must match those from the human analysis with an accuracy of no less than 90%, and the use of technical terminology must be precise. 2. An accuracy assessment is conducted by two independent researchers who cross-verify the core keywords output by GPT against the results of manual close reading, calculate the error rate (error rate < 5%), and label the specific types of errors (e.g., terminology omission, association errors, etc.). 3. The author is responsible for iterative optimization and logical calibration. Semantic associations from GPT that do not align with the research context are manually revised. In addition, for deviations in the interpretation of technical terms, authoritative domain-specific materials must be supplemented as training data to prompt GPT to conduct multiple rounds of deep learning inference until the output fully aligns with the technical context and research logic of earthen site surface conservation, ultimately meeting the predefined evaluation criteria.

Ultimately, under the guidance of the researchers, the evolution of research trends and their underlying logic are determined. The innovations of GPT-assisted semantic mining are reflected in the following aspects: 1. this overcomes the limitations of traditional quantitative tools by enabling a shift from data statistics to in-depth analysis through deep semantic association; 2. it establishes a closed-loop process of "manual close reading–GPT extraction–error calibration" (with an error rate of <5%), addressing semantic deviations inherent in purely machine-based analysis; 3. it efficiently integrates multi-source data, significantly improving efficiency compared with traditional manual analysis, enabling the rapid identification of core themes and trend inference. As illustrated in Figure 2, the research methodology integrates two major academic databases to reveal key driving factors, identify existing limitations, and trace emerging technological trends. This approach employs four strategies—quantitative statistics, causal analysis, theoretical exploration, and technological forecasting—to overcome the limitations of traditional bibliometric analysis and provide multidimensional references for future research.

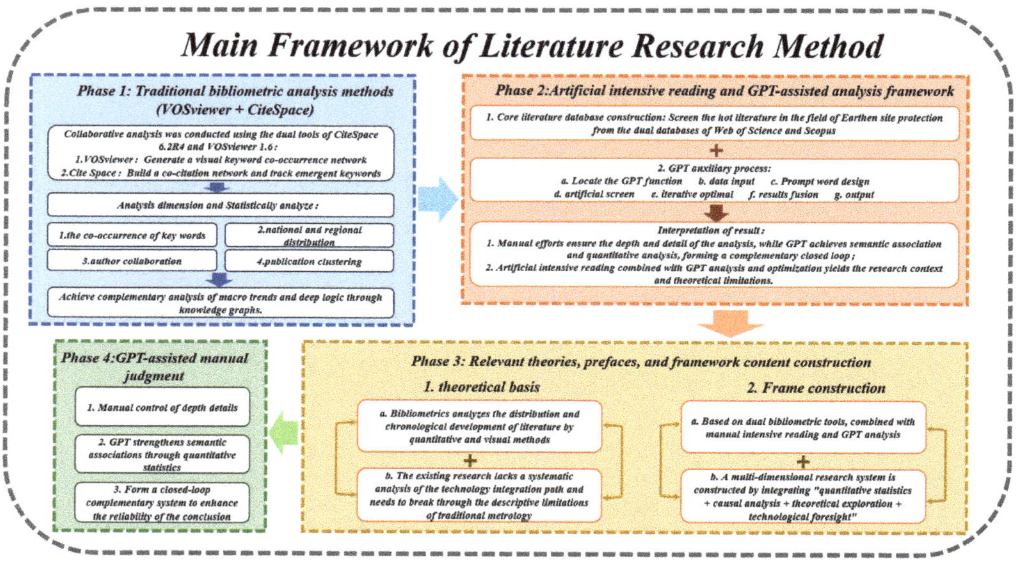

Figure 2. Framework of the literature research methodology (source: created by the authors).

3. Results

3.1. Analysis of Annual Publication Trends

This study analyzes the publication trends in the field of earthen site surface conservation from 2000 to 2025, covering a total of 967 articles. As shown in Figure 3, the research output has increased significantly over the last two decades and can be divided into three main stages: the initial stage, the transitional stage, and the rapid growth stage.

This study divides the development of research on earthen site surface conservation into three stages, with the starting point of the growth phase being appropriately advanced to reflect the increasing volume of publications.

Initial Phase (2000–2007): During this phase, the number of journal publications was low, indicating that the field was in its infancy with a relatively stable cumulative growth rate. The overall publication volume remained limited, reflecting a slow development trend. Research efforts were still exploring and developing foundational theories and methodologies. At this point, there was no mature research framework or comprehensive theoretical system, and studies primarily focused on the basic issues of earthen site surface conservation with limited interdisciplinary integration.

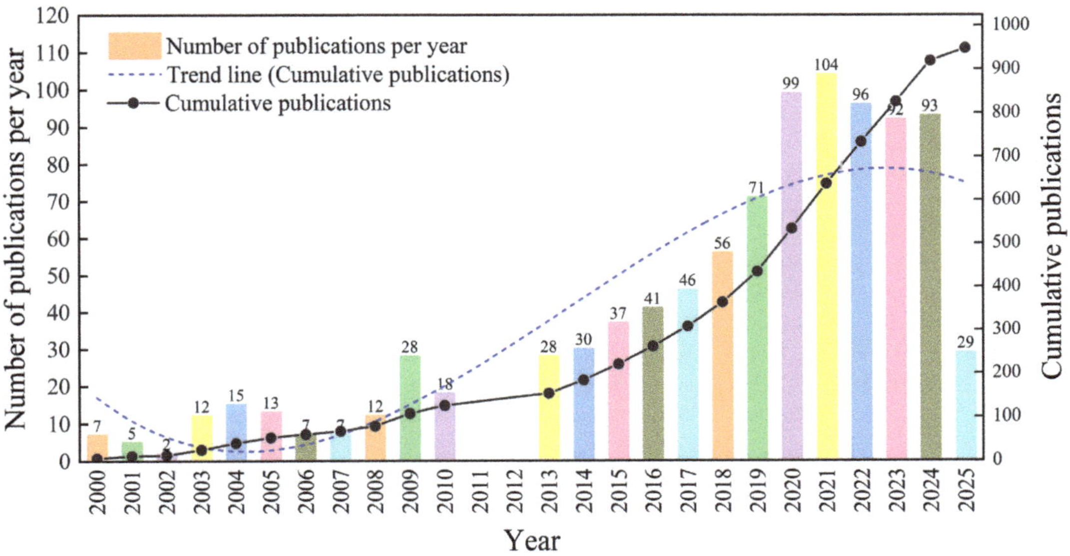

Figure 3. Publication trend analysis (source: created by the authors).

Transitional Phase (2008–2014): This period witnessed moderate fluctuations in publication output, showing a trend of staged growth. By 2014, the cumulative number of publications had reached approximately 450. The increasing application of spatial technologies such as GIS and RS to earthen site surface conservation [25,52], the gradual introduction of heritage conservation policies, and the development of interdisciplinary research all contributed to the accelerated progress during this phase. It is worth noting that there was no significant publication output in 2011–2012. Researchers speculate that this gap may be due to a shift in focus; the early years (2008–2010) emphasized reinforcement techniques [8], while the later years (2013–2014) focused on deterioration mechanisms [15,53]. This two-year gap might represent a period of technical validation. Moreover, the widespread adoption of GIS and RS for conservation monitoring purposes occurred after 2014 [54]; in contrast, the years 2011–2012 were still dominated by single-discipline studies (e.g., archaeology, material science), resulting in fragmented publications without prominent research hotspots.

Rapid Growth Phase (2015–present): Since 2015, the number of annual publications has consistently increased, peaking at 104 in 2021. The cumulative number of publications has shown an exponential upward trend, reflecting the growing attention to this field. The surge in research can be attributed to the rise of digital conservation [55–57], the growing emphasis on sustainable conservation practices, and the promotion of international cooperative projects (e.g., UNESCO's Asia–Pacific heritage conservation initiatives [58]). These developments highlight the fluctuating nature of research activities over time and validate the delineation of the three stages.

Under the implementation of global heritage digitization strategies [59], advances in artificial intelligence [60], and especially the promotion of conservation work at earthen sites along the Belt and Road Initiative [15], research on earthen site surface conservation is expected to yield more innovations in technology and sustainability. This field will likely remain a research hotspot for a considerable time, with increasing scholarly output in the form of papers and monographs.

3.2. Active Countries/Regions

The analysis of publication volume by country/region reveals the global distribution of research in the field, highlighting the necessity of strengthening international cooperation through scientific collaboration platforms. Such cooperation facilitates the global flow and exchange of knowledge and technology, ultimately enhancing the overall quality of research. As shown in Figure 4, the global publication distribution map generated using MapBox 2025 v11.13.0 (https://www.mapbox.com/ (accessed on 17 July 2025)) illustrates the number of publications by country through color coding and numerical labels. It can be observed that countries in North America, East Asia, Europe, and Oceania stand out with relatively high publication volumes, indicating strong research capacities. In contrast, countries in Africa, Central Asia, and South America generally show weaker research performance in this field, with some countries lacking data or having zero publications, resulting in blank areas on the map. And the publication gap between the United States (304 articles) and China (185 articles) may be influenced by the database's preference for English-language literature, rather than reflecting an actual disparity in research capacity. In addition, data gaps in regions such as Africa are not examined in terms of whether they result from the limited distribution of earthen sites or a lack of research resources.

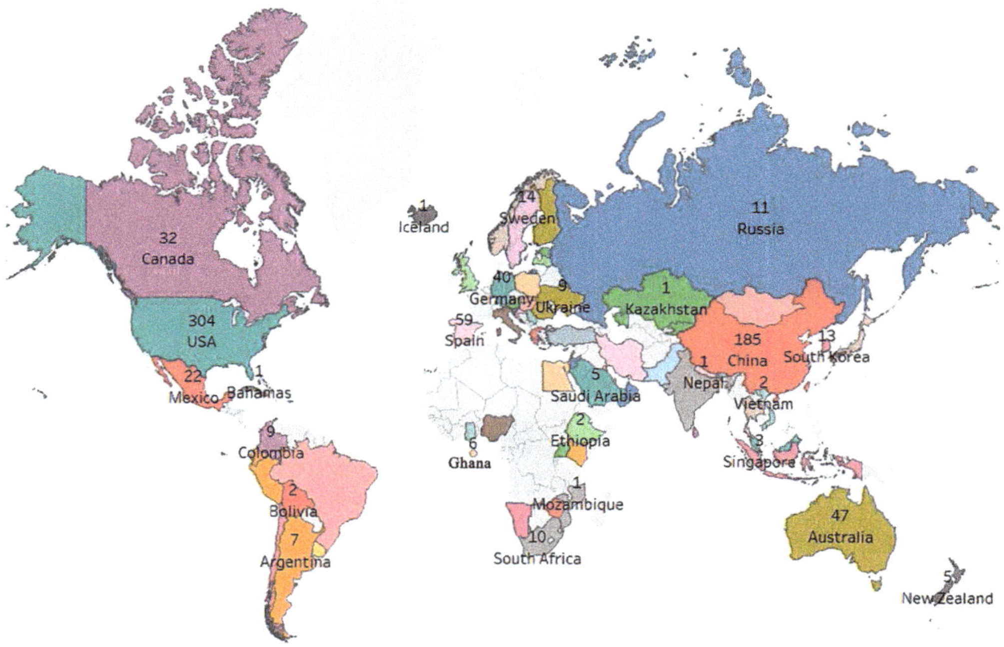

Figure 4. Global publication landscape in the field of earthen site surface conservation (source: created by the authors).

Figure 5 illustrates the top 15 countries/regions with the most significant citation bursts in the field of spatial conservation of earthen sites between 2000 and 2025. Citation burst analysis helps identify countries or regions whose publications have been cited intensively within a specific period, thereby indicating their rising influence in the field. Using CiteSpace, the analysis was conducted by setting the parameter to "Burst terms" to generate a burst detection map. A total of 15 countries were identified as having significant bursts. In this figure, Year indicates the first year that a country experienced a burst, while Strength denotes the intensity of the citation burst [61]. The light blue bars represent the full time span (2000–2025), the red segments indicate the duration of the burst period, and years without significant changes are filled in dark blue. The analysis reveals that most citation bursts occurred between 2000 and 2015, reflecting a growing interest in surface conservation research during that period. In recent years, a new wave of burst countries has emerged—including Colombia, the United Arab Emirates, China, and Denmark—suggesting renewed global attention to the field. Among these countries, China experienced a burst from 2023 to 2025, with the highest strength value of 16.88, indicating a rapid ascent in global influence and a growing leadership role in the field. Although China shows the highest burst intensity (16.88) during 2023–2025, there is no comparison of the citation impact of journals between China and the United States, making it insufficient to assess international recognition based solely on burst intensity. In contrast, the USA had an earlier burst (2014–2015) with a strength of 6.94, confirming its early central position in the field but suggesting a tapering of global attention in recent years. Other countries such as Turkey, Germany, Greece, and Belgium experienced notable citation bursts between 2013 and 2020, demonstrating Europe's early and solid research foundation in this area. Emerging research nations such as Colombia and the United Arab Emirates are gradually joining the ranks of high-impact contributors, indicating their potential to become major research forces in the near future.

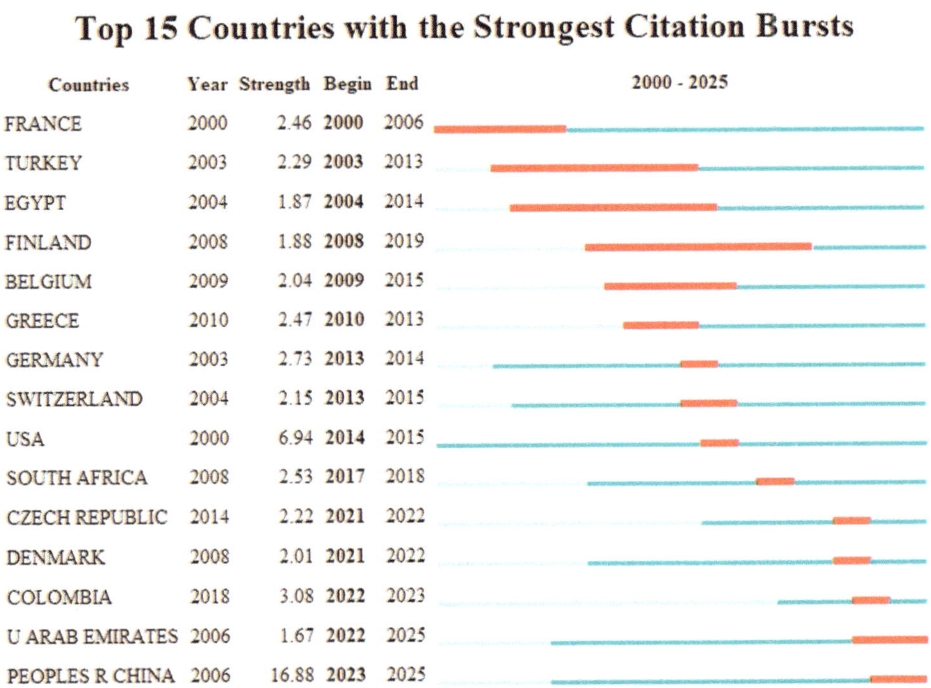

Figure 5. Top 15 countries with the strongest citation bursts (source: created by the authors).

Therefore, it is evident that China's academic influence in the field of earthen site surface conservation continues to expand globally. The United States, on the other hand, demonstrates both strong international citation impact and a relatively high frequency of international collaboration, thereby exerting substantial global influence. These differences reveal that various countries have distinct understandings of the roles and developmental paths related to earthen site surface conservation. Overall, this analysis indicates that

China is currently one of the countries conducting the most extensive research in this field, while also highlighting the academic interest, contributions, and recognition of research achievements from other nations. Future studies may further explore patterns of international collaboration and emerging research hotspots to support the sustainable development of this field.

3.3. Institutional Distribution Analysis

As shown in Figure 6, the VOSviewer software was used to analyze the research institutions active in the field of earthen site surface protection. A minimum threshold of two publications was set to ensure the clarity of the co-authorship network visualization. In the resulting visual map, each circle and text label represents an institution. The lines connecting the circles indicate co-authorship relationships, with the line thickness reflecting the strength of the collaboration. Gradient colors indicate the overall intensity of cooperation with other institutions, and the size of each circle is proportional to the number of publications produced by that institution [62]. Different colors represent distinct institutional clusters. For example, Cluster 1 includes the Chinese Academy of Sciences, Lanzhou University, and others. This cluster likely represents research forces centered on Chinese scientific research institutions and universities, mainly focusing on the theoretical foundations and material development for earthen site surface protection. Cluster 2 features institutions such as the University of Aveiro and the University of Minho, representing research groups from Portugal and other European countries that focus on techniques, methodologies, and protection management. Cluster 3, which includes several American universities, is likely oriented toward the use of advanced geographic information technologies and environmental monitoring techniques to observe and simulate the environmental conditions of earthen site surfaces. Connections between clusters show the presence of international collaborative research. For instance, the Chinese Academy of Sciences appears to collaborate with multiple international universities on research projects involving protective materials and monitoring technologies, including data and result sharing. In terms of institutional distribution, Chinese institutions dominate the field. The Chinese Academy of Sciences, Lanzhou University, Dunhuang Academy, and the University of Chinese Academy of Sciences are represented by relatively large circles, indicating a high volume of publications. The Dunhuang Academy, for example, has a high citation count (325 times) and an average citation rate of 17.1053, demonstrating significant academic impact. Lanzhou University has a relatively recent average publication year, indicating active and ongoing research in recent years. Domestically, institutions such as the Chinese Academy of Sciences and the University of Chinese Academy of Sciences, as well as Lanzhou University and Lanzhou University of Technology, are connected by thick lines, reflecting strong internal collaboration. Among international institutions, Michigan State University, Cornell University, and Stanford University have produced notable research outputs. However, their smaller circle sizes and thinner lines connecting them to Chinese institutions suggest a limited scale of international collaboration. The Chinese Academy of Sciences and Lanzhou University demonstrate close collaboration with dense linkages; however, the technological innovativeness and practical applicability of their joint research are not indicated. The sparse connections with overseas institutions are not analyzed in terms of whether they result from data-sharing barriers or insufficient international exchange. For example, Cornell University has an average publication year of 2012, indicating that its research activity peaked earlier and has decreased in recent years.

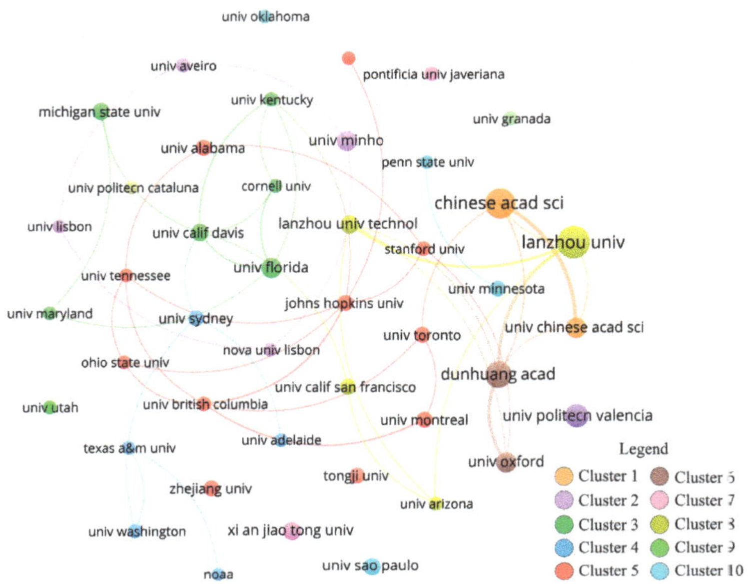

Figure 6. Institutional co-occurrence network map (source: created by the authors).

Figure 7 illustrates the top 20 most-cited institutions worldwide in the field of earthen site surface protection from 2000 to 2025. Notably, early research was predominantly led by US institutions—such as the United States Department of Agriculture and the University of California System—which laid the theoretical groundwork for the field. In the mid-term period, European institutions such as the University of Minho and Universitat Politècnica de València became prominent, focusing on the expansion and application of new technologies. In recent years, Chinese institutions have experienced a rapid surge in influence. Lanzhou University (2019–2025) stands out with a burst strength of 5.15, ranking first in the world in the field of earthen site surface protection, significantly ahead of other global institutions. Both Xi'an Jiaotong University (2022–2025) and the University of Chinese Academy of Sciences (2022–2025) also demonstrate strong intensity and sustained research activity. This trend indicates that China has shifted from merely participating in the field to playing a leading and demonstrative role, establishing a certain degree of international discourse power in the realm of earthen site surface protection. Lanzhou University (burst intensity: 5.15) shows a high burst, but its core technological focus is not specified (e.g., whether it concentrates on earthen sites in arid regions of Northwest China), making it difficult to infer the universality of its technologies. The burst of Xi'an Jiaotong University (2022–2025) does not indicate its collaborative relationships with other institutions.

Overall, Chinese research institutions have become a vital component of the global academic community, characterized by high research output and strong internal collaboration. However, international cooperation with institutions abroad still requires further strengthening. Moving forward, it is essential to promote a more integrated and interdisciplinary collaborative mechanism between domestic and international research bodies to advance the global development of research in this field.

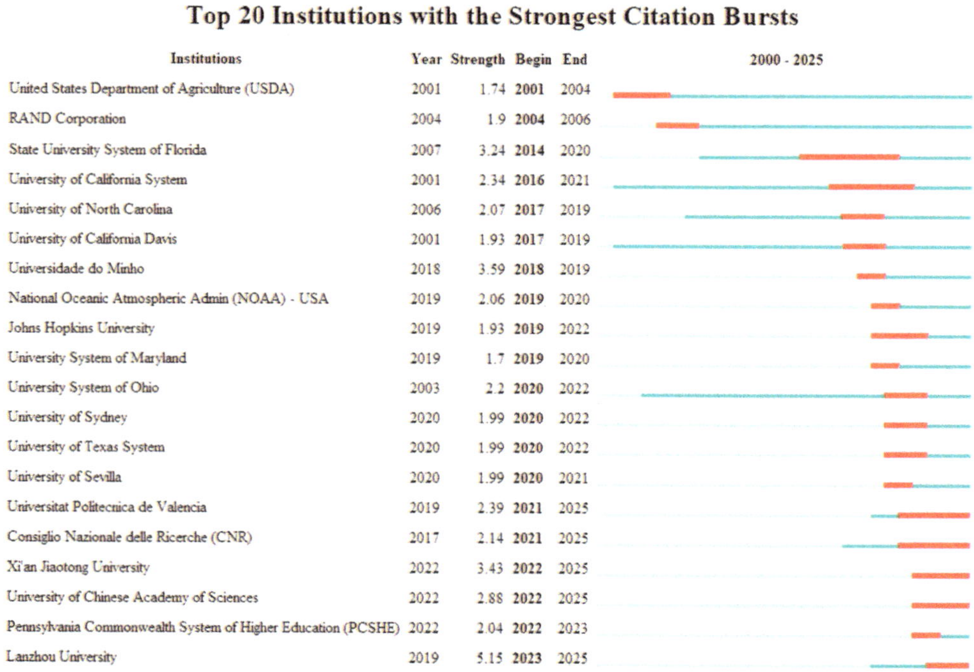

Figure 7. Top 20 institutions with the strongest citation bursts (source: created by the authors).

3.4. Analysis of Research Authors

To reveal the collaborative relationships among authors, a co-authorship analysis was conducted using VOSviewer, as shown in Figure 8, with the minimum publication threshold set at two papers per author. This analysis generated a co-authorship network map for researchers in the field of earthen site surface protection. Each circle in the map represents an individual author; the larger the circle, the more frequently the author appears or the more actively they are involved. Lines between circles indicate collaboration strength—thicker lines suggest stronger ties and more frequent co-authorships. Different colors represent distinct author clusters, indicating tighter intra-cluster relationships.

For example, Cluster 5 consists primarily of Chinese scholars whose research focuses on domestic earthen site surface protection, particularly in areas such as crack disease prevention and reinforcement materials for earthen sites in arid regions of northwest China. These researchers possess a strong foundation in local studies and practical experience. Cluster 3, by contrast, may include scholars from the UK, the US, and other Western countries who utilize advanced monitoring technologies and geographic information systems in their research, emphasizing environmental monitoring and data analysis of earthen site surfaces.

The lines connecting different clusters reflect instances of collaboration across groups, facilitating knowledge exchange. For instance, expertise in site management and planning from one cluster can inform strategies in others, while practical findings on disease prevention in earthen structures may offer useful insights to international peers. The visualization highlights that Chinese scholars form the core research cluster, with authors such as Chen Wenwu, Guo Qinglin, and Cui Kai appearing as large, densely connected nodes, indicating strong influence and leadership within the field. International scholars such as Lourenço Paulo B. and Heather Viles have also made significant contributions; however, their connections with Chinese scholars appear sparse, reflecting a relatively weak international collaboration network.

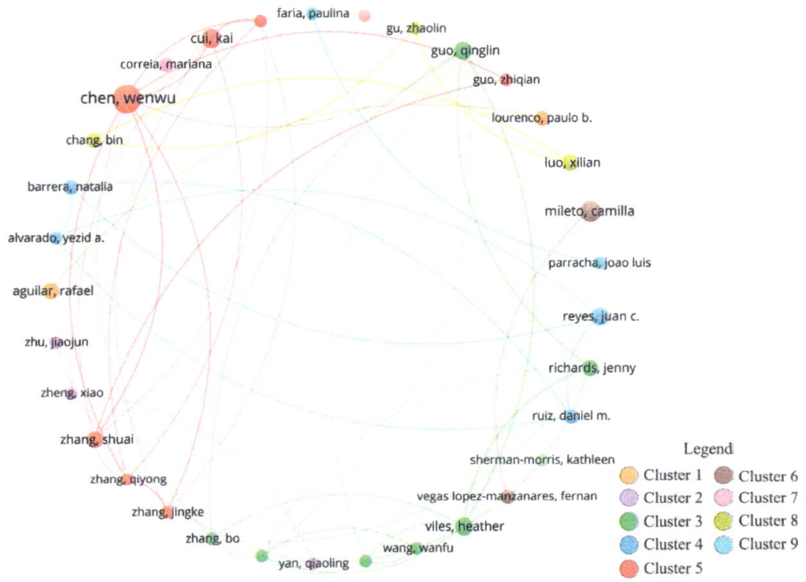

Figure 8. Author co-authorship network map (source: created by the authors).

Overall, while there remains substantial room to enhance global collaboration, the trend also reflects the rapid advancement and emergence of new academic forces in the field of earthen site surface protection worldwide.

Figure 9 displays the top 10 authors worldwide with the most significant citation bursts in the field from 2000 to 2025. The citation burst analysis identifies authors whose publication frequency or citation rate has increased markedly during a specific time period, helping to reveal the field's core contributors, emerging research trends, shifting focuses, and evolving hotspots. The author with the highest burst strength is Alvarado, Yezid A., whose citation burst began in 2022 and lasted through 2023, with a burst strength of 2.18. His research during this period gained substantial attention within the academic community. The author with the longest burst duration is Aguilar, Rafael, who maintained a high burst level from 2015 to 2019. His work has had a long-lasting influence and significant impact on subsequent studies. In recent years, Zhang, Shuai has also emerged as an influential scholar in the field, with growing recognition and impact for his recent research. These findings indicate that research in the field is undergoing a period of rapid development, marked by the emergence of a new generation of influential academic contributors.

3.5. Journal Co-Citation Analysis

In VOSviewer, a hotspot density map (or "cloud map") visually represents the distribution of research hotspots within a field by illustrating the co-occurrence frequency or association strength of entities such as keywords, documents, or authors through color gradients and density. Typically, the color scheme ranges from cool tones (e.g., blue) to warm tones (e.g., yellow), with warmer colors indicating higher co-occurrence frequencies or stronger associations, thus signifying core research hotspots in the field. Higher density—indicated by denser clusters of points—reflects more frequent co-occurrences, suggesting that the research themes in that area are more focused. In contrast, isolated keywords may represent peripheral topics or emerging directions. Clusters of varying sizes correspond to distinct thematic areas.

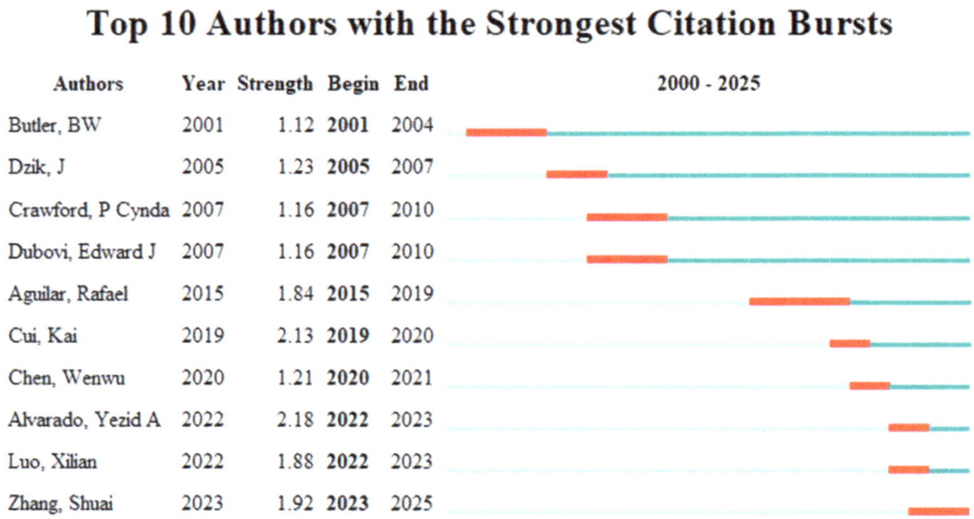

Figure 9. Top 10 authors with the strongest citation bursts (source: created by the authors).

Figure 10 presents a hotspot journal density map in the field of earthen site surface protection. Analysis reveals that the *Journal of Cultural Heritage*, *Construction and Building Materials*, *Nature*, and *Science* exhibit high cluster density and occupy the core of the hotspot map, marking them as authoritative and influential journals within the field. However, top-tier comprehensive journals such as *Nature* and *Science* may exhibit high clustering intensity due to interdisciplinary reviews or research, rather than systematic contributions to surface conservation technologies for earthen sites. In contrast, specialized journals such as the *Journal of Cultural Heritage* account for a larger proportion of publications on earthen site conservation, with a focus on core hotspots such as material development and monitoring technologies. Their role in technological leadership is weakened by the generalized representation of "hotspot clusters" in the diagram. Conversely, journals such as *Geophysical Research Letters* and *Journal of Bacteriology* exhibit lower cluster densities and are typically interdisciplinary in nature—related to geophysics or microbiology—with more indirect links to earthen site surface protection. *Heritage-Basel* appears in a bright color and forms an independent cluster, reflecting its focus on international heritage protection and signaling an emerging trend in the field. Core journals identified in the visualization can serve as key sources for literature review, while newly emerging journals often indicate potential research frontiers. Overall, the journal hotspot map offers a clear overview of journal distribution, disciplinary structure, and development trends in the field of earthen site surface protection, providing a valuable reference for future research planning.

Figure 11 presents the top 15 most-cited academic journals from 2000 to 2025. The data show that *Heritage-Basel* (6.86, 2023–2025) and *Heritage Science* (6.03, 2023–2025) exhibit the highest burst intensities, indicating that these two journals have had the greatest impact on academic research in their respective fields during this period. This suggests that research on earthen site surface protection has become a central focus in recent years. Additionally, journals such as *Remote Sensing-Basel* and *Geosciences* have shown strong citation trends since 2022, reflecting the growing application and influence of remote sensing and geophysical technologies in the field of earthen surface protection. Their rising attention in academic circles highlights significant advancements in related studies. The strong performance of these journals demonstrates their critical roles within their domains, and the duration and intensity of their citation bursts underscore their importance in driving academic research and the dissemination of knowledge.

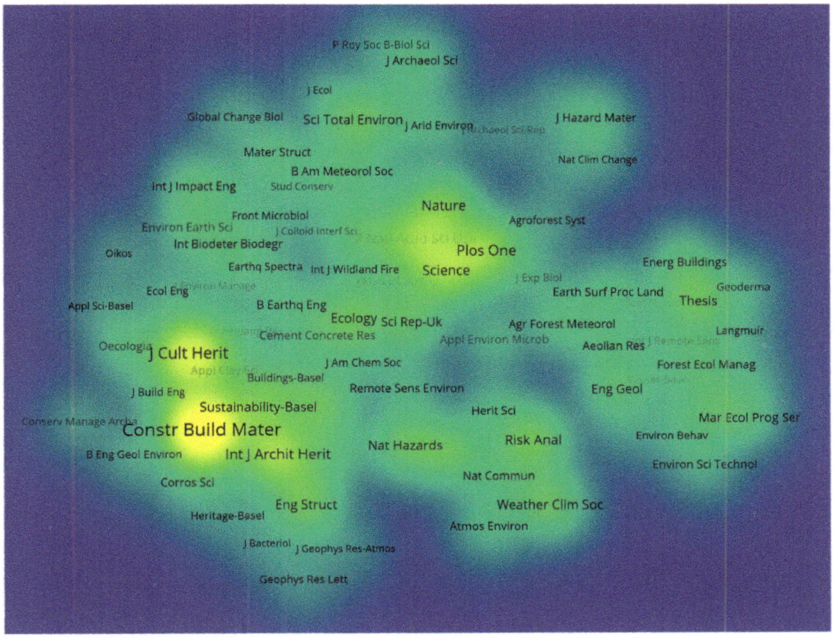

Figure 10. Hotspot journal visualization of earthen site surface protection (image source: drawn by the author).

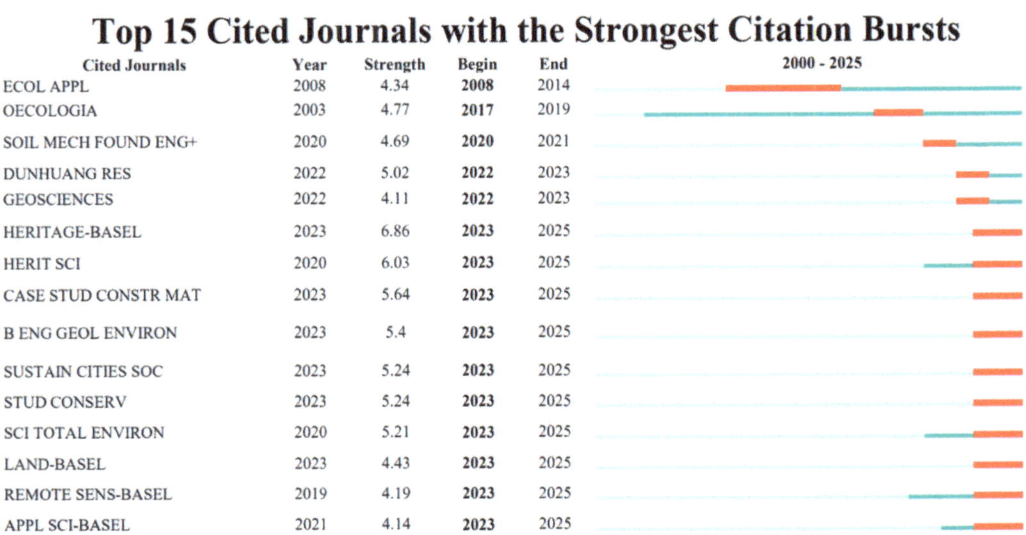

Top 15 Cited Journals with the Strongest Citation Bursts

Cited Journals	Year	Strength	Begin	End	2000 - 2025
ECOL APPL	2008	4.34	**2008**	2014	
OECOLOGIA	2003	4.77	**2017**	2019	
SOIL MECH FOUND ENG+	2020	4.69	**2020**	2021	
DUNHUANG RES	2022	5.02	**2022**	2023	
GEOSCIENCES	2022	4.11	**2022**	2023	
HERITAGE-BASEL	2023	6.86	**2023**	2025	
HERIT SCI	2020	6.03	**2023**	2025	
CASE STUD CONSTR MAT	2023	5.64	**2023**	2025	
B ENG GEOL ENVIRON	2023	5.4	**2023**	2025	
SUSTAIN CITIES SOC	2023	5.24	**2023**	2025	
STUD CONSERV	2023	5.24	**2023**	2025	
SCI TOTAL ENVIRON	2020	5.21	**2023**	2025	
LAND-BASEL	2023	4.43	**2023**	2025	
REMOTE SENS-BASEL	2019	4.19	**2023**	2025	
APPL SCI-BASEL	2021	4.14	**2023**	2025	

Figure 11. Top 15 most-cited journals with the strongest citation bursts (2000–2025) (source: created by the authors).

3.6. Document Co-Citation Analysis

Figure 12 presents an analysis of the citation frequency network based on the core literature published between 2000 and 2025. Using the CiteSpace software, the visualization illustrates the citation relationships among different publications, thereby helping to identify the most influential papers and scholars in the field. Each node represents a cited reference, and the larger the node, the more frequently the reference has been cited, indicating greater academic influence. The color gradient—from cool colors (blue, purple) to warm colors (yellow, orange)—represents the temporal span of citation activity. The connecting lines between nodes indicate citation relationships; the quantity and density of these lines reflect the closeness and academic relevance between references. Different colors of the lines correspond to citation relationships established during different time periods. Prominent nodes such as Richards, J. (2020) [63], Richards, J. (2019) [64], and Du, Y.M. (2017) [65] are frequently cited, demonstrating their high impact and foundational role in the field of surface protection of earthen sites. Moreover, the connections among these highly cited papers suggest that they are often co-cited in the same references, indicating thematic similarities. This further confirms the significant influence that these works have exerted on the research domain of earthen site surface conservation.

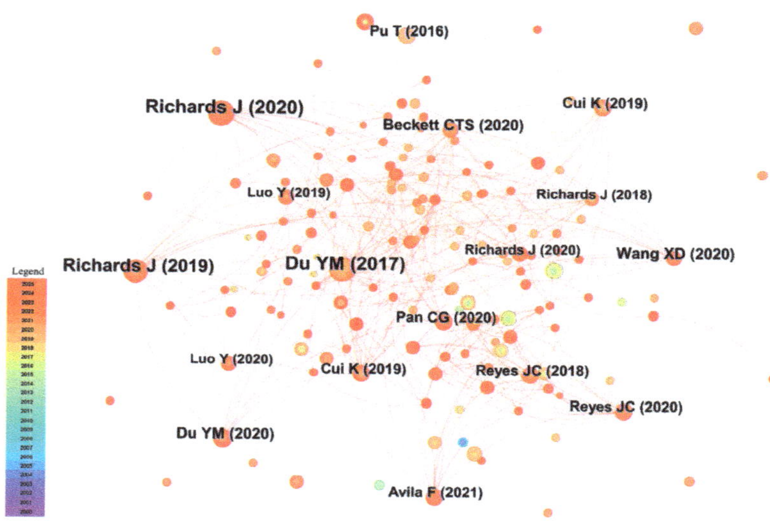

Figure 12. Document co-citation network analysis (source: created by the authors).

3.7. Keyword Co-Occurrence Cluster Analysis

Keyword clustering reveals the distribution and interrelations of various research hotspots, thereby reflecting the primary thematic focuses in the field of surface protection of earthen sites. As illustrated in Figure 13, this study conducted a keyword co-occurrence cluster analysis using the VOSviewer software. By setting the minimum occurrence threshold to 30, a total of 155 keywords were extracted from 18,467 keywords, resulting in a clear and detailed visualization. In the map, each node consists of a circle and a label, where the size of the circle indicates the frequency of the keyword's occurrence, and the thickness of the connecting lines represents the strength of the relationships between keywords. Different colors represent distinct research emphases, forming four separate clusters. This approach effectively highlights the domain's key areas of interest and provides insight into the evolving academic trends and emerging themes in the surface protection of earthen sites.

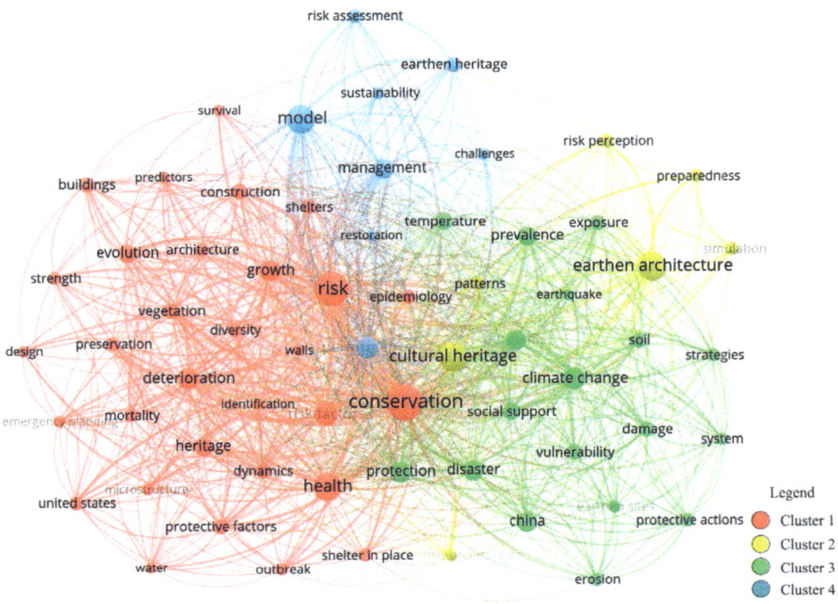

Figure 13. Cluster analysis of high-frequency keywords in the field of spatial conservation of earthen sites (source: created by the authors).

Cluster 1 (Red) focuses on risks and health-related conservation of earthen sites. Keywords include "risk," "health," "conservation," and "cultural heritage." This cluster centers on the various risks facing earthen site surfaces and explores how to ensure their healthy survival [33]. Emphasis is placed on the attributes of cultural heritage during protection processes, discussing how to balance risk response and heritage conservation, preserve traditional characteristics, and incorporate contemporary conservation approaches.

Cluster 2 (Green) centers on the natural environmental impacts on earthen site surfaces and corresponding conservation strategies. Keywords include "earthen architecture," "climate change," "soil," and "erosion." This cluster analyzes environmental factors such as climate change, soil conditions, and erosion that affect the surface of earthen sites and explores nature-based conservation strategies and interventions.

Cluster 3 (Blue) highlights the development of models and management systems for the surface protection of earthen sites, with keywords such as "model," "management," "sustainability," and "earthen heritage." This research cluster delves into the construction of quantitative models for surface protection and the establishment of scientific management systems, analyzing the driving mechanisms behind sustainable development and their implications for managing the surface conservation of earthen sites [21].

Cluster 4 (Yellow) addresses risk perception and response strategies related to earthen site surfaces. Notable keywords include "risk perception," "preparedness," "simulation," and "challenges." This cluster emphasizes a multidisciplinary approach to understanding public and professional perceptions of risks to earthen site surfaces. It examines the challenges through simulation methods and explores preparedness strategies, offering comprehensive approaches to enhance the effectiveness of surface protection for earthen sites.

However, there is a lack of representation for cross-disciplinary technological domains. For instance, "climate change" in the green cluster and "sustainability" in the blue cluster are connected by a relatively thin line in the diagram, which fails to reflect the strong causal relationship that exists in the surface conservation of earthen sites. Some cutting-edge technology keywords, such as DT, are absent from high-frequency clusters, resulting in a disconnect with the conclusions on "intelligentization" and "sustainabilization." This may be due to the keyword filtering threshold (a minimum of 30 occurrences), which excludes emerging terms within the field and hinders the clustering results from capturing the trends of technological integration.

The keyword temporal frequency map (Figure 14) illustrates the frequency and temporal distribution of keywords over time. The relationships among keywords help reveal thematic correlations within the research field and uncover the dynamics of academic development and emerging hotspots. In this visualization, each node consists of a circle and a label. The size of the circle is proportional to the frequency of keyword occurrence—larger circles and fonts represent higher frequencies. Lines between nodes indicate co-occurrence relationships, with line thickness reflecting the strength of those connections. The color of each circle, as indicated by the gradient legend in the bottom-right corner, represents the average year of occurrence. Blue indicates earlier appearance, while orange indicates more recent usage. For example, keywords such as "deterioration" and "heritage" are concentrated in the earlier, blue-colored region, reflecting early research interests. In contrast, keywords such as "sustainability" and "risk assessment" appear predominantly in the more recent, orange-colored region, indicating that they have become major focal points in the field in recent years.

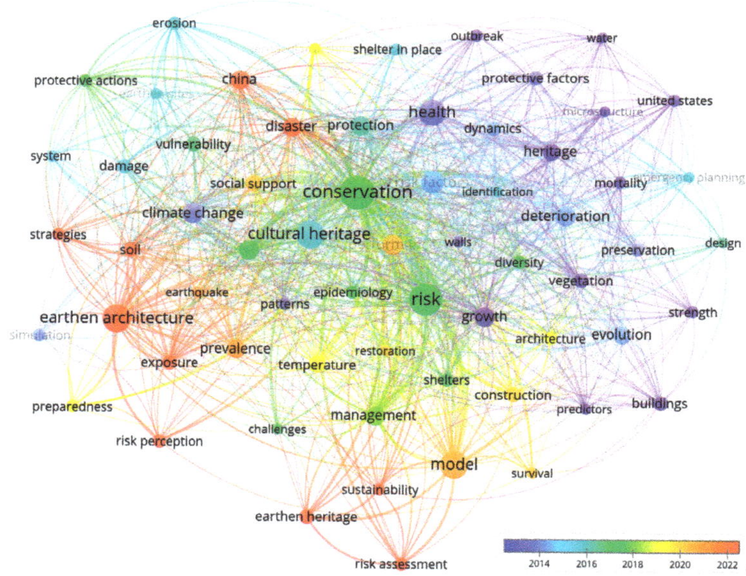

Figure 14. Temporal frequency variation of hot keywords in earthen site protection research (source: drawn by the authors).

This study reviews nearly 25 years of research on earthen site surface protection and proposes a five-phase model of the literature's evolution based on the keyword co-occurrence frequency and the temporal distribution of emerging keywords. Although the time intervals of each phase differ, the distinct characteristics and varying rates of development effectively reflect the corresponding stages of research progress. The first phase (2000–2014), lasting 15 years, primarily focused on the attributes and characteristics of earthen sites themselves and the establishment of theoretical frameworks for surface protection through case studies. This phase marks the initial stage of research on earthen site surface protection, which is characterized by relatively slow development. In contrast, the third phase (2018–2020) lasted only 3 years, coinciding with a period of technological integration. The widespread application of GIS and RS technologies greatly accelerated the advancement of earthen site surface protection, making it more intelligent and precise.

Phase 1 (2000–2014): Key hotspots during this period include "Heritage," "Cultural Heritage," and "Risk Factors." Research emphasized the cultural heritage value of earthen sites [16], deepening the understanding of their significance from a broader heritage perspective. The foundational frameworks for protection and risk response were developed [66], establishing theoretical underpinnings such as protection goals and risk classification, which later facilitated digital-technology-based research [67].

Phase 2 (2014–2016): Keywords such as "Earthen Sites," "Climate Change," and "Vulnerability" emerged [68]. This phase focused on refined analyses of earthen surface types and their correlations with climate change and vulnerability. The increasing attention to vulnerability mechanisms under environmental change marked a shift toward natural-environment-centered approaches to discovering new methods of surface protection [69,70].

Phase 3 (2016–2018): Hot keywords included "Conservation," "Durability," and "Management." The visualization results show that "Conservation" remained central in the keyword network, continuing the core agenda from earlier stages. Protection during this phase emphasized scientific and systematic approaches, shifting from conceptual frameworks to assessments of practical effectiveness and long-term feasibility. The prominence of "Durability" reflects growing concern with material weatherability and damage resistance [71]. The close linkage between "Management" and "Conservation" signals the transformation of theory into operational management systems, such as the development of risk assessment models or the incorporation of sustainable strategies addressing ecological, economic, and social dimensions [72,73].

Phase 4 (2018–2020): The key terms were "Model," "Construction," and "Architecture Evolution." This phase signaled a transition toward practical implementation, with emphasis on applying surface protection technologies to real-world contexts. The emergence of "Model" indicated efforts to develop and apply quantitative models for dynamic simulations and data-driven decision making [74]. "Construction" highlighted the use of new materials and techniques for surface repair and reconstruction [75]. "Architecture Evolution" reflected studies on how natural and anthropogenic forces alter earthen site surfaces, while architectural and mechanical knowledge was leveraged to enhance conservation strategies, such as in situ preservation [70,76].

Phase 5 (2020–2025): Emerging keywords include "Strategies," "Risk Perception," and "Sustainability." This phase centers on integrating various protection methods into comprehensive strategies. Greater attention is paid to understanding risk perceptions and advancing the sustainable development of earthen sites across ecological, economic, and social domains [77]. The rise of "Sustainability" as a core keyword reveals a heightened emphasis on the synergy between long-term development and surface protection, echoing the field's alignment with broader societal demands for ecological and socioeconomic sustainability [78] and promoting the dual goals of cultural heritage preservation and environmental conservation [79].

Figure 15 illustrates the top 15 most frequently cited keywords in the field of earthen site surface protection research between 2000 and 2005. Keywords such as "risk factors" and "cultural heritage" represent the academic focal points of that specific period. This analysis not only reveals the dynamic evolution of research hotspots across different stages but also reflects the shifting frontiers and focal themes within the field of earthen site surface protection. Among them, "cultural heritage" exhibits the highest burst intensity, with a rate of 4.7 during the burst interval of 2015–2021, indicating significant scholarly attention to topics related to cultural heritage in recent years. It is worth noting that there is an issue of term generalization regarding core concepts. For example, "cultural heritage" exhibited the highest burst intensity of 4.7 during 2015–2021, but it did not differentiate its proportion between earthen sites and other heritage types (such as stone heritage). As a distinctive type of heritage, earthen sites require targeted surface conservation technologies (e.g., rammed earth reinforcement, wind erosion control). The generalization of this keyword may obscure the lack of specialized technological research on earthen sites, leading to biased judgments on core issues within the field. In addition, keywords such as "earthen architecture" (2008–2025, burst intensity: 3.25) and "resilience" (2010–2020, burst intensity: 3.25) also show strong burst trends, highlighting major academic interest and activity within the domain. This trend suggests a global academic shift from the early focus on traditional risk factors to broader concerns regarding the relationship between large-scale sites and cultural heritage, as well as the development of long-term protection mechanisms. The emergence of these keywords not only underscores current research frontiers and hotspots but also helps scholars identify potentially significant future directions, such as in-depth studies around emerging keywords such as "heritage conservation" (2023–2025) and "earthen site" (2023–2025).

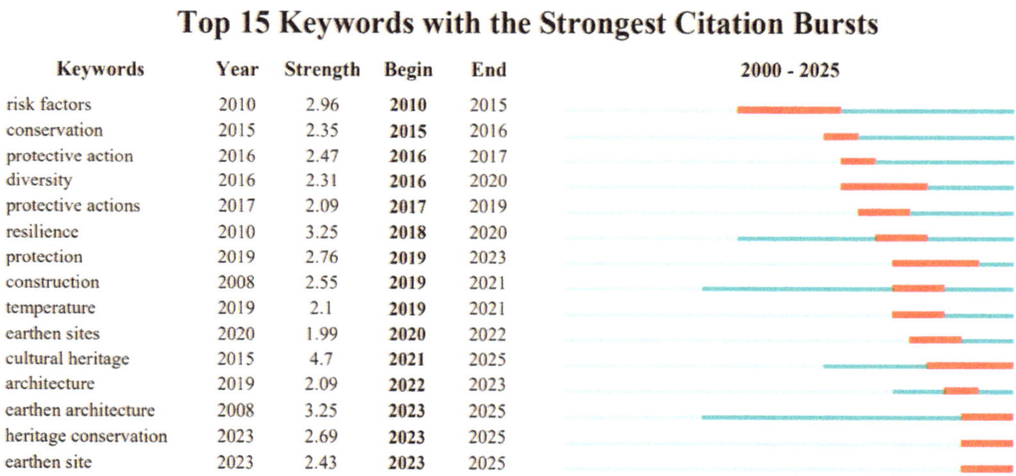

Figure 15. Top 15 journals with the strongest citation bursts in earthen site protection research (source: drawn by the authors).

Based on the above visualization results, the research trajectory of surface protection for earthen sites can be interpreted through a dual-dimensional lens of temporal distribution and burst intensity. In the early stage, foundational concepts such as "Cultural Heritage" appeared early and maintained a strong citation burst (4.7), consistently occupying a dominant position. With theoretical advances and technological progress, keywords such as "Resilience" emerged in the transitional zone of the keyword temporal frequency map, accompanied by a notable citation burst, marking a shift from theory-oriented to practice-driven research. In the later stage, the term "Sustainability" becomes prominent in the frequency timeline, while "Heritage Conservation" and "Earthen Site"—which exhibit strong citation bursts from 2023 to 2025—act as complementary indicators. Together, these patterns suggest a developmental trajectory that moves from foundational studies toward a sustainability-oriented research paradigm.

4. Discussion

4.1. Problem Revelation and Method Innovation

A review of the research on the surface protection of earthen sites over the last 25 years reveals a clear trend toward digital transformation. However, notable deficiencies remain in capturing surface deterioration details, enabling real-time monitoring, and ensuring long-term effectiveness. Although promising progress has been made in the application of digital technologies and interdisciplinary collaboration, certain limitations and contradictions persist, hindering the evolution of protection paradigms. As shown in Table 1, research hotspots currently focus on the application of advanced digital technologies, yet the limited depth of technological integration constrains the development of dynamic protection capabilities. In terms of keyword trends, there has been a shift from basic surface protection in earlier studies toward sustainable surface protection in recent years. This shift highlights a gap in interdisciplinary theoretical collaboration, as a systematic protection framework has yet to be fully established. Regarding research trends, while there have been explorations in constructing digital technology repositories and integrating them with city information modeling (CIM), the lack of interdisciplinary theoretical support often results in a disconnect between technological applications and practical protection work.

Table 1. Summary of research hotspots, keyword evolution, and future trends (source: compiled by the authors).

	Question 1	**Question 2**	**Question 3**
Question Content	What are the research hotspots of the article?	How have the hotspot keywords evolved?	What are the future research trends and technical priorities?
Brief Answer	Prioritizing the application of digital technologies, multi-tech collaborative monitoring, integration with sustainable development, and global interdisciplinary collaboration	Based on development stages, the evolution trend shifts from theory-driven to intelligence-driven, emphasizing sustainable protection	Building a digital technology base for earthen site protection, applying CIM, promoting smart and sustainable conservation, and emphasizing interdisciplinary and international cooperation
Objective	Exploring technological integration and long-term protection mechanisms	Clarifying the correlation, evolutionary logic, and driving mechanisms of keyword change	Promoting intelligent prediction, smart management, and sustainability in protection
Research Contribution	Revealing the evolution path of surface protection technologies, verifying the rationality of each phase, and providing innovative methodological guidance	Constructing a keyword evolution stage model and identifying key policy and technical drivers	Proposing a multidimensional technology application framework and clarifying scientific breakthroughs
Research Limitation	Technology integration is still fragmented, and a systematic technical framework is lacking, making it difficult to meet the needs of automated surface protection of earthen sites [80]	Early protection theories and keyword extraction lack coherence; analysis is insufficient, leading to weak theoretical consistency [22]	Data sources lack systematic integration, technical applicability is limited, and CIM theoretical support is weak, leaving the framework underdeveloped [81]

Based on the issues discussed above, it is suggested that a digital technology database for earthen sites should be developed [11,82]. This would allow for the integration of currently fragmented technologies into a systematic structure that connects digital technology theory with practical applications. Furthermore, on this foundation, an interdisciplinary framework integrating the theory of CIM [83] should be established, incorporating insights from archaeology, computer science, and other disciplines to address the current theoretical gaps. With the digital technology database as a foundation, tailored solutions can be proposed for different heritage, ecological, and community protection needs across varying timeframes. This will enable the iterative updating of conservation methods throughout both short-term protection and long-term evolution processes, advancing the development of a CIM-based coordinated conservation system.

In terms of methodology, this study does not rely solely on visualized outputs from VOSviewer and CiteSpace for bibliometric analysis. Instead, it incorporates in-depth manual reading of the key literature and employs GPT-assisted analysis to go beyond the traditional surface-level interpretations of bibliometric data visualizations, thereby uncovering trends in cutting-edge technologies for surface protection of earthen sites [84]. The value of GPT-assisted semantic mining in the methodology of this study is reflected in three aspects. 1. It compensates for the limitation of traditional quantitative methods that prioritize data over logical reasoning. In this study, semantic association is used to reveal the internal driving mechanism of the "digitization–intelligentization–sustainabilization" technological integration. 2. It addresses the challenge of identifying theoretical discontinuities in interdisciplinary research. For example, by using GPT-assisted analysis to examine the correlation between meteorological data and studies on the performance of reinforcement materials, theoretical gaps are identified, enabling the construction of interdisciplinary deterioration prediction models. 3. It enhances the credibility of the conclusions by quantitatively extracting the core points of the literature and validating the rationality of the three-stage technological evolution (single-point monitoring–spatial modeling–intelligent integration). GPT was used as a preliminary tool to extract information and key points from the literature, while the researchers applied professional judgment to review, revise, and supplement the preliminary outputs, performing in-depth analysis and making final decisions. This approach ensured the accuracy and reliability of critical information in the literature [85]. The analysis thus established a multi-level framework combining a qualitative close reading with quantitative technical analysis, revealing the underlying logic of theoretical innovation and technological advancement, and guiding the field toward deeper and more comprehensive development.

4.2. Analysis and Interpretation of Hot Literature

This study conducts a comprehensive analysis of the research content of hotspot articles to summarize the current state of research on surface protection of earthen sites. As shown in Table 2, a multi-technology collaborative monitoring system has gradually taken shape, with increasing emphasis on the coupling of natural and human-induced factors. The trend toward interdisciplinary integration is evident, and emerging technologies such as DT are being applied to enable preventive protection. Looking ahead, further research is needed on multi-physical field coupling, the development of standardized technical systems, the promotion of AI and DT applications, the expansion of VR-/AR-based site presentation, and the enhancement of international cooperation and cross-disciplinary integration, with greater attention to the synergy between heritage conservation and sustainable development.

Table 2. Analysis of research content in the key literature (source: compiled by the author).

No.	Author(s)	Year	Innovation Approach	Core Technology	Core Findings
1	Yao et al. [3]	2021	Integrated multidisciplinary research, established a quantitative evaluation method for surface damage vulnerability based on disaster risk theory	Used GIS to simulate RS-based data and proposed a surface risk vulnerability index system	Hard to quantify multi-source factors; cloud cover leads to RS data gaps; 70% of the evaluation is based on manual simulation
2	J. Richards et al. [63]	2019	Proposed a surface erosion simulation framework, focusing on prediction of erosion impact and classification of damage levels	Used PWESD and PFESD erosion simulation models; proposed an early warning system for surface stability	Limited applicability to different sites; difficult to validate and calibrate data in field conditions
3	Chen et al. [86]	2020	Established an ecological protection zoning system and quantitative damage classification index system	Used InSAR and NDVI surface vegetation indices to analyze slope and erosion risk	Vegetation indices have low correlation with fine structural diseases; natural restoration ability is difficult to predict
4	Guo et al. [33]	2022	Constructed a surface multi-type disease diversity diagnosis system, proposed a method for evaluating the impact of environmental factors	Used InSAR and NDVI remote sensing indices; proposed a comprehensive vulnerability index system	Difficult to determine interaction between factors; affected by regional climate and surface reflection effects
5	Li et al. [15]	2011	Built a PS monitoring model and used it to monitor the structural disease process of large-scale earthen sites	Used PS to extract deformation velocity and estimated the surface structural deformation rate	PS is suitable only for slow deformation zones; it cannot detect microcracks and surface morphology
6	Sánchez-Calvillo et al. [34]	2024	Based on SCOPUS 2004–2023 data, applied VOSviewer and Bibliometrix for the mapping of scientific research hotspots	Used big data bibliometric technology; visualized hotspot keyword clusters	Unable to distinguish the practical implementation of the keywords; lacks application relevance
7	Liu et al. [56]	2024	Proposed a surface VR digital twin damage simulation platform and evaluated its effectiveness through multi-angle verification	Built a VR/AR multi-scene interaction model to simulate disease evolution and preventive decision making	Simulation lacks physical feedback; difficult to replicate material reaction processes
8	Guo et al. [87]	2022	Combined LiDAR and UAV oblique photogrammetry; improved 3D model detail resolution	Used SFM reconstruction technology and the ICP algorithm to optimize point cloud data	Restricted by scanning coverage; limited by the number of control points and terrain occlusion
9	Li et al. [8]	2011	Built a PS monitoring model; explored the relationship between damage location and spatial structure	Used PS time-series analysis to extract damage zones; built surface displacement evolution models	PS has low resolution; unable to reflect surface deterioration process at the micro level
10	E Puertas et al. [57]	2024	Used multiple data sources such as RS, UAV, and 3D scanning; created a multidimensional protection system integrating ecology, disaster, and cultural value	Combined decision making models and data fusion models; proposed a multi-criteria collaborative surface protection system	Lack of unified standard model; difficulty in integrating data from different formats

Current research on the surface protection of earthen sites builds upon the foundational understanding that surface deterioration is primarily driven by environmental factors (e.g., freeze–thaw cycles, wind erosion). However, by conducting multi-field coupling experiments that accelerate the deterioration process, researchers have proposed the concept of a "deterioration chain," thereby expanding the traditional view of single-factor influence. The application of SBAS-InSAR technology in conjunction with LiDAR and unmanned aerial vehicle (UAV) collaborative modeling has further advanced earlier approaches that relied on RS and GIS for surface spatial analysis. This combination of LiDAR and UAV addresses the limitations of using single-source RS data and the inefficiencies of labor-intensive field methods (such as manual comparative experiments), enhancing overall efficiency and mitigating data deficiencies [64]. Moving forward, it remains essential to deepen the investigation of multi-field coupled deterioration mechanisms, establish standardized protection technology systems, and further apply cutting-edge digital technologies in practice. In particular, greater emphasis should be placed on exploring synergistic pathways between the surface protection of earthen sites and sustainable development.

4.3. Research Hotspot Transformation and Development Trends

The current challenges facing the surface protection of earthen sites stem from the dual impact of climate change and anthropogenic damage. Traditional protective technologies are inadequate for addressing the synergy between micro-scale deterioration and macro-environmental shifts, necessitating a shift toward digital preservation and enhanced international cooperation. The theoretical rationale for this trend is as follows: (1) as illustrated in the keyword co-occurrence map, terms such as "climate change," "disaster," and "risk factors" demonstrate strong interconnectivity, highlighting the global risks and challenges posed by climate change and disasters to the surface protection of earthen sites [69,88]. The integration of digital technologies enables global collaborative monitoring through tools such as the IoT for disaster detection [89,90] and artificial intelligence (AI) for simulating climate change impacts [91]. These technologies facilitate comprehensive risk monitoring and early warning, as well as the sharing of monitoring data and coordinated protection measures to address global climate-related risks [92]. (2) The burst keywords "cultural heritage" and "earthen architecture" underscore the global significance of earthen site surface protection. Accordingly, this study advocates for the development of a global deterioration database for earthen sites and the establishment of a unified protection technology model. This would facilitate a transition toward precise, sustainable, and internationalized protection, supporting the formation of a world heritage conservation system [7,93].

By analyzing the evolution of hotspot keywords in the field of earthen site surface protection from 2000 to 2025 (as shown in Table 3), it becomes evident that the shift has been driven by a combination of policy influence, climate change, technological advancement, international collaboration, and conceptual innovation. This has led to a transition from single-category structural protection toward a multi-category, full-life-cycle protection paradigm.

Table 3. Analysis of factors driving keyword evolution (source: created by the authors).

Policy/Background	Phase	Nature	Goal Orientation	Keywords	References
Earthen site surfaces suffer from natural weathering and human damage	2000–2010	Independent research by academic bodies	Identifying causes of surface damage and risk factors, determining material science mechanisms, and clarifying structural stability	Surface damage factors, risk assessment, architectural structure, mechanical properties, etc.	Shabani et al. (2020) [94], Nazari et al. (2017) [95], Aven (2009) [96], Fabbri et al. (2018) [97]
Inclusion of earthen sites into the cultural heritage system	2011–2015	Policy-driven preventive research	Establishing surface protection systems for earthen sites, exploring minimally invasive restoration techniques	Cultural heritage, surface protection, restoration, earthen structures, etc.	Kono (2014) [98], Liu et al. (2022) [99], Zhang et al. (2021) [100], Rosado Correia et al. (2014) [101]

Table 3. *Cont.*

Policy/Background	Phase	Nature	Goal Orientation	Keywords	References
Intensified impacts of climate and temperature on earthen sites; emergence of emergency management systems	2016–2018	Multidisciplinary and cross-sectoral research	Constructing climate-adaptive protection technologies and emergency response systems to improve the resilience of earthen site surfaces	Climate change, conservation action, emergency planning, durability, etc.	Sardella et al. (2020) [102], Zhang (2023) [103], Han et al. (2022) [104], Shi et al. (2020) [105]
The Paris Agreement promotes green transformation of heritage conservation	2019–2020	Policy-driven interdisciplinary research	Integrating ecological needs, exploring sustainable renewal models for urban heritage sites, enhancing resilience	Resilience, adaptive management, sustainability, temperature, etc.	Peters et al. (2017) [106], Zuazo et al. (2008) [107], Qu et al. (2025) [108]
Popularization of digital technologies and the "Belt and Road" Initiative driving transboundary earthen site protection cooperation	2021–2023	Technology-driven international cooperation	Achieving precise protection, building cross-border heritage conservation networks	Earthen sites, digital surface protection, international cooperation, added value, etc.	Barton (2009) [109], Lercari (2019) [110], Guo et al. (2018) [111]
Dynamic conservation of earthen sites to address long-term risks	2024–2025	Future-oriented and community-guided research	Constructing frameworks of "Conservation–Transmission–Development", promoting local knowledge integration	Sustainable development, dynamic conservation, long-term risks, knowledge systematization, etc.	Pereira Roders et al. (2011) [112], Zhong et al. (2021) [113]

Based on the evolution of hotspot keywords, this study attempts to summarize the developmental pattern of earthen site surface protection into six stages after several rounds of selection.

(1) Basic Protection Stage (2000–2010): Due to severe natural disasters and human-induced damage, earthen sites were in urgent need of protection during this period. Archaeologists began to explore the fundamental scientific issues related to earthen sites. The main focus was on the architecture and material mechanics of earthen sites, using simulation experiments to identify weak points in site walls and earthen structures [94,95]. A pioneering "risk factor" quantitative assessment framework was proposed [96], and univariate analyses on material mechanical properties were conducted [97]. However, a comprehensive protection system had yet to be established, and policy interventions were largely absent.

(2) Policy Foundation Stage (2011–2015): This stage followed China's endorsement of *The Nara Document on Authenticity* (Asia–Pacific region) [98]. During this time, the principle of authenticity was officially established, and earthen sites were incorporated into the legal framework of "cultural heritage" protection [99]. The research focus shifted from individual surface restoration to systematic protection. Emphasis was placed on minimal intervention techniques based on the theoretical principles of authenticity and reversibility [100,101]. Efforts were made to build international standard protection procedures and cooperation models, although the field was still in the early stage of multidisciplinary integration.

(3) Transformation and Response Stage (2016–2018): With increasingly extreme global weather, earthen sites faced intensified damage from water erosion, salt corrosion, and thermal expansion and contraction [102–104]. "Climate change," "water," and "temperature" became key research topics in surface protection. Disaster modeling was introduced [105], and weather-resistant reinforcement materials were developed alongside on-site emergency response plans. The research emphasis shifted from static protection to dynamic prevention, following a multidisciplinary research path.

(4) Conceptual Expansion Stage (2019–2020): The *Paris Agreement* emphasized synergizing carbon reduction with local development [106]. The keyword "resilience" surged in 2018 and became a central topic. This stage marked a shift beyond traditional

protection concepts, incorporating approaches such as vegetation-based protection and microclimate regulation to enhance system robustness [107,108].

(5) Innovation and Cooperation Stage (2021–2023): Advanced digital technologies rapidly developed during this period, with 3D laser scanning and BIM technology being widely applied in the surface protection of earthen sites [109,110]. At the same time, the *Belt and Road Initiative* facilitated international cooperation in the protection of earthen site surfaces [111], promoting a shift from traditional material improvement to intelligent monitoring enabled by digital technology.

(6) Integrated Protection Stage (2024–2025): Following advances in frontier technologies and updates in conservation philosophy, earthen site surface protection has centered on "sustainable strategies," "living conservation," and "dynamic monitoring" [112]. The focus has turned to integrated approaches combining conservation, heritage transmission, and development, with tools such as VR virtual displays enabling the continuation of living heritage and transformation of its value [113].

In conclusion, the current research trend in earthen site surface protection is evolving from qualitative to quantitative approaches and from single-method studies to multi-method interdisciplinary collaboration. With the continuous development of frontier digital technologies, significant progress has been made in the scientific understanding and practical implementation of earthen site surface protection.

4.4. Summary of Digital Technology Methods and Application Scenarios

Since the 21st century, surface protection technologies for earthen sites have developed into a comprehensive system encompassing digital modeling, intelligent monitoring, and sustainable management [4], resulting in the digital technology application framework presented in Table 4, which spans macro-, meso-, and micro-scales, the problems to be addressed, and the corresponding methods.

Table 4. Advanced technological application framework for the surface spatial protection of earthen sites (source: compiled by the authors).

Research Scale	Digital Technology	Problems Addressed	Methods	Function	Technical Limitations
Macro-Scale	DT	Monitoring dynamic changes in site clusters and conservation planning	Multi-source sensing data modeling and IoT data collection	Scenario simulation and risk prediction	Difficult to capture fine surface details; modeling precision needs improvement
	BIM	Spatial distribution of site clusters and environmental impact assessment	3D model integration and environmental zoning analysis	Coordination of protection and development	Data accuracy is low; environmental zoning precision is insufficient
	IoT	Large-scale environmental risk forecasting	Sensor networks for regional risk monitoring	Regional risk alerting and real-time monitoring	Sensor coverage is incomplete; maintenance costs are high
	CIM	Coordination of site and city planning, environmental synergy evaluation	Integration of spatial datasets and multi-scenario simulation modeling	Assists macro-scale heritage protection and policy formulation	Data integration is difficult; simulation accuracy needs improvement
Meso-Scale	LiDAR	3D modeling of single site structures	Ground-based scanning and point cloud processing	Diagnosis of site surface structure and disease	Complex data processing; low efficiency for surface detail representation
	BIM	Life-cycle information management of site surfaces	Structural restoration modeling	Supports site-level conservation and resource integration	Data freshness is low; surface data feedback lag
	AR	Visualization of restoration plans	Virtual plan demonstration and verification	Assists decision making with immersive interaction	Low realism and precision; limited interactive display capability
Micro-Scale	LiDAR	Microstructural degradation analysis of site surfaces	Detailed scanning and micro-geometry extraction	Accurate quantification of surface degradation level	Scanning period is long; surface fine detail loss is possible
	VR	Immersive damage diagnosis and risk research	Interactive VR scene construction	Helps managers simulate decision making and training	VR equipment is expensive; user experience varies

At present, cutting-edge conservation technologies are advancing into a new era characterized by digitalization, intelligentization, and sustainability. As illustrated in Figure 16, future digital system integration must incorporate multiscale frameworks and leverage interdisciplinary tools such as LiDAR for data acquisition, BIM and GIS for modeling, IoT for real-time monitoring, and AI for diagnostics. These should be integrated with CIM to address the limitations of insufficient multidisciplinary collaboration and the constraints of isolated technologies in the field of earthen site surface conservation. This will facilitate the full-process, efficient integration of digital twin systems for earthen site surface protection, covering data acquisition, model construction, dynamic monitoring, and management decision making, thereby enhancing conservation effectiveness and promoting a shift from passive to proactive protection. For instance, the "Digital Dunhuang" initiative by the Dunhuang Academy applies LiDAR scanning technology to document mural surfaces and establish a three-dimensional disease archive. By integrating multi-source data with real-time information from IoT sensors, it forms a three-tier system ranging from regional to individual site and micro-structural levels. Specifically, a GIS is used to assess ecological impacts, LiDAR captures surface crack distributions, and IoT monitors temperature and humidity variations, thereby enabling real-time supervision of both macro and micro dimensions across the entire conservation chain [114]. Similarly, the Western Xia Imperial Tombs in China, as a typical earthen site in a semi-arid region, have adopted technologies such as RS, GIS, and the IoT to establish a multi-indicator monitoring system, achieving multiscale monitoring of the earthen site surface [115]. This has verified the application value of technologies such as LiDAR and GIS at the meso- and macro-scales. However, from a critical perspective, although the monitoring system has integrated multiple technologies, there is weak interoperability among the data from different technologies. Moreover, it lacks community participation and quantitative evaluation of ecological benefits, reflecting that current digital conservation of earthen site surfaces still needs breakthroughs in technological integration depth, interdisciplinary collaboration, and sustainability assessment [22].

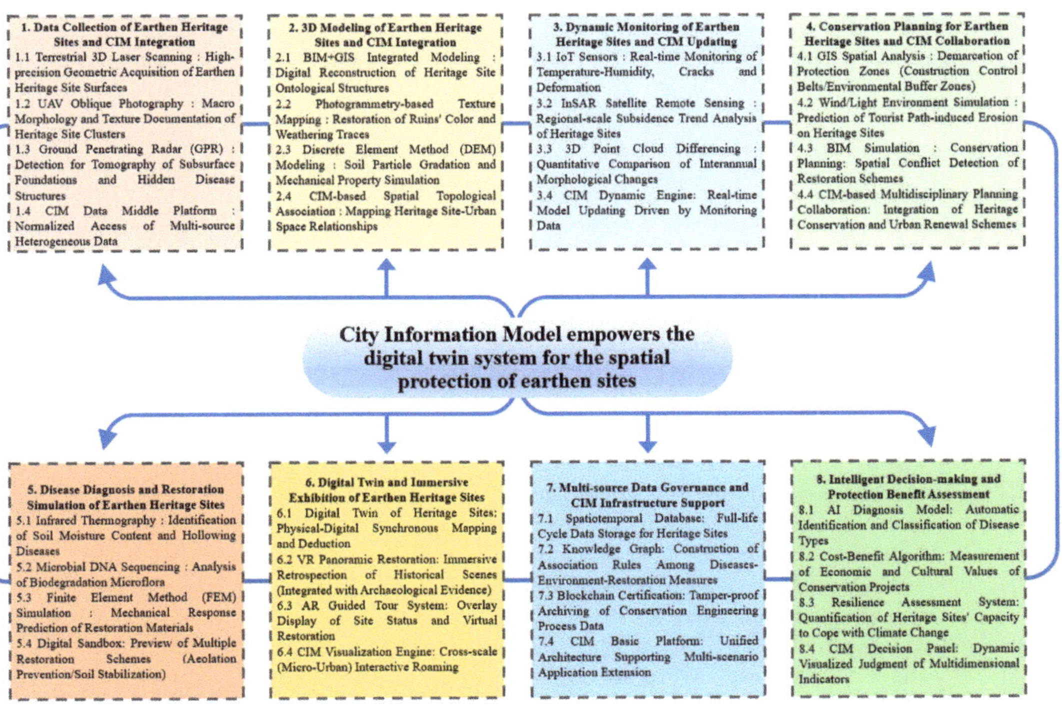

Figure 16. CIM-driven digital system integration technology roadmap (image source: created by the authors).

With the advancement of digitalization, intelligent technologies, and sustainability-focused approaches, the use of technologies such as DT and LiDAR has broken through the limitations of traditional protection methods, making proactive prevention for earthen site surface conservation possible. However, due to the diversity in surface materials, preservation conditions, and spatial scales of earthen sites, the application of a single technology is insufficient to meet comprehensive surface protection demands. This highlights the urgency and necessity of selecting appropriate technologies for surface protection, as illustrated in Table 5. By integrating multiple technologies, the conservation system can be upgraded to achieve efficient integration from data acquisition to governance decision making, thereby improving protection efficiency, avoiding technology misuse, and providing effective safeguards for the sustainable preservation of world cultural heritage.

Table 5. Technical selection table for the surface spatial protection of earthen sites (source: created by the authors).

Protection Needs	Technology Types	Selection Basis	Technical Limitations
3D Modeling	LiDAR, UAV Photogrammetry	For high-precision modeling, choose LiDAR; for large-area surveys, use UAV photogrammetry.	Precision loss with complex terrain; data stitching errors
Environmental Monitoring	IoT Environmental Sensors	To obtain long-term and real-time data, choose low-power and wireless-transmission IoT sensors.	Response delay, durability differences, and limited data accuracy
Pathology Diagnosis	Computer Vision (AI Image Recognition), Spectral Analysis	For the rapid identification of deterioration, choose high-resolution models of computer vision and multi-spectral image/spectral analysis.	Surface image recognition accuracy affected by lighting; bias in data processing
Material Selection	Nanomaterial Compatibility Testing	To match the chemical properties of earthen materials, conduct spectral analysis first, then select appropriate protective agents.	Compatibility issues; long-term effects difficult to predict

4.5. Future Research Directions and Development Trends

The development trend of hotspot keywords indicates that future research on the surface protection and utilization of earthen sites will increasingly focus on smart, digital, and sustainable approaches [116]. As illustrated in Figure 17, digital technologies are currently being applied across multiple scales, offering valuable support for both the protection and presentation of site surfaces. Looking ahead, the establishment of a cloud-based surface protection platform for earthen sites—integrating GIS, the IoT, and digital technologies—could help uncover the spatiotemporal evolution patterns of sites, thereby providing a robust technical foundation for future conservation efforts [4].

As shown in Table 6, the future direction of surface protection for earthen sites will emphasize multidimensional integration, as well as digital and intelligent conservation. This approach highlights the importance of interdisciplinary collaboration and fusion to enhance scientific rigor while promoting the widespread application and accessibility of advanced technologies. In practice, intelligent prediction models may be employed to enable proactive risk warnings, while real-time monitoring platforms can be built using multi-source sensors and IoT technologies [117]. CIM can be utilized to integrate data on a site and its surrounding environment, thereby optimizing macro-level conservation planning [118]. Moreover, blockchain technology offers innovative solutions for heritage authentication and theft prevention. For example, the blockchain monitoring project for the Chan Chan archaeological site in Peru leverages blockchain's immutability to store detailed data—such as geolocation and material composition—on-chain, thereby granting each site a unique digital identity and forming a relatively complete traceability system to enhance site security [119]. Finally, the development of digital science communication platforms can increase public awareness and engagement in site preservation, ultimately fostering a long-term protection model that integrates prevention, monitoring, restoration, security, and public participation, balancing site conservation with regional development.

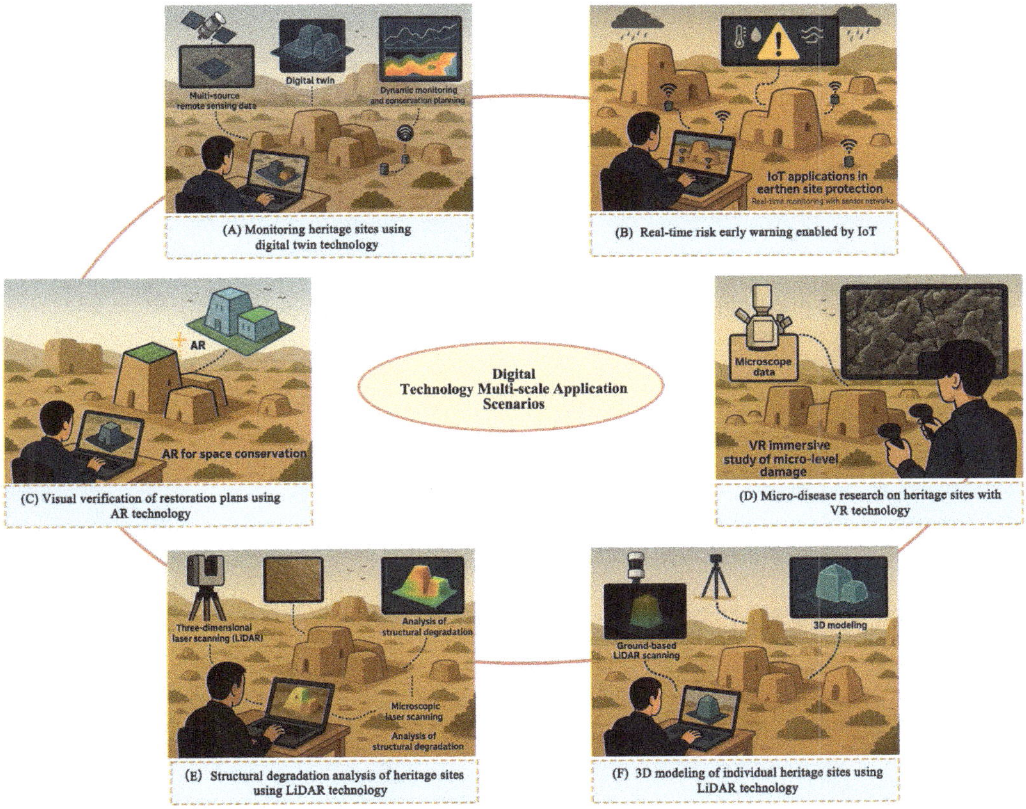

Figure 17. Multiscale application scenarios of digital technologies for earthen heritage surface protection (image source: drawn by the authors).

Table 6. Multidimensional technological applications and innovative developments in the surface spatial protection of earthen sites (source: created by the authors).

Protection Dimension	Research Technologies	Application Scenarios	Problems Solved	Technical Advantages
Preventive protection	Digital twin and intelligent prediction modeling	Predictive protection and risk warning	Traditional preventive measures lack precision	Difficult to predict micro-damage on surfaces
Real-time monitoring	Multi-source sensor networks	Real-time non-contact monitoring of displacement, humidity, and other data	Monitoring methods are limited, and data acquisition is insufficient	Surface data feedback is delayed

Table 6. *Cont.*

Protection Dimension	Research Technologies	Application Scenarios	Problems Solved	Technical Advantages
Damage warning	Digital restoration technologies	Assisting in the development of restoration plans and simulation of restoration processes and effects	Traditional restoration relies heavily on experience, making it hard to estimate results	Surface restoration effects are hard to estimate
Anti-theft measures	Blockchain-based cultural relic authentication platform	Digital certification for earthen heritage; anti-counterfeiting using blockchain	Theft and illegal trafficking of artifacts	Insufficient protection of surface information
Public awareness of site protection	Blockchain combined with VR navigation	VR-based immersive experiences and interactive exhibitions; science education	Weak awareness and limited participation in the valuation and recognition of cultural heritage	Large gap between virtual and actual surface

At present, the research trend of spatial surface protection for earthen sites, as shown in Figure 18, is entering an era of digitization, precision, and sustainability [120]. It is evident that current research hotspots, themes, and trends in this field are largely driven by digital technologies such as digital twins, blockchain, and geospatial analysis (GI) [121]. These technologies support the development of systems for virtual mirroring of sites and restoration traceability [122,123], transforming protection practices from experience-based approaches to data-driven decision making. With IoT sensors providing comprehensive site coverage and deep integration of AI algorithms [124], intelligent early warnings for site deterioration and personalized, closed-loop restoration management are becoming possible [125].

Future surface conservation of earthen sites demonstrates clear directions across five key dimensions—disease prevention and control, monitoring methods, material technologies, conservation models, and technological integration—highlighting potential research breakthroughs and challenging the limitations of traditional approaches (see Table 7).

Table 7. Future breakthrough points in earthen site conservation (source: created by the authors).

Dimension	Traditional Methods	Future Directions	Key Breakthrough Point
Disease Prevention and Treatment	Chemical disinfection	Biological crusting, microbial mineralization	Eco-friendly and long-lasting approaches
Monitoring Methods	Manual patrol, single-point measurement	AI visual recognition, wireless sensor networks (WSNs)	All-factor, real-time monitoring
Material Technologies	Cement-based materials	Modified traditional materials, bio-based materials	Compatibility and reversibility
Conservation Models	Passive restoration	Multi-scenario simulation, automatic risk prediction	Precision and proactive prevention
Technological Integration	Single-discipline approach	Interdisciplinary integration (e.g., computing, mechanics, archaeology, etc.)	Cross-domain collaborative innovation

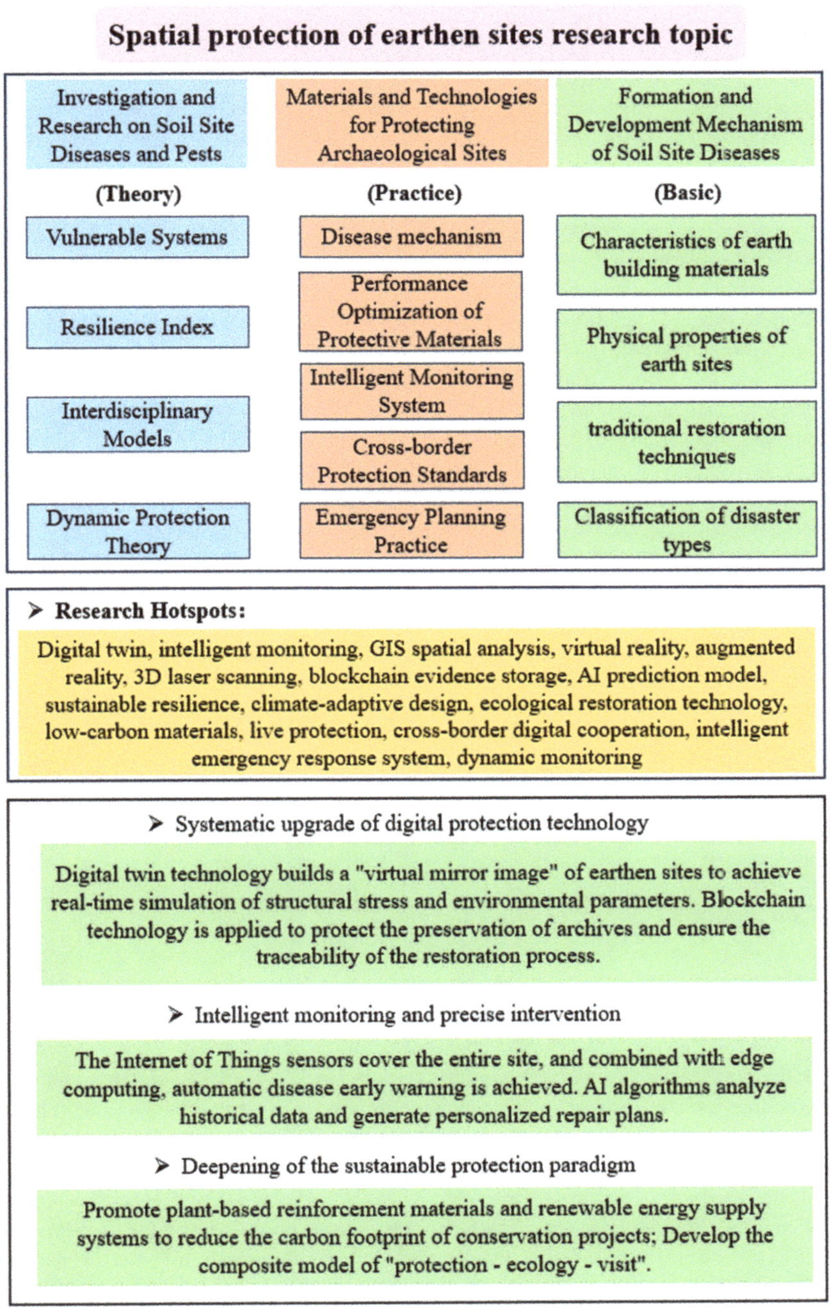

Figure 18. Schematic diagram of research trends in the spatial surface protection of earthen sites (source: drawn by the authors).

The use of biological soil crusts [126,127] and microbial mineralization [128] as alternatives to chemical sterilization harnesses the metabolic and degradative capabilities of microorganisms to achieve reinforcement and conservation. This approach not only protects the site itself but also helps maintain a healthy ecological state, contributing to the creation of long-term, ecologically friendly historical and cultural heritage sites and ensuring long-term stability [129]. To address the limitations of traditional manual on-site monitoring—such as for fragile sections, unstable collapsed supports, and voids at the top of sites—AI-based visual recognition [130], wireless sensor networks (WSNs), and robotic monitoring systems [131] are being increasingly employed for high-risk zones. Meanwhile, the use of physicochemical modification techniques and bio-based materials (which match the physical and chemical properties of the soil and are also removable after application) is improving the performance of site materials [132]. The protection model has evolved from passive restoration to preventive and precision-based conservation, supported by multi-field coupling simulation and risk prediction. For example, in the Mogao Grottoes multi-field coupling laboratory, various extreme climate conditions are simulated to improve traditional protective measures [33]. Emphasis is placed on anticipatory and precise intervention. Furthermore, the integration of multiple technologies is breaking through the constraints of single disciplines. Cross-disciplinary innovations now combine biology (e.g., microbial remediation [125]), computer science (e.g., digital twin technology [121]), and mechanics (e.g., structural mechanical analysis of earthen materials [133]). In the protection of the Sanxingdui site, technologies from computer science, archaeology, and history have been integrated to achieve 3D modeling and historical reconstruction [87].

In conclusion, the future of spatial surface protection for earthen sites lies in building a digital technology repository to serve as a foundational support system. This will deepen fundamental research, foster interdisciplinary collaboration, enhance scientific protection strategies, and promote technological implementation. Based on this repository, intelligent prediction, full-data monitoring, and early-warning systems can be established. Urban information modeling (CIM) can be used to improve the planning of surrounding areas. In parallel, blockchain-based anti-theft systems will enhance site security, while efforts to raise public awareness will encourage broader societal participation. Together, these elements form a long-term, cooperative protection mechanism that balances conservation and development. Ultimately, this will result in a protection model that is digitally driven, intelligent, precise, and sustainable, with interdisciplinary collaboration at its core [35].

5. Conclusions

At present, the surface protection of earthen sites is shifting toward a preventive paradigm, yet there remain significant gaps in data integration, the application of cutting-edge technologies, and the coordination between ecological and cultural dimensions. This study systematically analyzes the technological evolution of earthen site surface protection since the 21st century through bibliometric analysis and GPT-assisted semantic mining. The findings reveal a three-phase transition pattern in technological development: from 2000 to 2007, single-point monitoring technologies (RS, GPS) were predominant; between 2008 and 2014, the focus shifted to spatial modeling (LiDAR, GIS), enabling a transition from passive repair to proactive early warning; since 2015, the field has entered an era of intelligent integration, with emerging technologies such as the IoT and DT contributing to 62% of the annual publication growth, indicating the maturation of the preventive protection paradigm. However, critical deficiencies persist in technical applications; while foundational technologies (GIS, BIM) boast coverage rates exceeding 78%, cross-platform integration remains low at just 23%; frontier technologies (e.g., blockchain, AI-based diagnostics) are largely confined to laboratory testing (application rate < 7%), and an integrated ecological–cultural assessment framework is absent (only 9% of studies involve carbon footprint evaluation).

This study proposes three major innovations. (1) Methodologically, it introduces a pioneering "dual-tool synergy + GPT semantic mining" framework, in which VOSviewer is used to identify hotspot clusters (e.g., a linkage strength of 0.68 between "risk perception" and "sustainability strategies"), CiteSpace captures emergent technologies (e.g., a 320% three-year growth in "blockchain-based preservation"), and GPT enhances the semantic extraction efficiency by 40% (error rate < 5%). (2) Theoretically, it constructs a triadic framework of "smart monitoring–technology integration–application synergy", clarifying the role of DTs in risk simulation and that of VR in enabling public participation. (3) Practically, the framework is validated using the Dunhuang case, through which a "demand–technology–evaluation" decision matrix is developed; at the micro level, bio-based materials are prioritized for site repair; at the macro level, AI-driven vision systems are combined with drone networks for comprehensive monitoring.

Two limitations of this study remain. First, the data sample is dominated by English-language journals (though Chinese CNKI data were included), which may underrepresent region-specific technologies. Second, long-term evolutionary mechanisms are insufficiently explored due to the lack of data predating the 21st century. Meanwhile, the GPT-assisted semantic mining method used in this study has broad applicability in other fields. For example, in the field of urban planning, GPT-assisted semantic mining can be employed to analyze the correlation patterns between sustainable development and technological integration, providing solutions to bottlenecks in multi-technology integration. In the field of rural revitalization, it can be used to explore the relationships between industrial integration and cultural heritage, offering strategies for the development of rural characteristics. Additionally, in interdisciplinary research, this analytical method can help identify theoretical discontinuities and technological gaps, thereby promoting a shift from fragmented studies toward the construction of systematic frameworks. Future research should focus on three key directions: (1) developing a DT-driven multi-field coupled simulation platform to quantify the climate–structure interaction mechanisms in surface protection; (2) establishing the "Belt and Road Heritage Protection Alliance" to advance blockchain-based cross-border preservation and standard co-construction; (3) formulating a three-dimensional "ecological–economic–social" benefit index system aligned with SDG 11.4 (sustainable heritage protection) and SDG 13 (climate action), thereby achieving a smart and sustainable transformation in the surface protection of earthen sites.

Author Contributions: Conceptualization, Y.X. and Y.C.; methodology, Y.X.; software, Y.X.; validation, Y.X., Y.C. and Y.H.; formal analysis, Y.X., Y.C. and Y.H.; investigation, Y.X.; resources, Y.H.; data curation, Y.X.; writing—original draft preparation, Y.X., Y.C. and Y.H.; writing—review and editing, Y.X., Y.H. and Y.Y.; visualization, Y.X.; supervision, Y.H.; project administration, Y.Y.; funding acquisition, Y.H. All authors have read and agreed to the published version of the manuscript.

Funding: This research was funded by the following: (1) the General Research Project of the Zhejiang Provincial Department of Education: "The Impact of Local Identity on Rural Settlements in the Middle Reaches of the Nanxi River and Optimization Strategies" (grant number: Y202353008); (2) the 2024 Zhejiang Provincial Philosophy and Social Science Planning "Provincial-City Cooperation" Project: "The Distribution Characteristics, Spatiotemporal Evolution, and Influencing Factors of Rural Settlements in the Nanxi River Basin from 1990 to 2020" (grant number: 24SSHZ085YB).

Institutional Review Board Statement: Not applicable.

Informed Consent Statement: Not applicable.

Data Availability Statement: Please contact Yingzhi Xiao at lennon@xauat.edu.cn for further information.

Conflicts of Interest: The authors declare no conflicts of interest.

References

1. Guo, H.; Chen, F.; Tang, Y.; Ding, Y.; Chen, M.; Zhou, W.; Zhu, M.; Gao, S.; Yang, R.; Zheng, W. Progress toward the sustainable development of world cultural heritage sites facing land-cover changes. *Innovation* **2023**, *4*, 100496. [CrossRef] [PubMed]
2. Jokilehto, J. The context of the Venice Charter (1964). *Conserv. Manag. Archaeol. Sites* **1998**, *2*, 229–233. [CrossRef]
3. Yao, Y.; Wang, X.; Lu, L.; Liu, C.; Wu, Q.; Ren, H.; Yang, S.; Sun, R.; Luo, L.; Wu, K. Proportionated distributions in spatiotemporal structure of the world cultural heritage sites: Analysis and countermeasures. *Sustainability* **2021**, *13*, 2148. [CrossRef]
4. Laohaviraphap, N.; Waroonkun, T. Integrating Artificial Intelligence and the Internet of Things in Cultural Heritage Preservation: A Systematic Review of Risk Management and Environmental Monitoring Strategies. *Buildings* **2024**, *14*, 3979. [CrossRef]
5. Li, J.; Pei, Y.; Zhao, S.; Xiao, R.; Sang, X.; Zhang, C. A review of remote sensing for environmental monitoring in China. *Remote Sens.* **2020**, *12*, 1130. [CrossRef]
6. Li, Y.; Zhao, L.; Chen, Y.; Zhang, N.; Fan, H.; Zhang, Z. 3D LiDAR and multi-technology collaboration for preservation of built heritage in China: A review. *Int. J. Appl. Earth Obs. Geoinf.* **2023**, *116*, 103156. [CrossRef]
7. Xiao, W.; Mills, J.; Guidi, G.; Rodríguez-Gonzálvez, P.; Barsanti, S.G.; González-Aguilera, D. Geoinformatics for the conservation and promotion of cultural heritage in support of the UN Sustainable Development Goals. *ISPRS J. Photogramm. Remote Sens.* **2018**, *142*, 389–406. [CrossRef]
8. Li, Z.; Wang, X.; Sun, M.; Chen, W.; Guo, Q.; Zhang, H. Conservation of Jiaohe ancient earthen site in China. *J. Rock Mech. Geotech. Eng.* **2011**, *3*, 270–281. [CrossRef]
9. Luo, L.; Liu, J.; Cigna, F.; Evans, D.; Hernandez, M.; Tapete, D.; Shadie, P.; Agapiou, A.; Elfadaly, A.; Chen, M. Space technology: A powerful tool for safeguarding world heritage. *Innovation* **2023**, *4*, 100420. [CrossRef] [PubMed]
10. Holtorf, C. Embracing change: How cultural resilience is increased through cultural heritage. *World Archaeol.* **2018**, *50*, 639–650. [CrossRef]
11. Poulopoulos, V.; Wallace, M. Digital technologies and the role of data in cultural heritage: The past, the present, and the future. *Big Data Cogn. Comput.* **2022**, *6*, 73. [CrossRef]
12. Bernard, L.; Van't Veer, R.; Huvila, I.; Moreau, A.; Verhagen, P.; Cooper, A.; Green, C.; van Manen, N.; Mckeague, P. Mapping our heritage: Towards a sustainable future for digital spatial information and technologies in European archaeological heritage management. *J. Comput. Appl. Archaeol.* **2019**, *2*, 89–104. [CrossRef]
13. Axelsson, R.; Angelstam, P.; Elbakidze, M.; Stryamets, N.; Johansson, K.-E. Sustainable development and sustainability: Landscape approach as a practical interpretation of principles and implementation concepts. *J. Landsc. Ecol.* **2011**, *4*, 5–30. [CrossRef]
14. Zhang, L.; Lu, P.; Lau, R.; Yan, L.; Li, X.; Yang, R.; Leung, H.H.; Chen, P.; Wang, X. Unique traditional villages on the Loess Plateau of China: Historic evolution and challenges to sustainable development of silo-caves. *Herit. Sci.* **2021**, *9*, 118. [CrossRef]
15. Li, L.; Shao, M.; Wang, S.; Li, Z. Preservation of earthen heritage sites on the Silk Road, northwest China from the impact of the environment. *Environ. Earth Sci.* **2011**, *64*, 1625–1639. [CrossRef]
16. Mileto, C.; Vegas López-Manzanares, F. Earthen architectural heritage in the international context: Values, threats, conservation principles and strategies. *J. Cult. Herit. Manag. Sustain. Dev.* **2022**, *12*, 192–205. [CrossRef]
17. Lennon, J. The Venice charter down under: Its legacy in landscape preservation. *Change Over Time* **2014**, *4*, 322–336. [CrossRef]
18. Taylor, K. Cultural heritage management: A possible role for charters and principles in Asia. *Int. J. Herit. Stud.* **2004**, *10*, 417–433. [CrossRef]
19. Hua, S. World heritage classification and related issues—A case study of the "convention concerning the protection of the World cultural and natural heritage". *Proc. Soc. Behav. Sci.* **2010**, *2*, 6954–6961. [CrossRef]
20. Li, W.; Xie, Q.; Ao, J.; Lin, H.; Ji, S.; Yang, M.; Sun, J. Systematic review: A scientometric analysis of the status, trends and challenges in the application of digital technology to cultural heritage conservation (2019–2024). *npj Herit. Sci.* **2025**, *13*, 90. [CrossRef]
21. Lu, Y.; Ahmad, Y. Heritage protection perspective of sustainable development of traditional villages in Guangxi, China. *Sustainability* **2023**, *15*, 3387. [CrossRef]
22. Lian, Y.; Xie, J. The evolution of digital cultural heritage research: Identifying key trends, hotspots, and challenges through bibliometric analysis. *Sustainability* **2024**, *16*, 7125. [CrossRef]
23. Ashworth, G. Preservation, conservation and heritage: Approaches to the past in the present through the built environment. *As. Anth.* **2011**, *10*, 1–18. [CrossRef]
24. King, J.; Brown, C. Environmental flows: Striking the balance between development and resource protection. *Ecol. Soc.* **2006**, *11*, 2. [CrossRef]
25. Mao, F.; Liu, Z.; Zhou, W.; Huang, J.; Li, Q. The research and application of spatial information technology in cultural heritage conservation—Case study on Grand Canal of China. *Int. Arch. Photogramm. Remote Sens. Spat. Inf. Sci* **2008**, *38*, 999–1006. [CrossRef]
26. Lasaponara, R.; Masini, N. Satellite remote sensing in archaeology: Past, present and future perspectives. *J. Archaeol. Sci.* **2011**, *38*, 1995–2002. [CrossRef]

27. Yao, Y.; Wang, X.; Luo, L.; Wan, H.; Ren, H. An overview of GIS-RS applications for archaeological and cultural heritage under the DBAR-heritage mission. *Remote Sens.* **2023**, *15*, 5766. [CrossRef]
28. Zhang, C.; Chen, J.; Li, P.; Han, S.; Xu, J. Integrated high-precision real scene 3D modeling of karst cave landscape based on laser scanning and photogrammetry. *Sci. Rep.* **2024**, *14*, 20485. [CrossRef] [PubMed]
29. Limongiello, M.; Musmeci, D.; Radaelli, L.; Chiumiento, A.; Di Filippo, A.; Limongiello, I. Parametric GIS and HBIM for Archaeological Site Management and Historic Reconstruction Through 3D Survey Integration. *Remote Sens.* **2025**, *17*, 984. [CrossRef]
30. Buckley, S.J.; Howell, J.; Enge, H.; Kurz, T. Terrestrial laser scanning in geology: Data acquisition, processing and accuracy considerations. *J. Geol. Soc.* **2008**, *165*, 625–638. [CrossRef]
31. Sun, L.; Shang, Z.; Xia, Y.; Bhowmick, S.; Nagarajaiah, S. Review of bridge structural health monitoring aided by big data and artificial intelligence: From condition assessment to damage detection. *J. Struct. Eng.* **2020**, *146*, 04020073. [CrossRef]
32. Fang, S.; Da Xu, L.; Zhu, Y.; Ahati, J.; Pei, H.; Yan, J.; Liu, Z. An integrated system for regional environmental monitoring and management based on internet of things. *IEEE Trans. Ind. Inf.* **2014**, *10*, 1596–1605. [CrossRef]
33. Guo, Q.; Wang, Y.; Chen, W.; Pei, Q.; Sun, M.; Yang, S.; Zhang, J.; Du, Y. Key Issues and Research Progress on the Deterioration Processes and Protection Technology of Earthen Sites under Multi-Field Coupling. *Coatings* **2022**, *12*, 1677. [CrossRef]
34. Sánchez-Calvillo, A.; Rincón, L.; Hamard, E.; Faria, P. Bibliometric Analysis on Earthen Building: Approaches from the Scientific Literature and Future Trends. *Buildings* **2024**, *14*, 3870. [CrossRef]
35. Li, Y.; Ip, K.H. Exploring Preventive Conservation Strategies for the Preservation of Rammed Earth Architectural Sites. *Acad. J. Archit. Geotech. Eng.* **2023**, *5*, 10–17. [CrossRef]
36. Fan, Y.; Chao, W.; Haiting, S.; Mu, J. Construction characteristics and reinforcement and maintenance technology of native earth buildings in Liling area. *J. Asian Archit. Build. Eng.* **2024**, *13*, 1–17. [CrossRef]
37. Zheng, Y.; Liu, S.; Zhang, Y.; Du, Z.; Li, M.; Luo, H.; Zhu, J. Effect of In Situ Polymerization of Super Absorbent Polymers on the Protection of Earthen Heritage Sites in Semi-Arid Regions. *Materials* **2024**, *17*, 4839. [CrossRef] [PubMed]
38. Meskell, L. States of conservation: Protection, politics, and pacting within UNESCO's World Heritage Committee. *Anthrop. Q.* **2014**, *87*, 217–243. [CrossRef]
39. Gao, Y.; Lin, R.; Lu, Y. A visualized analysis of the research current hotspots and trends on innovation chain based on the knowledge map. *Sustainability* **2022**, *14*, 1708. [CrossRef]
40. Huang, Y. Bibliometric analysis of GIS applications in heritage studies based on Web of Science from 1994 to 2023. *Herit. Sci.* **2024**, *12*, 57. [CrossRef]
41. Burnham, J.F. Scopus database: A review. *Biomed. Digit. Libr.* **2006**, *3*, 1. [CrossRef] [PubMed]
42. Yan, L.; Zhiping, W. Mapping the literature on academic publishing: A bibliometric analysis on WOS. *Sage Open Eng.* **2023**, *13*, 21582440231158562. [CrossRef]
43. Baghini, M.S.; Mohammadi, M.; Norouzkhani, N. Usability Testing: A Bibliometric Analysis Based on WoS Data. *J. Scientom. Res.* **2024**, *13*, 9–24. [CrossRef]
44. Zhu, J.; Liu, W. A tale of two databases: The use of Web of Science and Scopus in academic papers. *Scientometrics* **2020**, *123*, 321–335. [CrossRef]
45. Huang, Y.; Xia, J.; Chen, Y.; Hong, J. Global research progress on rural settlement spatial patterns from 2000–2024: A comprehensive review of hotspots, frontiers, and emerging trends. *Front. Earth Sci.* **2024**, *12*, 1505570. [CrossRef]
46. Lawani, S.M. Bibliometrics: Its theoretical foundations, methods and applications. *Libri* **1981**, *31*, 294–315. [CrossRef]
47. Gutiérrez-Salcedo, M.; Martínez, M.Á.; Moral-Munoz, J.A.; Herrera-Viedma, E.; Cobo, M.J. Some bibliometric procedures for analyzing and evaluating research fields. *Appl. Intell.* **2018**, *48*, 1275–1287. [CrossRef]
48. Ramachandran, R.; Bugbee, K.; Murphy, K. From open data to open science. *Earth Space Sci.* **2021**, *8*, e2020EA001562. [CrossRef]
49. Fang, Q.; Inoue, T.; Li, D.; Liu, Q.; Ma, J. Transit-Oriented Development and Sustainable Cities: A Visual Analysis of the Literature Based on CiteSpace and VOSviewer. *Sustainability* **2023**, *15*, 8223. [CrossRef]
50. Jentner, W.; Lindholz, G.; Hauptmann, H.; El-Assady, M.; Ma, K.-L.; Keim, D. Visual analytics of co-occurrences to discover subspaces in structured data. *ACM Trans. Interact. Intell. Syst.* **2023**, *13*, 1–49. [CrossRef]
51. Meng, L.; Wen, K.-H.; Brewin, R.; Wu, Q. Knowledge atlas on the relationship between urban street space and residents' health—A bibliometric analysis based on VOSviewer and CiteSpace. *Sustainability* **2020**, *12*, 2384. [CrossRef]
52. Liu, B.; Wu, C.; Xu, W.; Shen, Y.; Tang, F. Emerging trends in GIS application on cultural heritage conservation: A review. *Herit. Sci.* **2024**, *12*, 139. [CrossRef]
53. Shao, M.; Li, L.; Wang, S.; Wang, E.; Li, Z. Deterioration mechanisms of building materials of Jiaohe ruins in China. *J. Cult. Herit.* **2013**, *14*, 38–44. [CrossRef]
54. Igbokwe, J.; Akinyede, J.; Dang, B.; Alaga, T.; Ono, M.; Nnodu, V.; Anike, L. Mapping and monitoring of the impact of gully erosion in Southeastern Nigeria with satellite remote sensing and Geographic Information System. *Int. Arch. Photogramm. Remote Sens. Spat. Inf. Sci.* **2008**, *37*, 865–872. [CrossRef]

55. Campiani, A.; Lingle, A.; Lercari, N. Spatial analysis and heritage conservation: Leveraging 3-D data and GIS for monitoring earthen architecture. *J. Cult. Herit.* **2019**, *39*, 166–176. [CrossRef]
56. Liu, Z.; Wang, J. Protection and Utilization of Historical Sites Using Digital Twins. *Buildings* **2024**, *14*, 1019. [CrossRef]
57. Puertas, E.; Ávila, F.; García-Macías, E.; Gallego, R. Preventive Preservation of Rammed Earth Historical Heritage Through Continuous Monitoring, Architectural Inspections, and Data Fusion. *Buildings* **2024**, *14*, 3294. [CrossRef]
58. Akagawa, N.; Sirisrisak, T. Cultural landscapes in Asia and the Pacific: Implications of the World Heritage Convention. *Int. J. Herit. Stud.* **2008**, *14*, 176–191. [CrossRef]
59. Zhang, R.; Zhang, C. Classification and application of digital technologies in landscape heritage protection. *Land* **2022**, *11*, 1699. [CrossRef]
60. Fan, J.; Chen, Y.; Zheng, L. Artificial Intelligence for Routine Heritage Monitoring and Sustainable Planning of the Conservation of Historic Districts: A Case Study on Fujian Earthen Houses (Tulou). *Buildings* **2024**, *14*, 1915. [CrossRef]
61. Guo, Y.; Geng, X.; Chen, D.; Chen, Y. Sustainable building design development knowledge map: A visual analysis using CiteSpace. *Buildings* **2022**, *12*, 969. [CrossRef]
62. McAllister, J.T.; Lennertz, L.; Atencio Mojica, Z. Mapping a discipline: A guide to using VOSviewer for bibliometric and visual analysis. *Sci. Technol. Libr.* **2022**, *41*, 319–348. [CrossRef]
63. Richards, J.; Viles, H.; Guo, Q. The importance of wind as a driver of earthen heritage deterioration in dryland environments. *Geomorphology* **2020**, *369*, 107363. [CrossRef]
64. Richards, J.; Zhao, G.; Zhang, H.; Viles, H. A controlled field experiment to investigate the deterioration of earthen heritage by wind and rain. *Herit. Sci.* **2019**, *7*, 51. [CrossRef]
65. Du, Y.; Chen, W.; Cui, K.; Gong, S.; Pu, T.; Fu, X. A model characterizing deterioration at earthen sites of the Ming Great Wall in Qinghai Province, China. *Soil Mech. Found. Eng.* **2017**, *53*, 426–434. [CrossRef]
66. Li, L.; Feng, R.; Xi, J. Ecological risk assessment and protection zone identification for linear cultural heritage: A case study of the Ming Great Wall. *Int. J. Environ. Res. Public Health* **2021**, *18*, 11605. [CrossRef] [PubMed]
67. Tsakiridis, N.L.; Samarinas, N.; Kalopesa, E.; Zalidis, G.C. Cognitive soil digital twin for monitoring the soil ecosystem: A conceptual framework. *Soil Syst.* **2023**, *7*, 88. [CrossRef]
68. Adger, W.N. Vulnerability. *Glob. Environ. Change* **2006**, *16*, 268–281. [CrossRef]
69. Nakhaei Ashtari, M.; Correia, M. Assessment of vulnerability and site adaptive capacity to the risk of climate change: The case of Tchogha Zanbil World Heritage earthen site in Iran. *J. Cult. Herit. Manag. Sustain. Dev.* **2022**, *12*, 107–125. [CrossRef]
70. Pan, C.; Chen, K.; Chen, D.; Xi, S.; Geng, J.; Zhu, D. Research progress on in-situ protection status and technology of earthen sites in moisty environment. *Constr. Build. Mater.* **2020**, *253*, 119219. [CrossRef]
71. Beckett, C.; Jaquin, P.; Morel, J.-C. Weathering the storm: A framework to assess the resistance of earthen structures to water damage. *Constr. Build. Mater.* **2020**, *242*, 118098. [CrossRef]
72. Bonazza, A.; Sardella, A. Climate change and cultural heritage: Methods and approaches for damage and risk assessment addressed to a practical application. *Heritage* **2023**, *6*, 3578–3589. [CrossRef]
73. He, D.; Hu, J.; Zhang, J. Assessment of sustainable development suitability in linear cultural heritage—A case of Beijing great wall cultural belt. *Land* **2023**, *12*, 1761. [CrossRef]
74. Ning, X.; Qi, J.; Wu, C. A quantitative safety risk assessment model for construction site layout planning. *Saf. Sci.* **2018**, *104*, 246–259. [CrossRef]
75. Kumar, V.; Singh, R.; Ahuja, I.; Hashmi, M.J. On technological solutions for repair and rehabilitation of heritage sites: A review. *Adv. Mater. Process. Technol.* **2020**, *6*, 146–166. [CrossRef]
76. Manzano-Fernández, S.; Mileto, C.; Vegas López-Manzanares, F.; Cristini, V. Conservation and in situ enhancement of Earthen Architecture in Archaeological sites: Social and anthropic risks in the Case studies of the Iberian Peninsula. *Heritage* **2024**, *7*, 2239–2264. [CrossRef]
77. Crowley, K.; Jackson, R.; O'connell, S.; Karunarthna, D.; Anantasari, E.; Retnowati, A.; Niemand, D. Cultural heritage and risk assessments: Gaps, challenges, and future research directions for the inclusion of heritage within climate change adaptation and disaster management. *Clim. Resil. Sustain.* **2022**, *1*, e45. [CrossRef]
78. Qin, R.J.; Leung, H.H. Becoming a traditional village: Heritage protection and livelihood transformation of a Chinese village. *Sustainability* **2021**, *13*, 2331. [CrossRef]
79. Oriaghe, T. Exploration on the Win-win Cooperation between Cultural Heritage and Natural Environment Protection Based on the Theory of Cooperation and Competition. *Nat. Environ. Prot.* **2023**, *4*, 49–59. [CrossRef]
80. Mendoza, M.A.D.; De La Hoz Franco, E.; Gómez, J.E.G. Technologies for the preservation of cultural heritage—A systematic review of the literature. *Sustainability* **2023**, *15*, 1059. [CrossRef]
81. Gil, J. City information modelling: A conceptual framework for research and practice in digital urban planning. *Built Environ.* **2020**, *46*, 501–527. [CrossRef]

82. Siliutina, I.; Tytar, O.; Barbash, M.; Petrenko, N.; Yepyk, L. Cultural preservation and digital heritage: Challenges and opportunities. *Amazon. Invest.* **2024**, *13*, 262–273. [CrossRef]
83. Okonta, E.D.; Israel, G.C.; Okeke, F.O. Scientometric review and visualisation of City Information Modeling (CIM) trends and future directions. *Discov. Cities* **2025**, *2*, 28. [CrossRef]
84. Lim, W.M.; Kumar, S. Guidelines for interpreting the results of bibliometric analysis: A sensemaking approach. *Glob. Bus. Organ. Excell.* **2024**, *43*, 17–26. [CrossRef]
85. Megawati, R.; Listiani, H.; Pranoto, N.W.; Akobiarek, M. The role of GPT chat in writing scientific articles: A systematic literature review. *J. Penel. Pend. IPA* **2023**, *9*, 1078–1084. [CrossRef]
86. Chen, F.; Zhou, W.; Xu, H.; Parcharidis, I.; Lin, H.; Fang, C. Space technology facilitates the preventive monitoring and preservation of the Great Wall of the Ming dynasty: A comparative study of the Qingtongxia and Zhangjiakou sections in China. *IEEE J. Sel. Top. Appl. Earth Obs. Remote Sens.* **2020**, *13*, 5719–5729. [CrossRef]
87. Guo, M.; Fu, Z.; Pan, D.; Zhou, Y.; Huang, M.; Guo, K. 3D Digital protection and representation of burial ruins based on LiDAR and UAV survey. *Meas. Control* **2022**, *55*, 555–566. [CrossRef]
88. Prosser, C.D.; Burek, C.V.; Evans, D.H.; Gordon, J.E.; Kirkbride, V.B.; Rennie, A.F.; Walmsley, C.A. Conserving geodiversity sites in a changing climate: Management challenges and responses. *Geoheritage* **2010**, *2*, 123–136. [CrossRef]
89. Ray, P.P.; Mukherjee, M.; Shu, L. Internet of things for disaster management: State-of-the-art and prospects. *IEEE Access* **2017**, *5*, 18818–18835. [CrossRef]
90. Sharma, K.; Anand, D.; Sabharwal, M.; Tiwari, P.K.; Cheikhrouhou, O.; Frikha, T. A disaster management framework using Internet of Things-based interconnected devices. *Math. Probl. Eng.* **2021**, *2021*, 9916440. [CrossRef]
91. Luccioni, A.; Schmidt, V.; Vardanyan, V.; Bengio, Y. Using artificial intelligence to visualize the impacts of climate change. *IEEE Comput. Graph. Appl.* **2021**, *41*, 8–14. [CrossRef] [PubMed]
92. Tsatsaris, A.; Kalogeropoulos, K.; Stathopoulos, N.; Louka, P.; Tsanakas, K.; Tsesmelis, D.E.; Krassanakis, V.; Petropoulos, G.P.; Pappas, V.; Chalkias, C. Geoinformation technologies in support of environmental hazards monitoring under climate change: An extensive review. *ISPRS Int. J. Geo-Inf.* **2021**, *10*, 94. [CrossRef]
93. Cirulis, A.; De Paolis, L.T.; Tutberidze, M. Virtualization of digitalized cultural heritage and use case scenario modeling for sustainability promotion of national identity. *Procedia Comput. Sci.* **2015**, *77*, 199–206. [CrossRef]
94. Shabani, A.; Kioumarsi, M.; Plevris, V.; Stamatopoulos, H. Structural vulnerability assessment of heritage timber buildings: A methodological proposal. *Forests* **2020**, *11*, 881. [CrossRef]
95. Nazari, Y.R.; Saatcioglu, M. Seismic vulnerability assessment of concrete shear wall buildings through fragility analysis. *J. Build. Eng.* **2017**, *12*, 202–209. [CrossRef]
96. Aven, T. A new scientific framework for quantitative risk assessments. *Int. J. Bus. Contin. Risk Manag.* **2009**, *1*, 67–77. [CrossRef]
97. Fabbri, A.; Morel, J.-C.; Gallipoli, D. Assessing the performance of earth building materials: A review of recent developments. *RILEM Tech. Lett.* **2018**, *3*, 46–58. [CrossRef]
98. Kono, T. Authenticity: Principles and notions. *Change Over Time* **2014**, *4*, 436–460. [CrossRef]
99. Liu, Y.; Wang, Y.; Dupre, K.; McIlwaine, C. The impacts of world cultural heritage site designation and heritage tourism on community livelihoods: A Chinese case study. *Tour. Manag. Perspect.* **2022**, *43*, 100994. [CrossRef]
100. Zhang, Y.; Dong, W. Determining minimum intervention in the preservation of heritage buildings. *Int. J. Archit. Herit.* **2021**, *15*, 698–712. [CrossRef]
101. Rosado Correia, M.R.A.; Walliman, N.S.R. Defining criteria for intervention in earthen-built heritage conservation. *Int. J. Archit. Herit.* **2014**, *8*, 581–601. [CrossRef]
102. Sardella, A.; Palazzi, E.; von Hardenberg, J.; Del Grande, C.; De Nuntiis, P.; Sabbioni, C.; Bonazza, A. Risk mapping for the sustainable protection of cultural heritage in extreme changing environments. *Atmosphere* **2020**, *11*, 700. [CrossRef]
103. Zhang, Y. Knowledge of earthen heritage deterioration in dry areas of China: Salinity effect on the formation of cracked surface crust. *Herit. Sci.* **2023**, *11*, 41. [CrossRef]
104. Han, W.; Gong, R.; Liu, Y.; Gao, Y. Influence mechanism of salt erosion on the earthen heritage site wall in Pianguan Bastion. *Case Stud. Constr. Mater.* **2022**, *17*, e01388. [CrossRef]
105. Shi, P.; Ye, T.; Wang, Y.; Zhou, T.; Xu, W.; Du, J.; Wang, J.a.; Li, N.; Huang, C.; Liu, L. Disaster risk science: A geographical perspective and a research framework. *Int. J. Disaster Risk Sci.* **2020**, *11*, 426–440. [CrossRef]
106. Peters, G.P.; Andrew, R.M.; Canadell, J.G.; Fuss, S.; Jackson, R.B.; Korsbakken, J.I.; Le Quéré, C.; Nakicenovic, N. Key indicators to track current progress and future ambition of the Paris Agreement. *Nat. Clim. Change* **2017**, *7*, 118–122. [CrossRef]
107. Zuazo, V.c.H.D.; Pleguezuelo, C.R.o.R. Soil-Erosion and Runoff Prevention by Plant Covers: A Review. *Agron. Sustain. Dev.* **2008**, *28*, 65–68. [CrossRef]
108. Qu, J.; Sun, M.; Wang, F.; Liu, K.; Miao, W. Analysis of microenvironment characteristics and the impact on the preservation of heritage sites: A case study of the Jinsha Earthen Site. *npj Herit. Sci.* **2025**, *13*, 109. [CrossRef]

109. Barton, J. 3D laser scanning and the conservation of earthen architecture: A case study at the UNESCO World Heritage Site Merv, Turkmenistan. *World Archaeol.* **2009**, *41*, 489–504. [CrossRef]
110. Lercari, N. Monitoring earthen archaeological heritage using multi-temporal terrestrial laser scanning and surface change detection. *J. Cult. Herit.* **2019**, *39*, 152–165. [CrossRef]
111. Guo, H.; Liu, J.; Qiu, Y.; Menenti, M.; Chen, F.; Uhlir, P.F.; Zhang, L.; van Genderen, J.; Liang, D.; Natarajan, I. The Digital Belt and Road program in support of regional sustainability. *Int. J. Digit. Earth* **2018**, *11*, 657–669. [CrossRef]
112. Pereira Roders, A.; Van Oers, R. Bridging cultural heritage and sustainable development. *J. Cult. Herit. Manag. Sustain. Dev.* **2011**, *1*, 5–14. [CrossRef]
113. Zhong, H.; Wang, L.; Zhang, H. The application of virtual reality technology in the digital preservation of cultural heritage. *Comput. Sci. Inf. Syst.* **2021**, *18*, 535–551. [CrossRef]
114. Yu, T.; Lin, C.; Zhang, S.; Wang, C.; Ding, X.; An, H.; Liu, X.; Qu, T.; Wan, L.; You, S. Artificial intelligence for Dunhuang cultural heritage protection: The project and the dataset. *Int. J. Comput. Vis.* **2022**, *130*, 2646–2673. [CrossRef]
115. Lercari, N.; Jaffke, D.; Campiani, A.; Guillem, A.; McAvoy, S.; Delgado, G.J.; Bevk Neeb, A. Building cultural heritage resilience through remote sensing: An integrated approach using multi-temporal site monitoring, datafication, and Web-GL visualization. *Remote Sens.* **2021**, *13*, 4130. [CrossRef]
116. Mazzetto, S. Integrating Emerging Technologies with Digital Twins for Heritage Building Conservation: An Interdisciplinary Approach with Expert Insights and Bibliometric Analysis. *Heritage* **2024**, *7*, 6432–6479. [CrossRef]
117. Malek, Y.N.; Kharbouch, A.; El Khoukhi, H.; Bakhouya, M.; De Florio, V.; El Ouadghiri, D.; Latré, S.; Blondia, C. On the use of IoT and big data technologies for real-time monitoring and data processing. *Procedia Comput. Sci.* **2017**, *113*, 429–434. [CrossRef]
118. Omrany, H.; Ghaffarianhoseini, A.; Ghaffarianhoseini, A.; Clements-Croome, D.J. The uptake of City Information Modelling (CIM): A comprehensive review of current implementations, challenges and future outlook. *Smart Sustain. Built Environ.* **2023**, *12*, 1090–1116. [CrossRef]
119. Trček, D. Cultural heritage preservation by using blockchain technologies. *Herit. Sci.* **2022**, *10*, 6. [CrossRef]
120. Chakraborty, S.; Ji, S. Analyzing spatial technology trends in urban heritage preservation: A bibliometrics study. *Int. J. Archit. Comput.* **2024**, *23*, 548–566. [CrossRef]
121. Dang, X.; Liu, W.; Hong, Q.; Wang, Y.; Chen, X. Digital twin applications on cultural world heritage sites in China: A state-of-the-art overview. *J. Cult. Herit.* **2023**, *64*, 228–243. [CrossRef]
122. Qiao, C.; Zhang, W.; Gong, D.; Gong, Y. In situ virtual restoration of artifacts by imaging technology. *Herit. Sci.* **2020**, *8*, 110. [CrossRef]
123. Pietroni, E.; Ferdani, D. Virtual restoration and virtual reconstruction in cultural heritage: Terminology, methodologies, visual representation techniques and cognitive models. *Information* **2021**, *12*, 167. [CrossRef]
124. Casillo, M.; Colace, F.; Gaeta, R.; Lorusso, A.; Santaniello, D.; Valentino, C. Revolutionizing cultural heritage preservation: An innovative IoT-based framework for protecting historical buildings. *Evol. Intell.* **2024**, *17*, 3815–3831. [CrossRef]
125. Wang, C.; Hou, L.; Jiang, N.; Wang, Y.; Mao, X.; Zhou, P.; Xia, Y.; Wang, Y.; Chen, C.; Yang, X. Microbial diversity in earthen site of exhibition Hall of pit no. 1 at the terracotta warriors Museum in Emperor Qinshihuang's mausoleum site museum and its correlation with environmental factors. *Front. Microbiol.* **2024**, *15*, 1378180. [CrossRef] [PubMed]
126. Bowker, M.A. Biological soil crust rehabilitation in theory and practice: An underexploited opportunity. *Restor. Ecol.* **2007**, *15*, 13–23. [CrossRef]
127. Eldridge, D. Ecology and management of biological soil crusts: Recent developments and future challenges. *Bryologist* **2000**, *103*, 742–747. [CrossRef]
128. Benzerara, K.; Miot, J.; Morin, G.; Ona-Nguema, G.; Skouri-Panet, F.; Férard, C. Significance, mechanisms and environmental implications of microbial biomineralization. *C. R. Geosci.* **2011**, *343*, 160–167. [CrossRef]
129. Wang, T.; Guo, Q.; Pei, Q.; Chen, W.; Wang, Y.; Zhang, B.; Yu, J. Destruction or protection? Experimental studies on the mechanism of biological soil crusts on the surfaces of earthen sites. *Catena* **2023**, *227*, 107096. [CrossRef]
130. Mansuri, L.E.; Patel, D. Artificial intelligence-based automatic visual inspection system for built heritage. *Smart Sustain. Built Environ.* **2022**, *11*, 622–646. [CrossRef]
131. Ahmed, S.A.; Popov, V.L.; Topalov, A.V.; Shakev, N.G. Environmental monitoring using a robotized wireless sensor network. *AI Soc.* **2018**, *33*, 207–214. [CrossRef]
132. Jha, M.K.; Joshi, S.; Sharma, R.K.; Kim, A.A.; Pant, B.; Park, M.; Pant, H.R. Surface modified activated carbons: Sustainable bio-based materials for environmental remediation. *Nanomaterials* **2021**, *11*, 3140. [CrossRef] [PubMed]
133. Kong, D.; Wan, R.; Chen, J.; Jing, Y.; Huang, W.; Wang, Y. The study on engineering characteristics and compression mechanisms of typical historical earthen site soil. *Constr. Build. Mater.* **2019**, *213*, 386–403. [CrossRef]

coatings

MDPI

Article

Microchemical Analysis of Rammed Earth Residential Walls Surface in Xiaochikan Village, Guangdong

Liang Zheng, Qingnian Deng, Jingwei Liang, Zekai Guo, Yufei Zhu, Wei Liu and Yile Chen *

Faculty of Humanities and Arts, Macau University of Science and Technology, Avenida Wai Long, Taipa, Macau 999078, China; zhengliang@must.edu.mo (L.Z.); 2250000398@student.must.edu.mo (Q.D.); 2250000402@student.must.edu.mo (J.L.); 2250018776@student.must.edu.mo (Z.G.); 2250013581@student.must.edu.mo (Y.Z.); 2250026607@student.must.edu.mo (W.L.)

* Correspondence: chenyile@must.edu.mo

Abstract: Xiaochikan Village, located in Guangdong Province in South China, is one of the few remaining traditional rammed earth dwellings of the Cantonese ethnic group in the Lingnan region. However, the influence of Zhuhai's subtropical maritime monsoon climate has led to continuous physical and chemical erosion of the rammed earth walls. For example, cracking occurs due to high temperatures and heavy rain, accelerated weathering occurs due to salt spray deposition, and biological erosion occurs due to high humidity and high temperatures. Therefore, two experimental analysis techniques, X-ray diffraction (XRD) and scanning electron microscopy-energy dispersive spectrometer (SEM-EDS), were used to explore the structural anti-erosion mechanism of the ancient, rammed earth buildings in Xiaochikan Village. The results show that (1) the morphological characteristics of the east and west walls of the rammed earth dwellings in Xiaochikan Village are more similar. The particles on the east wall are regular spherical or polygonal, small, and evenly distributed, while the particles on the west wall are mainly spherical and elliptical, with consistent size and less agglomeration. The surfaces of the particles on both walls are relatively smooth and flat. (2) The core element bases of the four wall samples are consistent, with C, Si, Al, Ca, and Fe as the core, accounting for more than 93%, reflecting the base characteristics of the local alluvial soil "silicate skeleton–carbonate cementation–organic matter residue" and reflecting the "local material" attribute of rammed earth. Except for the south wall sample, the Cl content of the remaining samples exceeds 1%. In the thermal map, Cl shows "pore/interstitial enrichment", which confirms that the salinization process of marine aerosols with rainwater infiltration and evaporation residue is a common influence of marine climate. (3) The rammed earth walls in Xiaochikan Village consist of three main minerals: Quartz (SiO_2, including alpha-type SiO_2), Calcite ($CaCO_3$, including synthetic calcite), and Gibbsite ($Al(OH)_3$).

Keywords: wall surface; SEM-EDS analysis; XRD analysis; rammed earth walls; rammed earth dwellings; Ming and Qing dynasties; South China

1. Introduction

1.1. Research Background

As one of the world's oldest building types, earthen buildings, with their ecological characteristics of locally sourced materials and low carbon footprint, have become a crucial carrier of regional climate adaptation wisdom and local cultural memories [1,2]. Located in the subtropical monsoon climate zone, earthen dwellings in Guangdong's Pearl River Delta region are subject to long-term environmental stresses such as high temperatures and humidity, typhoons and heavy rains, and microbial erosion [3,4]. The microstructural stability of their wall surfaces (the direct interface between the building and the external environment) directly determines the durability of the buildings and their ability to survive as a heritage site. Xiaochikan Village, a typical traditional Cantonese village in Doumen District, Zhuhai, boasts over 60 extant earthen dwellings dating from the Ming and Qing dynasties (A.D. 1368–1912) to the Republic of China (A.D. 1912–1949) [5]. Their walls utilize a traditional blend of alluvial soil from the Pearl River Estuary, plant ash, and glutinous rice paste. A slurry-coated finish creates a unique, dense protective layer. This layer retains the core advantages of Cantonese earthen architecture: breathability and moisture resistance, while exhibiting distinct micromorphological characteristics distinct from loess cave dwellings in northern China and rammed earth walls in northwest China. It serves as a prime example for studying the surface protective mechanisms of earthen buildings in the humid and hot regions of Lingnan.

The earth dwellings of Xiaochikan Village face dual demands for preservation and utilization as rural revitalization and traditional village preservation strategies advance. On the one hand, field investigations revealed that over half of the village's earth dwellings have suffered from disrepair, with surface wall deterioration exhibiting microscopic signs of increased porosity, weathering of clay minerals, and microbial attachment, posing a direct threat to structural safety. On the other hand, as important showcases of traditional architectural techniques and intangible cultural heritage, earth dwellings must preserve their historical character and craftsmanship during restoration to avoid the distortion of their cultural value caused by modern material substitution. Against this backdrop, it is crucial to conduct a microstructural analysis of the surface wall structures of the earth dwellings in Xiaochikan Village.

1.2. Literature Review

As a traditional village on the west bank of the Pearl River Estuary, the rammed earth buildings in Xiaochikan Village, Zhuhai, which date back to the Ming and Qing Dynasties, are important physical examples of rammed earth technology in the Lingnan region [6]. Currently, due to environmental factors such as high temperature, high humidity, and typhoons, the rammed earth walls in the village are generally experiencing a decrease in strength, and some buildings are even facing the risk of collapse. In rammed earth buildings, the particles are tightly connected to each other, which can generate greater friction, thereby promoting the stability of the wall structure [7]. Therefore, systematically sorting out the research results related to the microstructure of Lingnan rammed earth has important guiding significance for the protection and restoration of the rammed earth buildings in Xiaochikan Village. Traditional rammed earth in the Lingnan region is mostly based on "three-in-one soil" (local sandy soil, lime, and plant ash) and is often mixed with glutinous rice paste or brown sugar paste as an organic gelling agent [8]. The rammed earth buildings in Xiaochikan Village also follow this process. Oyster shells with a calcium carbonate content of over 90% are also added to some walls to further optimize the structural strength [9,10].

Some existing studies have shown that environmental factors have a multi-mechanism synergistic effect on the erosion of the microstructure of rammed earth dwellings in villages [11–13]. Physically, the temperature difference between day and night in the village can reach 15–20 °C and repeated thermal expansion and contraction can easily cause microscopic cracks between rammed earth particles. Chemically, salt spray from the offshore environment penetrates the pores of the wall, generating "salt expansion pressure" through a dissolution-crystallization cycle, causing particles to fall off and gradually increasing the porosity [8,14]. In addition to physical and chemical factors, biological erosion has a particularly significant impact on the north wall. Because the north wall is in a shaded environment for a long time, the humidity often exceeds 85%, providing suitable growth conditions for organisms such as mosses and actinomycetes. The acidic substances secreted by these organisms during metabolism will continuously decompose the mineral components and cementitious materials in the rammed earth, weakening the bonding force between particles and further exacerbating the looseness of the microstructure [15]. Some scholars have found that while the selection of soil for rammed earth is similar in the Mediterranean and China, the different soil compositions and materials constitute the key differences in reinforcing rammed earth [16].

Various experimental analysis techniques are also important ways to explore rammed earth architectural heritage. SEM (Scanning Electron Microscopy), EDS (Energy-Dispersive X-ray Spectroscopy), XRD (X-ray Diffraction), FTIR (Fourier-Transform Infrared Spectroscopy), XRF (X-ray Fluorescence), and other techniques are also very common in architectural heritage [17,18]. For example, infrared thermal imaging, rebound hammer, ultrasonic testing, thermocouple testing, and load tests on the roof frame and floor of Fujian Tulou [19] were used; drip tests on Fujian Tulou sites showed that the degree of erosion was greatest when the rainfall direction was in the range of 30° to 45°, which also provided a reference for the influence of wind and rain on the durability of rammed earth materials [20]. Cui et al. [21] analyzed the development of defects and quality evaluation results of 25 typical rammed earth sites in Gansu Province and conducted a global sensitivity analysis by using fuzzy hierarchical analysis (FAHP), the coefficient of variation method, the entropy method, and the CRITIC method to determine the comprehensive weight of evaluation indicators, thus verifying the applicability of the comprehensive weight-TOPSIS method in the quality evaluation of rammed earth sites. Despite the increasing diversity of analytical methods, the analysis of rammed earth heritage in the Lingnan region of China still requires a long period of exploration.

1.3. Problem Statement and Objectives

While existing research has revealed the fundamental structural characteristics and erosion patterns of rammed earth materials in the Lingnan region, studying the rammed earth structure in the specific area of Xiaochikan Village remains valuable. Future research should focus on rammed earth samples from Xiaochikan Village, conducting targeted analysis using techniques such as SEM-EDS and XRD. This will provide a reference for wall restoration and ultimately contribute to the preservation of traditional rammed earth architecture. This study attempts to explore and answer three main questions: (1) the main architectural structure of the rammed earth dwellings in Xiaochikan Village; (2) revealing the main mineral components of the surface of the rammed earth walls in Xiaochikan Village through XRD; and (3) revealing its main distributed elemental components through SEM-EDS.

2. Materials and Methods

2.1. Study Area: Xiaochikan Village in Guangdong

Xiaochikan Village, the study area, is located northeast of Doumen Town, Doumen District, Zhuhai City, Guangdong Province, China (geographical coordinates: 22°21′–22°24′ N, 113°11′–113°14′ E) [22,23]. It is located on the northwest side of Huangyang Mountain (Figure 1), bordering Huangyang Mountain to the southeast, Shangzhou Village (2.5 km away) to the west, Lianzhou Town Center (5 km away) to the north, and Hutiaomen Waterway (3 km away) to the west. It is a typical "mountain-river-sea" intersection landform with outstanding regional transportation accessibility. It is 55 km away from Zhuhai City, and the Lianzhou Exit of the Guangzhou-Zhuhai West Expressway is 10 km away from the village. It is worth noting that this area has a subtropical maritime monsoon climate, similar to the climate characteristics of Doumen, Zhuhai: the average annual temperature is 22.6 °C, with an extreme high of 38.5 °C and an extreme low of 2.5 °C. The average annual rainfall is 2080.6 mm, with the rainy season running from April to September, accounting for 84.7% of the annual total [24]. There is an average of 11 days of heavy rain per year, with a maximum daily rainfall of 393.7 mm. The average annual relative humidity is 80%, the average annual sunshine hours are 1876.3 h, and the annual accumulated temperature $\geq$ 10 °C is 8166.6 °C. Typhoons are most frequent from July to September [25], with one to two typhoons directly affecting the area each year and maximum wind speeds of 12–14. The high salt content of the air, with an average annual salt spray concentration of 0.05–0.12 mg/m^3, causes continuous physical and chemical erosion of the rammed earth walls. The walls are prone to cracking due to high temperatures and heavy rain, accelerated weathering due to salt spray deposition, and biological erosion due to the high humidity and temperature. Furthermore, due to the presence of Huangyang Mountain (its main peak is 583 m above sea level) to the southeast, the area is less affected by southeasterly winds, with northeasterly and westerly winds being the dominant wind directions (Figure 2).

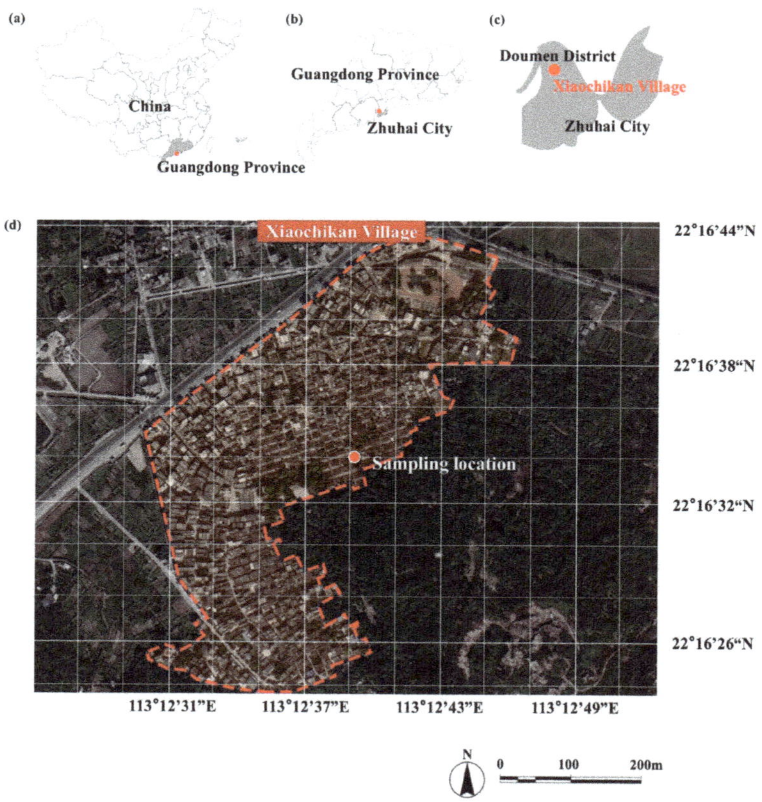

Figure 1. Location analysis map of Xiaochikan Village, Doumen Town, Zhuhai City. (**a**) Guangdong Province's location in China; (**b**) Zhuhai City's location within Guangdong Province; (**c**) Xiaochikan Village's location within Zhuhai City; (**d**) Xiaochikan Village's geographic area. Coordinates: 22°16′26–22°16′44″ N, 113°12′31″–113°12′49″ E; scale: 0–200 m. (Image source: drawn and annotated by the author).

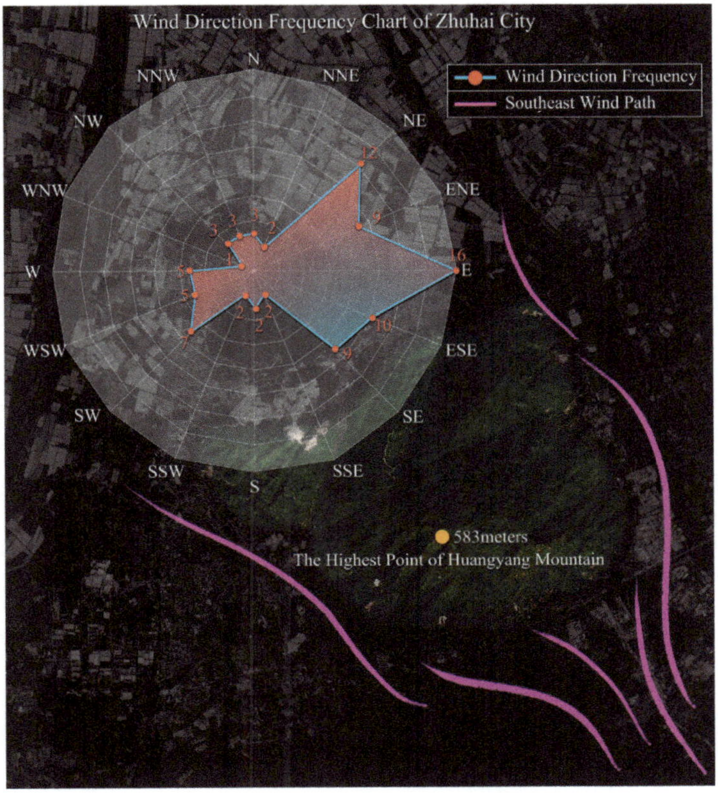

Figure 2. Wind direction frequency distribution in Zhuhai. (Image source: drawn and annotated by the author).

Xiaochikan Village is one of the best-preserved traditional rammed earth villages in Doumen District, Zhuhai City, with architectural history dating back to the Wanli period of the Ming Dynasty (roughly the late 16th to early 17th centuries) (Figure 3). Protected by relevant policies, Xiaochikan Village is one of the most ecologically well-preserved natural villages in Guangdong Province, offering the unique characteristics of a Lingnan market town known as "Little Guangzhou". Xiaochikan Village boasts numerous cultural relics and historical sites, including ancient Xiaochikan dwellings, Xiaochikan Stone Street, the Huanggong Ancestral Halls of Youkuan, Zhiwu, Yuting, and the Ma Nanbao Tomb. In 1931, donations from overseas Chinese in Xiaochikan paved a 1500 m-long, 1.6 m-wide granite street running east–west through the village. The pavement is constructed from four slabs, each approximately 0.4 m wide and varying in length. This stone street covers an area of 2400 m^2 and is the longest and best-preserved stone street in all administrative villages in Zhuhai City. In addition, there are more than 60 well-preserved ancient houses in Xiaochikan Village, all of which are a mixed structure of rammed earth, brick, and wood. They are composed of a gable roof, a dragon boat ridge, a blue brick front wall, a rammed earth gable, and a back wall. They have a distinct Lingnan architectural style from the Qing Dynasty [26–28]. Plate-building technology forms the rammed earth walls of the ancient houses in Xiaochikan Village [29]. After hundreds of years, the basic structure has remained intact. The microstructure of the material has a high historical value for studying the ancient residential buildings in the coastal areas of Guangdong. However, due to the typical environment of high temperature and humidity, frequent typhoons, and salt spray in the Pearl River Delta, the rammed earth walls of Xiaochikan Village are facing a serious threat of deterioration.

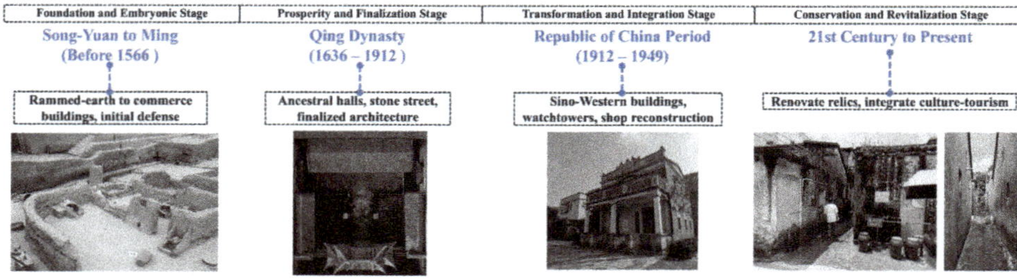

Figure 3. The evolution of architectural features in Xiaochikan Village. (Image source: drawn and annotated by the author).

2.2. Field Investigation and Sample Collection

This study conducted a field survey and sample collection at a typical rammed earth wall building in Xiaochikan Village, Doumen Town, Zhuhai City, from 10:00 AM to 1:00 PM on 29 June 2025. The target was a relatively well-preserved, representative single-unit rammed earth house in the village (Figure 4). According to on-site mapping, the building's floor plan is approximately 10.90 m × 10.70 m, with a height of approximately 5.87 m from the eaves to the ridge. It has a gabled, double-sloped roof covered with grayish-gray tiles. The exterior walls are original rammed earth, with some traces of later repairs (Figure 5).

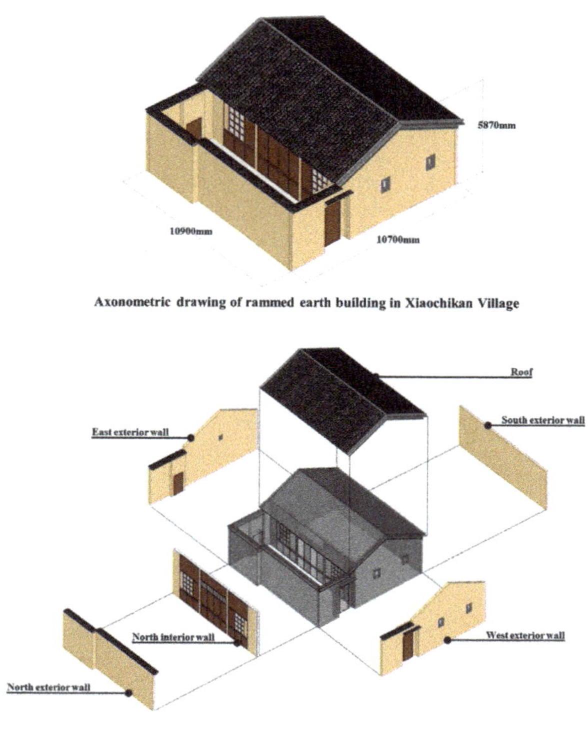

Figure 4. Axonometric drawing of rammed earth building in Xiaochikan Village and explosive drawing of rammed earth Xiaochikan Village. (Image source: drawn and annotated by the author).

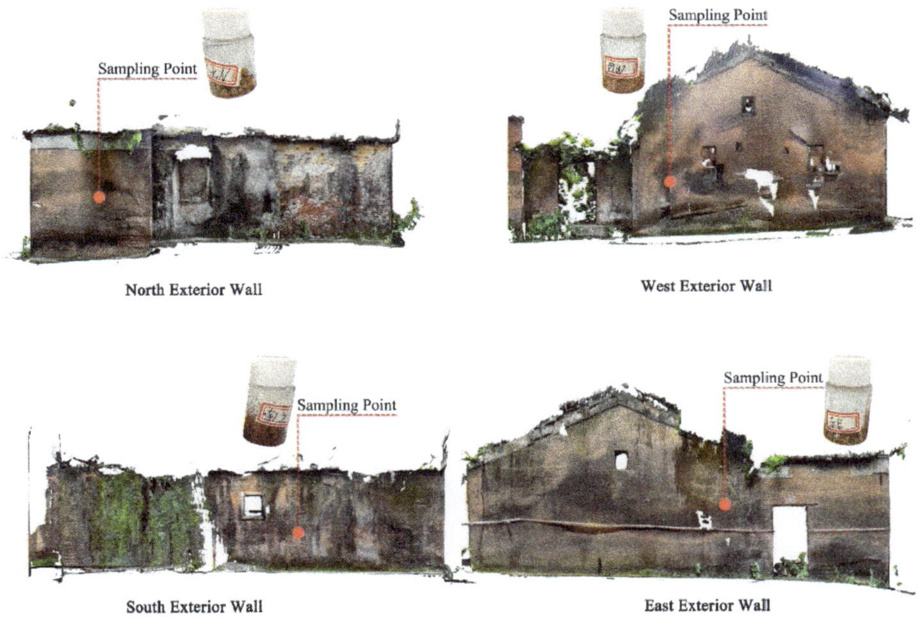

Figure 5. Walls and points for sample collection. (Image source: drawn and annotated by the author).

The sampling plan adhered to the principles of representativeness, comparability, minimal damage, and traceability and referenced technical specifications for sampling and sample management in the field of earthen site conservation and geotechnical testing. Given the objective of this study to interpret the composition and damage causes of rammed earth, only the surface and near-surface layers of the exterior walls were selected as comparison samples: one sampling point was set on each of the east, south, west, and north exterior walls, for a total of four sites (Figure 5). Each site avoided window openings, post-repair sites, and crack tips and was located in the middle of the wall section, 1.0 m or more from the corners, to minimize boundary effects. The sampling height was controlled at approximately 1.2 m ($\pm$0.2 m) above the ground to avoid significant areas of capillary water upwelling and splash erosion near the surface, while also ensuring that subsequent damage observations were not affected. At each site, a 2–3 mm layer of surface weathering/biological attachment was lightly removed with a soft brush. A stainless flat steel shovel was then used to scrape approximately 2–3 cm of near-surface rammed earth from the same location, with a single sample size of 10 g. Each sample is immediately placed in a 30 mL polypropylene sample bottle and sealed tightly. A self-adhesive label is attached to the outside of the bottle, and the location of the sampling point is recorded. At the same time, the point coordinates are marked on the on-site plan and elevation drawings to ensure that indoor testing and field records can be traced in both directions.

2.3. Analytical Techniques

This study combined X-ray diffraction (XRD) with scanning electron microscopy-energy dispersive spectroscopy (SEM-EDS) to analyze the structural corrosion resistance mechanism of the Xiaochikan Village ancient, rammed earth wall under a marine monsoon climate from a three-level perspective: mineral, structure, and elemental analysis. XRD was used to identify and (semi-)quantify the main and secondary mineral phases (such as quartz, calcite, and possible secondary salts), indicating chemical degradation pathways such as dissolution-redeposition, clay interlayer hydration, and salt crystallization through relative content and peak shape changes, and assessing the chemical corrosion resistance potential of the material from the perspective of mineral reactivity and stability. SEM-EDS obtained information such as particle contact relationships, pore throat scale, microcrack density, and elemental enrichment zones of Cl, S, Na, and Ca at the pore edges, characterizing the physical degradation processes such as particle shedding, skeleton disintegration, and salt crystallization expansion and cracking under the action of rain wet–dry cycles and salt spray. The two types of data are interpreted in conjunction at the same sampling point: when the structure exhibits "dense skeleton + low soluble salt + weak edge enrichment", it is inferred that the structure has high corrosion resistance; conversely, "high carbonate/secondary salt + interconnected pores + significant edge enrichment/microcracks" indicates reduced corrosion resistance. Thus, XRD provides mineralogical evidence of chemical stability and potential reactivity, while SEM-EDS reveals the structural weak points and stress concentration paths. Together, they support the mechanistic interpretation of the corrosion resistance mechanism of rammed earth wall structures and provide quantifiable indicators for protection strategies such as "salt barrier-moisture control-reinforcement".

2.3.1. X-Ray Diffraction Analysis

Systematic sampling was conducted along the east, south, west, and north directions of the rammed earth building site. Mineral composition analysis was performed using a Rigaku SmartLab high-resolution X-ray diffractometer (Rigaku Holdings Corporation, Tokyo, Japan). The experiment strictly adhered to the *X-ray Diffraction Analysis Method for Clay Minerals and Common Non-Clay Minerals in Sedimentary Rocks (SY/T 5163-2018)* [30]. Non-destructive testing was performed on bulk samples, each weighing approximately 10 g and precisely controlled to a thickness of 10 mm. Through detailed analysis of the characteristic XRD patterns, the phase composition of clay and non-clay minerals, including quartz, plagioclase, potassium feldspar, calcite, and montmorillonite, was identified. Their crystallinity parameters and semi-quantitative relative content were also determined. This microscopic study, based on mineral crystal field theory, provides core structural data for understanding the weathering mechanisms of rammed earth sites and developing targeted conservation plans.

2.3.2. SEM-EDS Analysis

Scanning electron microscopy-energy dispersive spectroscopy (SEM-EDS) was used to collaboratively characterize the microstructure and elemental composition of rammed earth samples oriented east, south, west, and north. SEM was used to observe differences in particle morphology, particle size distribution, and micropore structure. EDS was used to analyze the elemental composition (e.g., typical rammed earth elements such as Si, Al, Ca, Fe, and Mg) and their regional distribution. This approach further correlated the "structure-composition-erosion resistance" relationship and explored the causes of differences and anisotropic characteristics among samples oriented differently.

This experiment used a Czech TESCAN MIRA LMS (TESCAN, Brno, Czech Republic) as core equipment. The electron microscope's magnification was flexibly adjusted to 1.00 kX (for macroscopic pore observation), 5.00 kX (for observation of particle aggregates), and 10.00 kX (for observation of microscopic particle morphology), depending on observation requirements. Imaging was performed under a high vacuum environment to ensure clear images of the non-conductive rammed earth samples. EDS analysis was performed with a detection range of C (atomic number 6) to U (atomic number 92) to cover the major and trace elements in the rammed earth. Quantitative calibration was performed using Cu and Al alloy standard samples to ensure relative elemental content errors of less than 5%. The experimental operations sequentially carried out sample pretreatment, fixation and conductive coating, SEM micromorphology observation, EDS elemental analysis, and data processing and statistics. After the sample is fixed, a thin layer of Au conductive coating is applied to improve the imaging quality and signal-to-noise ratio of the non-conductive rammed earth. Then, SEM imaging and EDS analysis are carried out. The coating element signal is not included in the EDS quantification and normalization. The M-series peak of Au is shielded in the spectral fitting to ensure that the interpretation of light elements and halogen elements (such as C, S, and Cl) is not affected by the coating. The samples in four orientations were scanned, and secondary electron images at different magnifications were simultaneously recorded to capture microscopic features such as particle arrangement and pore morphology. EDS analysis was then carried out on typical characteristic areas such as pore edges, particle agglomerates, and abnormal mineral particles in the SEM images, and element distribution spectrum (mapping) were generated to intuitively present the uniformity of the elemental micro distribution. This study used images to qualitatively identify particle size distribution and pore morphology; however, it did not calculate the quantitative particle size distribution and porosity. The elemental data were normalized using the Oxford Instruments software AZtecTEM supporting EDS (https://nano.oxinst.com/products/aztec/aztectem#product-information-tabs, accessed on 4 October 2025), and the average relative contents of key elements such as Ca and Si in the four orientation samples were calculated.

3. Results

3.1. Mineral Composition Comparison Results

By analyzing XRD patterns on the rammed earth wall surface in four directions (east (E), south (S), west (W), and north (N)), we understood the mineralogical composition and stability of the surface material. Figure 6 shows three main minerals: Quartz (SiO_2, including alpha-SiO_2), Calcite ($CaCO_3$, including synthetic calcite), and Gibbsite ($Al(OH)_3$). Quartz is a very stable mineral with high hardness and durability, which significantly enhances the overall stability of the material. As can be seen in the figures, Quartz diffraction peaks are prominent in all samples, indicating that they account for a significant proportion of the material in all directions. The environment significantly affects the chemical stability of calcite (including synthetic calcite), with acidic environments being particularly sensitive. Its presence may make the material susceptible to dissolution under acidic conditions, affecting its structural stability. Gibbsite ($Al(OH)_3$) may decompose under certain temperature and humidity fluctuations, potentially affecting the long-term stability of the material.

The mineral phase groups of the four-directional samples are mainly quartz, calcite, and gibbsite, with similar overall peak shapes; the differences are mainly reflected in the relative peak intensity and peak shape. As shown in Figure 6, the quartz main peak in the east/west-oriented samples is more stable and relatively more prominent, indicating a stronger contribution from the silicate framework; the calcite-related peaks in the south/north-oriented samples are relatively more prominent, suggesting a higher proportion of carbonate cementation and greater sensitivity to acid rain and corrosion environments; gibbsite is identifiable in some samples but is a secondary phase. No typical sulfate/chloride diagnostic peaks were observed in any of the four-directional spectra, indicating that soluble salts exist mostly in trace or subcrystalline forms, and their influence is more evident in the pore edge enrichment and microstructure deterioration of SEM-EDS. The aforementioned mineralogical differences are consistent with the on-site environmental loading: the east/west direction is relatively mildly affected by heat, humidity, and rain, and the quartz-dominated "stable framework" characteristics are more obvious; the south/north direction is under strong sunlight/rain/high humidity conditions, the relative proportion of calcite increases, and it is easy to dissolve and redeposit in wet–dry and acidic environments, reducing the surface corrosion resistance stability.

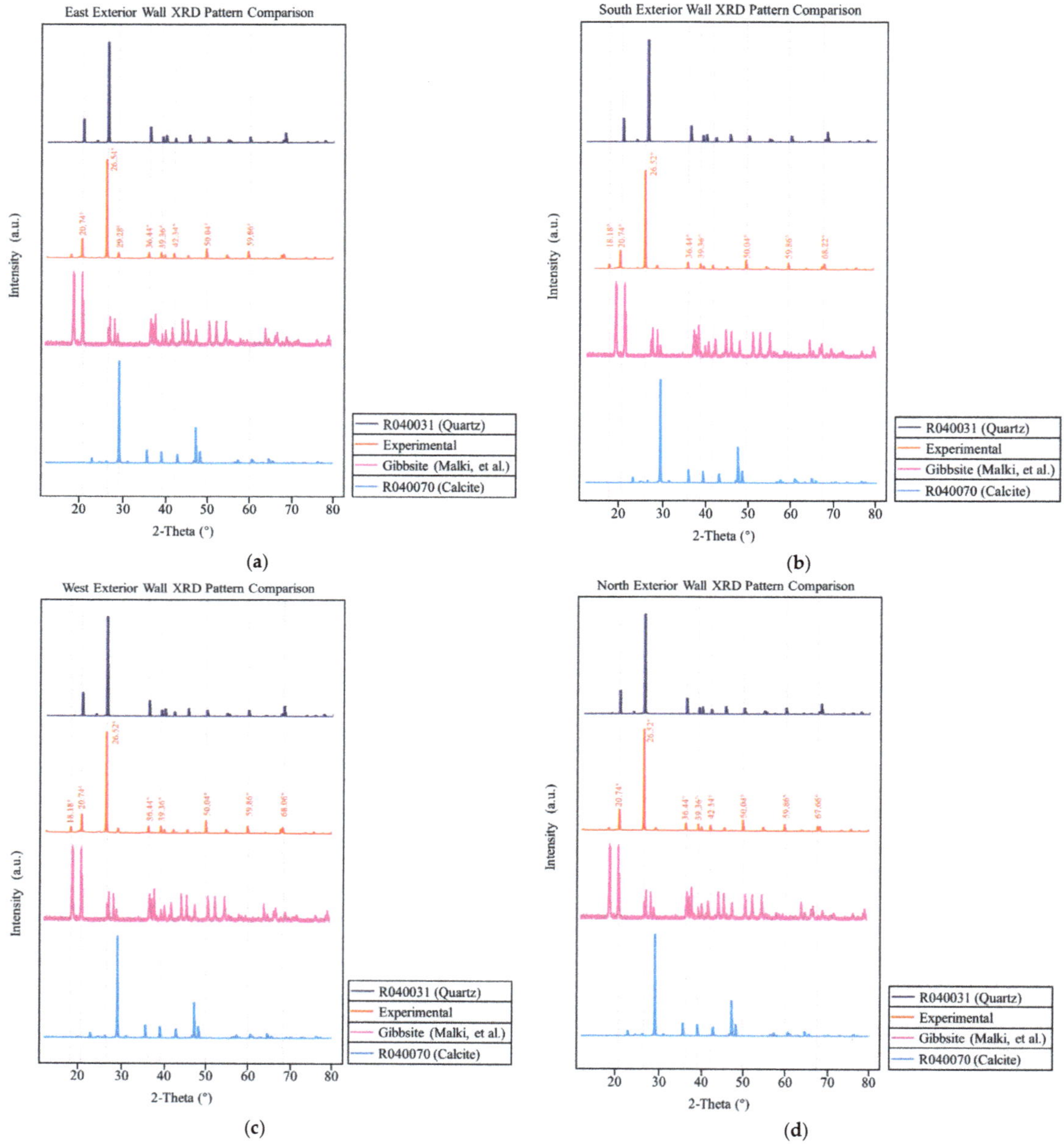

Figure 6. Comparative XRD results of four wall samples. (**a**) East Exterior Wall XRD Pattern Comparison; (**b**) South Exterior Wall XRD Pattern Comparison; (**c**) West Exterior Wall XRD Pattern Comparison; (**d**) North Exterior Wall XRD Pattern Comparison. (Image source: drawn and annotated by the author. The peak distribution spectrum of Gibbsite is from reference [31]).

3.2. Comparison Results of SEM Images

Through SEM imaging analysis of wall rammed earth samples facing east, south, west, and north, it was found that there were significant differences in their micromorphology, particle distribution, and surface characteristics (Figure 7). For other detailed SEM image results at different magnifications, please refer to Appendix A.

Figure 7. Comparative SEM results of four wall samples. (**a**) SEM morphology of the east wall sample particles; (**b**) SEM morphology of the south wall sample particles; (**c**) SEM morphology of the west wall sample particles; (**d**) SEM morphology of the north wall sample particles. (Image source: compiled by the author based on SEM analysis results).

In SEM images, the particles from the east wall samples (Figure 7a) are primarily regular in shape, either spherical or polygonal, small in size, evenly distributed with only slight agglomeration, and have a relatively smooth overall surface. This reflects the uniformity and stability of the growth or processing environment for these samples. The particles of the south wall samples (Figure 7b) exhibit significant irregularity, predominantly polygonal, with large variations in particle size, uneven distribution, and significant agglomeration. The surface is rough and accompanied by microcracks, suggesting that the samples in this direction may have been subject to significant stress during their formation.

The particles in the west wall sample (Figure 7c) are mainly nearly round to elliptical, with a medium size range, relatively weak agglomeration, and a relatively flat surface, indicating good particle uniformity and stable and uniform growth or processing conditions. The particles in the north wall samples (Figure 7d) are fragmented or flaky, with significant size variation, a wide distribution, and more prominent agglomeration. The surfaces are also rough, with visible cracks and pores, reflecting strong anisotropy in the microstructure of the samples in this direction, influenced by environmental or processing factors.

Overall, the SEM characteristics of the east and west walls of the rammed earth dwellings in Xiaochikan Village are more similar. The particles on the east wall are regular spherical or polygonal, small, and evenly distributed. The particles on the west wall are predominantly spherical and elliptical, uniform in size, and less agglomerated. The particle surfaces on both walls are relatively smooth and flat. This phenomenon is directly related to the more moderate environmental impacts on east- and west-facing walls under local climatic conditions. As the direction away from direct sunlight and most exposed to concentrated rainfall, these walls are shielded from both the prolonged high temperatures and heavy rainfall from the south and the persistent high humidity from the north. This results in a more intact rammed earth particle structure and a more regular microscopic morphology. However, the microscopic characteristics of the south and north walls differ significantly, and both exhibit signs of structural damage. The south wall exhibits irregular particles, uneven size, and a rough, cracked surface. This feature is due to its position as the surface exposed to intense summer sunlight and the primary rain during the rainy season. The alternating high temperatures, dryness, and wetness, along with rainfall, lead to a breakdown of rammed earth particle agglomerations and surface cracking. The fragmented or flaky particles and pronounced porosity on the north wall indicate that the north direction has the shortest sunlight exposure and relatively high humidity. This long-term humid environment causes the rammed earth particles to flake, leading to the development of a porous structure. These phenomena indicate that the microstructural differences among the four walls of this rammed earth dwelling are essentially due to the differential effects of local natural environmental factors such as sunlight, precipitation, and humidity on the walls facing different directions, ultimately resulting in significant microstructural anisotropy.

3.3. Comparison Results of EDS Element

3.3.1. East Wall Sample

The Ca spectrum signal in the east wall sample is forceful and concentrated, making it the most prominent feature (Figure 8). This is particularly true in the inter-grain cementation region, where the Ca signal appears as continuous bands or flakes, demarcating it from the grain framework (the Si and Al signal regions). The C signal is widely distributed, overlapping with the Ca signal in some areas and scattered across the grain surfaces in others. The signal intensity is moderate and uneven. The Si and Al signals are concentrated in the core of grain aggregates, forming strong, massive signals with clear grain boundaries, distinguishable from the surrounding Ca cement regions. The Fe signal is moderate and evenly distributed at the edges of grain aggregates or within the cement, with occasional strong, dot-like signals at unusual mineral grains. It is speculated that Fe is dispersed within the framework and cement as iron oxides. The Cl signal is slightly enriched at the pore edges, with lower intensity than that of the south wall sample and higher than that of the northwest wall sample. The signal distribution is relatively limited, primarily concentrated near the surface pores, likely due to small amounts of soluble salts remaining after rainwater infiltration.

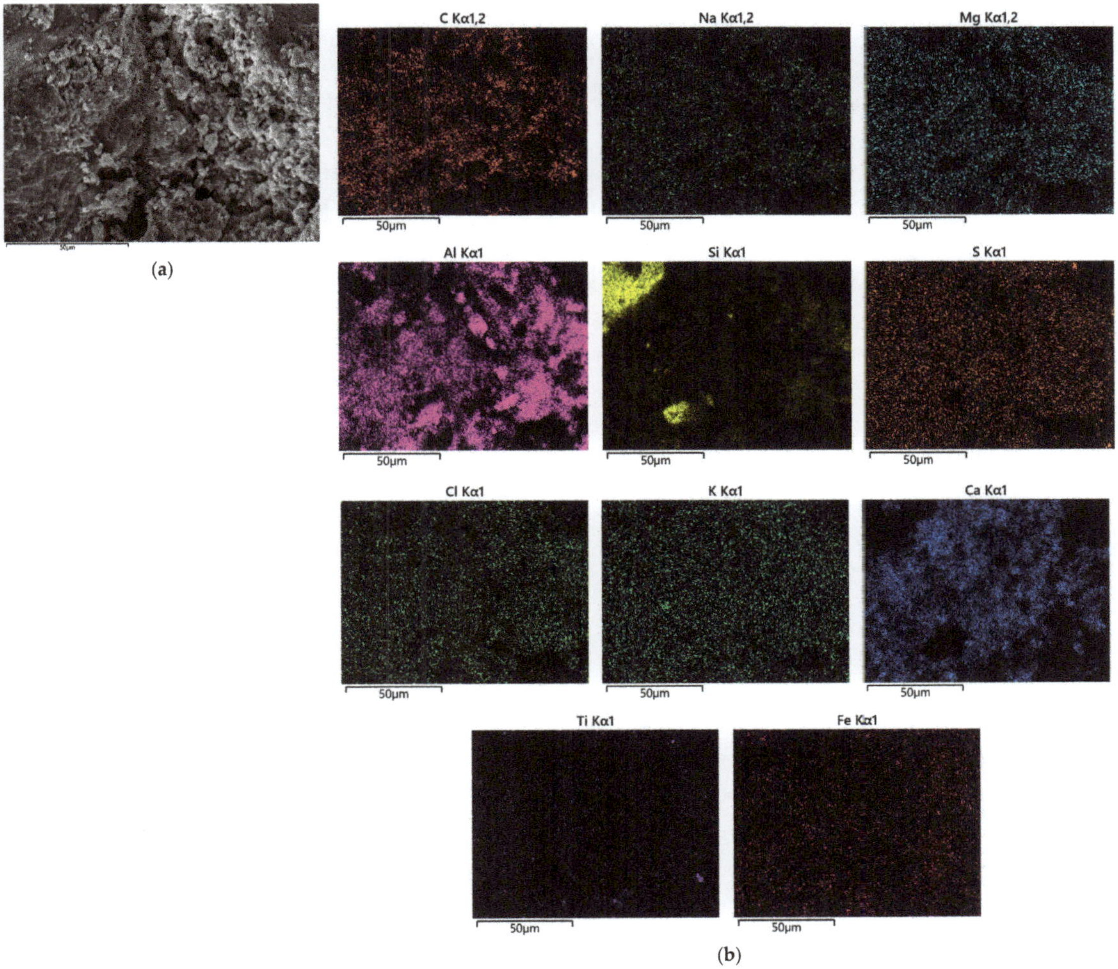

Figure 8. Element mapping results of the east wall rammed earth sample. (**a**) Investigated area; (**b**) imaging results. (Image source: compiled by the author based on SEM analysis results).

Furthermore, the east wall's Ca content, at 30.36% by weight and 15.68% by atomic percentage, is the highest among the four samples (Figure 9, Table 1). This suggests that large amounts of lime or carbonate-rich soil were added during preparation, or that the soil itself is rich in carbonate minerals, resulting in an abundance of cementing material. Theoretically, the rammed earth's integrity and compressive strength should be higher than those of the other samples. The C content, at 35.10% by weight, likely originates not only from organic matter but also from a significant amount of carbonate carbon (due to the extremely high Ca content, the proportion of carbonate carbon is likely even higher), accounting for over 60% by atomic percentage, indicating that organic carbon and carbonates together form the basis of the high carbon content. The Al content is only slightly higher than that of the north wall sample, with a moderate clay mineral content. The Si content is the lowest among the four samples, with a low proportion of sandy skeletons. Other trace elements indicate moderate impurity content.

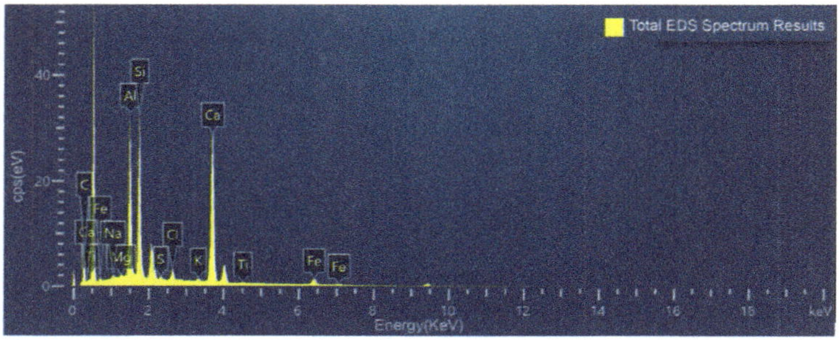

Figure 9. EDS spectrum at east wall sample. In the energy spectrum, the *X*-axis represents the energy (in keV) or wavelength of the X-rays; the *Y*-axis represents the number of X-rays detected at a specific energy/wavelength, i.e., the intensity or count. (Image source: compiled by the author based on SEM analysis results).

Table 1. The energy-dispersive spectrometer component analysis results for east wall sample.

Element	Line Type	Wt%	Wt% Sigma	At%
C	K-series	35.1	0.63	60.48
Na	K-series	0.4	0.03	0.36
Mg	K-series	0.14	0.03	0.12
Al	K-series	12.76	0.14	9.78
Si	K-series	14.35	0.15	10.57
S	K-series	0.22	0.03	0.14
Cl	K-series	1.42	0.04	0.83
K	K-series	0.45	0.03	0.24
Ca	K-series	30.36	0.31	15.68
Ti	K-series	0.29	0.05	0.12
Fe	K-series	4.51	0.11	1.68

Source: Author's statistics.

3.3.2. South Wall Sample

The spectrum shows continuous, high-intensity, uniform signals for Si and Al in the south wall sample, primarily concentrated in the particle aggregates (Figure 10). At the 50 μm scale, the Si and Al signals almost completely overlap, forming blocky or strip-like areas of a strong signal. These signals correspond to silicate mineral particles within the rammed earth. These particle aggregates form the soil's skeleton. The continuous signal indicates a relatively uniform distribution of the skeleton particles, with no obvious localized absence or over-concentration, indicating excellent structural integrity. The spectrum signal for Ca is moderate in intensity and primarily distributed in the interstices between particle aggregates, exhibiting a "filling" distribution. For example, at the boundaries between Si-Al framework particles, the Ca signal is significantly enhanced, indicating that Ca fills the interstices between the framework particles as a cement, exerting a cementing effect. The relatively continuous Ca signal distribution indicates that the cement fills the interstices uniformly, without the presence of excessive porosity caused by localized absence of cement. The spectrum signal for C is high in intensity but unevenly distributed, displaying a "scattered dot pattern with localized banding". Scattered point signals may correspond to tiny plant debris particles, while localized band signals may represent the remnants of fibrous organic matter. This uneven distribution is consistent with the natural retention of organic matter in rammed earth. Furthermore, the C signal is not concentrated in specific areas (such as pore edges), indicating that organic matter has not undergone significant migration due to environmental factors. The Fe spectrum signal intensity is moderate and relatively evenly distributed, occurring in both skeletal particles and cementing areas, with occasional localized point signal enhancements. This distribution indicates that Fe is uniformly dispersed in the rammed earth matrix as a trace impurity, with no significant enrichment or loss. The auxiliary cementation effect of iron oxides is evident throughout the sample.

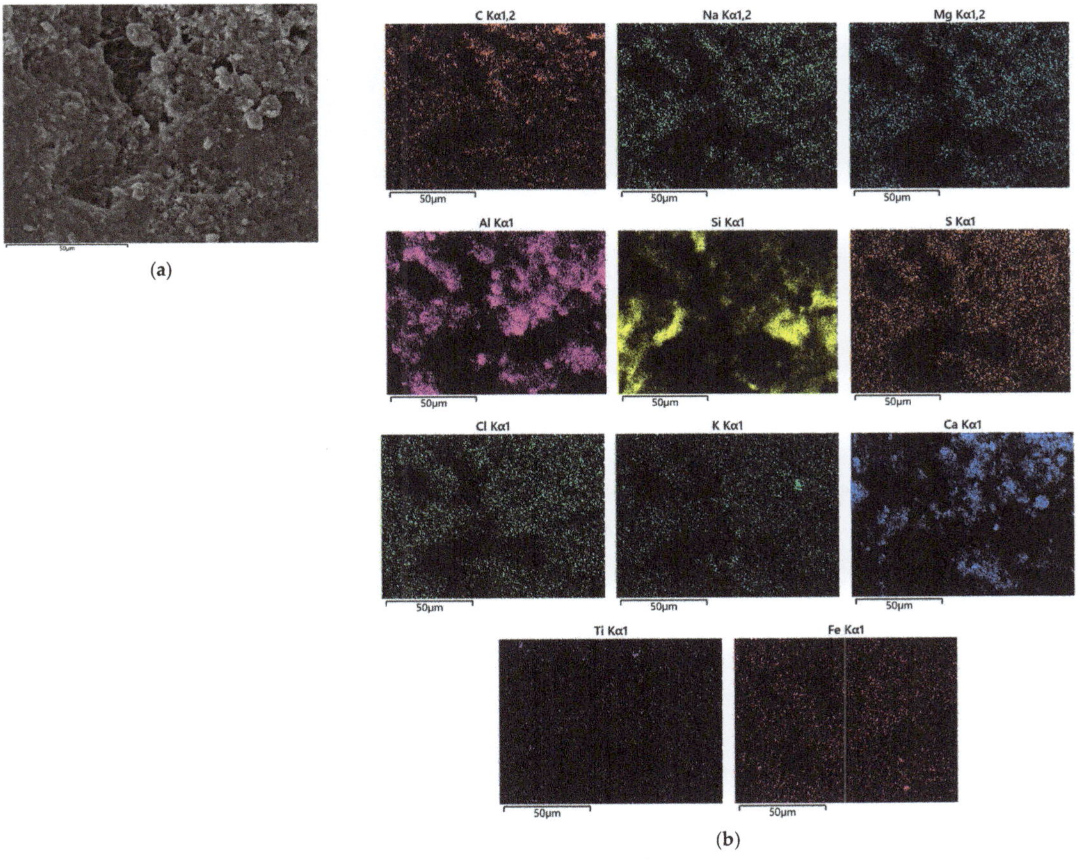

Figure 10. Element mapping results of the south wall rammed earth sample. (**a**) Investigated area; (**b**) imaging results. (Image source: compiled by the author based on SEM analysis results).

C, at 40.77% by weight and 64.70% by atomic percentage, is the most abundant element in the sample (Figure 11, Table 2). This high C content may be due to two factors: first, residual organic matter such as plant debris and fibers mixed into the rammed earth during preparation; second, carbonaceous components in the soil itself or carbon deposited in the environment later. From an atomic perspective, C accounts for over 60% of the atoms in the microstructure, making it a key element influencing the organic matter content of rammed earth. Si, at 23.39% by weight and 15.87% by atomic percentage, is the second most abundant element. Si is a core component of silicate minerals, forming the "skeleton" of rammed earth and fundamentally determining its mechanical strength. Its high content indicates a high proportion of sandy or silty minerals in the sample, resulting in a more stable skeleton structure. Ca, primarily derived from carbonate minerals (such as calcite $CaCO_3$) in the rammed earth or from lime added during preparation (which forms a calcium carbonate cement upon hydration), is a crucial "binder" for rammed earth, filling the gaps between mineral particles, strengthening the bonds between particles, and enhancing the integrity of the rammed earth.

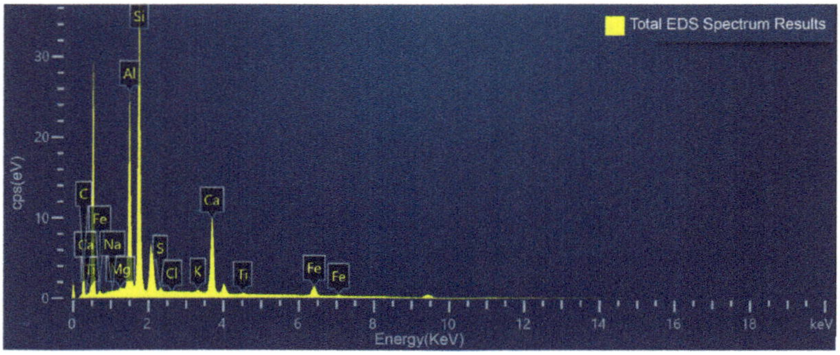

Figure 11. EDS spectrum at south wall sample. In the energy spectrum, the *X*-axis represents the energy (in keV) or wavelength of the X-rays; the *Y*-axis represents the number of X-rays detected at a specific energy/wavelength, i.e., the intensity or count. (Image source: compiled by the author based on SEM analysis results).

Table 2. The energy-dispersive spectrometer component analysis results for south wall sample.

Element	Line Type	Wt%	Wt % Sigma	At%
C	K-series	40.77	0.74	64.7
Na	K-series	0.07	0.03	0.06
Mg	K-series	0.09	0.03	0.07
Al	K-series	13.81	0.19	9.75
Si	K-series	23.39	0.31	15.87
S	K-series	0.31	0.04	0.18
Cl	K-series	0.09	0.03	0.05
K	K-series	0.48	0.04	0.23
Ca	K-series	14.15	0.2	6.73
Ti	K-series	0.3	0.06	0.12
Fe	K-series	6.56	0.16	2.24

Source: Author's statistics.

3.3.3. West Wall Sample

The Si and Al spectrum signals in the west wall sample are forceful, dominating the image and exhibiting a "large, continuous, blocky" distribution that covers most of the sample's microscopic area (Figure 12). The Si and Al signals completely overlap, forming densely distributed, large-scale framework aggregates (mostly larger than 20 μm) with narrow inter-aggregate gaps. This indicates a high content of sandy and clay mineral particles, a dense framework structure, and the absence of significant macropores or framework vacancies. This distribution perfectly matches the high Si and Al content and is typical of sandy rammed earth microstructures. The Fe signal is moderate in intensity and relatively uniformly distributed, primarily accompanying the Si and Al signals within framework aggregates, with occasional localized, dotted areas of enhancement. Due to the densely packed framework particles, the Fe distribution is widespread, helping to strengthen the internal bonding of the framework particles. The signal intensity of Cl is moderate, and its distribution characteristic is "framework interstitial enrichment"—it is mainly concentrated in the narrow gaps between Si-Al framework aggregates. Because the gaps are narrow and there is no large amount of Ca cement filling, Cl-containing water is retained here and evaporates and remains, forming interstitial enrichment. However, due to the narrow gaps, the enrichment area is distributed linearly, rather than being enriched in sheets at the pore edges, like in the north wall sample. The result shows that although there are salt residues in this sample, the pore space is limited, and the salt enrichment range is relatively limited.

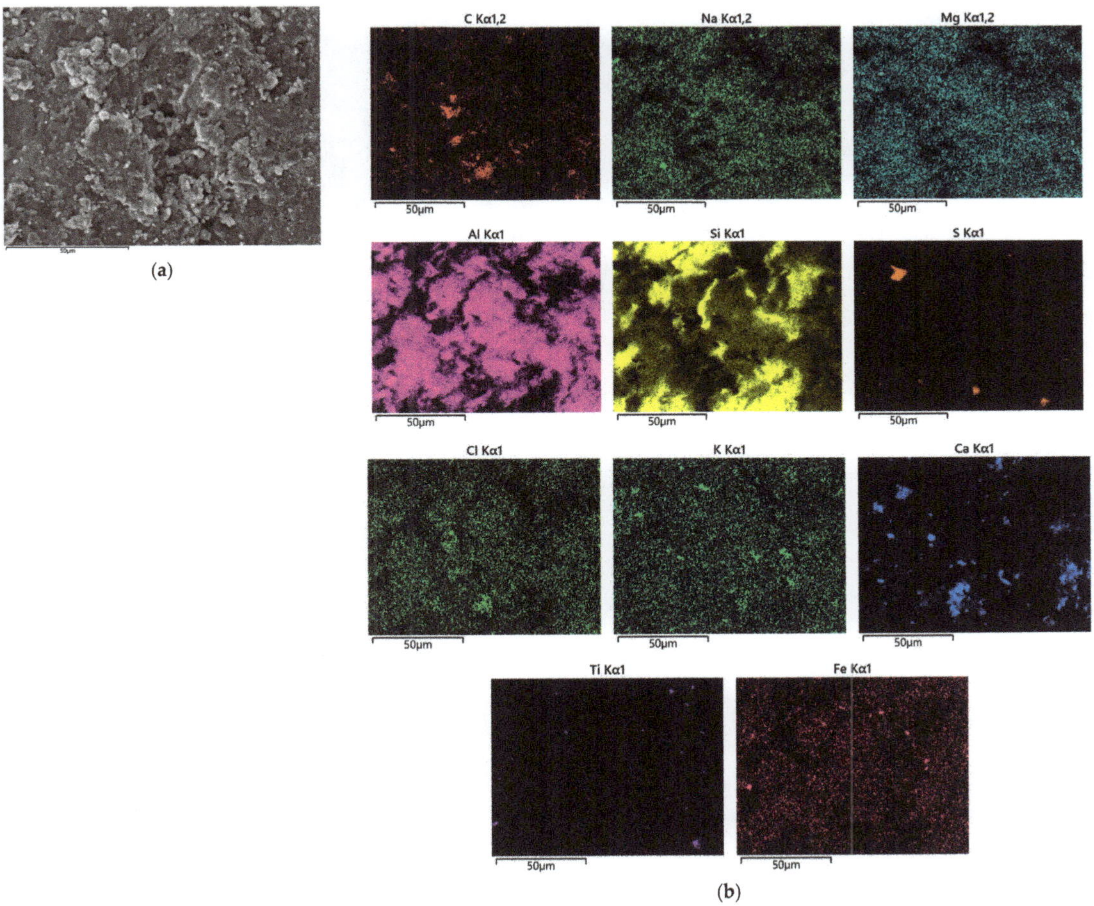

Figure 12. Element mapping results of the west wall rammed earth sample. (**a**) Investigated area; (**b**) imaging results. (Image source: compiled by the author based on SEM analysis results).

Si, at 38.03% by weight and 28.03% by atom, is the highest of the four samples and a core characteristic element (Figure 13, Table 3). This high Si content indicates a high proportion of sandy silicate minerals in this sample, a typical "sandy skeleton-dominated" rammed earth, with a strong mechanical stability and deformation resistance. Al, at 18.79% by weight and 14.42% by atom, is also the highest of the four samples and, together with Si, forms the basis for the high silicate mineral content. This high Al content indicates a rich content of clay or feldspar minerals. Clay minerals enhance the plasticity of rammed earth, making it easier for sandy particles to agglomerate, while feldspar minerals strengthen the skeleton. The carbon content, at 29.53% by weight and 50.89% by atom, is the lowest among the four samples, indicating low organic matter and carbonated carbon content. These findings may be due to the wall's more intense weathering or leaching, which resulted in organic carbon loss. Furthermore, the cement content is low, resulting in a low proportion of carbonate. The calcium content, at 5.63% by weight and 2.91% by atom, is the lowest among the four samples, indicating minimal carbonate cement or lime addition, indicating a lack of cementing material. The integrity of the rammed earth relies primarily on the viscosity of clay minerals and interparticle friction, rather than chemical cementation.

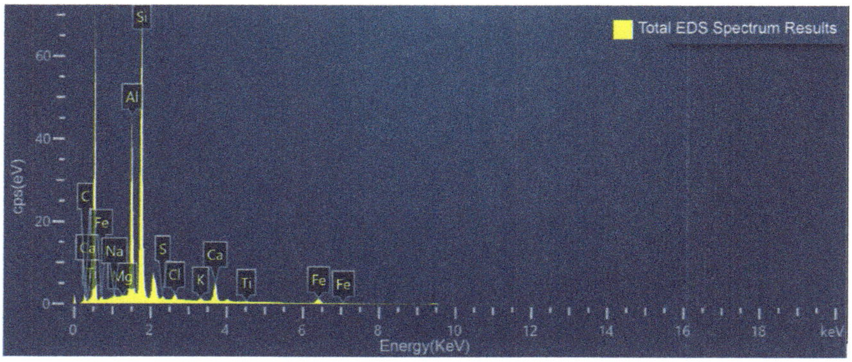

Figure 13. EDS spectrum at west wall sample. In the energy spectrum, the *X*-axis represents the energy (in keV) or wavelength of the X-rays; the *Y*-axis represents the number of X-rays detected at a specific energy/wavelength, i.e., the intensity or count. (Image source: compiled by the author based on SEM analysis results).

Table 3. The energy-dispersive spectrometer component analysis results for west wall sample.

Element	Line Type	wt%	Wt % Sigma	At%
C	K-series	29.53	0.58	50.89
Na	K-series	0.38	0.02	0.34
Mg	K-series	0.13	0.02	0.11
Al	K-series	18.79	0.16	14.42
Si	K-series	38.03	0.32	28.03
S	K-series	0.48	0.02	0.31
Cl	K-series	1.33	0.03	0.77
K	K-series	0.55	0.02	0.29
Ca	K-series	5.63	0.06	2.91
Ti	K-series	0.24	0.03	0.1
Fe	K-series	4.92	0.08	1.83

Source: Author's statistics.

3.3.4. North Wall Sample

The carbon spectrum signal in the north wall sample is high and evenly distributed (Figure 14). Unlike the uneven distribution in the south wall sample, the carbon signal in this sample is diffuse, with no distinct concentrated areas, such as dots or stripes. It is speculated that the organic matter may have undergone a certain degree of decomposition or dispersion and was evenly mixed in the rammed earth matrix in the form of finer particles, without any local residual accumulation. The Si signal intensity is high and concentrated in the particle agglomerate area, forming a blocky strong signal area. However, the size of the agglomerates is slightly smaller than the Si-Al agglomerates in the south wall sample, and the gaps between the agglomerates are wider, corresponding to the slightly lower Si content. The skeleton particles are still evenly distributed without obvious vacancies. The Ca signal intensity is high, and it is the strongest signal in the thermodynamic diagram after C and Si. It has a wide distribution range and fills the gaps between Si aggregates and slightly overlaps with the Si signal in some areas. This indicates that the Ca content is sufficient. The cement not only fills the gaps but also may form a thin cementing layer on the surface of the particles, enhancing the bonding force between the particles and the cement. The signal is continuous without obvious breakpoints, and the cementation uniformity is excellent. The Cl heatmap signal intensity is moderate, making it the clearest among the low-content elements. Its distribution is notable—it is primarily concentrated at pore edges and on the outer surfaces of particle aggregates, exhibiting an "edge enrichment" phenomenon. At a 50 μm scale, the Cl signal around pores is significantly stronger than in the matrix. This attribute is due to Zhuhai's high rainfall and humidity. When Cl-containing rainwater or moisture penetrates the rammed earth, soluble chloride salts are left behind in the pores due to evaporation, resulting in edge enrichment. This characteristic is consistent with the salinization trend caused by the maritime climate.

C, at 39.98% by weight and 64.35% by atom, is second only to the south wall sample and remains the most abundant element in the sample (Figure 15, Table 4). The C origin is consistent with that of the south wall, primarily organic and carbonaceous components, accounting for over 60% of the total atomic content. This value indicates that this sample also contains a high organic matter content, similar to that of the south wall. Si, at 20.40% by weight and 14.04% by atom, is the second most abundant element, slightly lower than the Si content in the south wall sample. Ca, at 20.19% by weight and 9.74% by atom, is significantly higher than the Ca content in the south wall sample (14.15%) and is a prominent core element characteristic of this sample. The high Ca content indicates a higher amount of carbonate cement or lime added to this sample, resulting in a more robust cementing base and potentially greater integrity. S, at 2.77% by weight and 1.67% by atom, is the highest of the four samples and the most prominent secondary component characteristic of this sample. The high sulfur content may be due to two factors: first, the high content of sulfate minerals in the local soil; and second, Zhuhai's proximity to the ocean. Sulfur compounds in the atmosphere (such as sulfate in marine aerosols) settle and penetrate the rammed earth, leading to sulfur enrichment. The Cl content, at 1.69% by weight and 0.92% by atomic percentage, is also the highest among the four samples, significantly higher than the 0.09% found in the south wall sample. Cl is a typical soluble salt element, and its high Cl content is closely related to Zhuhai's maritime climate. Sodium chloride (sea salt) in the marine atmosphere penetrates the rammed earth with rainwater or moisture, forming soluble salt residues.

Overall, the core element composition of the four samples is consistent, with C, Si, Al, Ca, and Fe constituting over 93% of the total. This reflects the fundamental characteristics of the local alluvial soil, described as "silicate skeleton-carbonate cementation-organic matter residue" [32], and highlights the "local material" attribute of rammed earth. Except for the south wall sample, the Cl content of the remaining samples exceeds 1%. In the thermal map, Cl shows "pore/interstitial enrichment", confirming that the salinization process of marine aerosols with rainwater infiltration and evaporation residues is a common influence of marine climate.

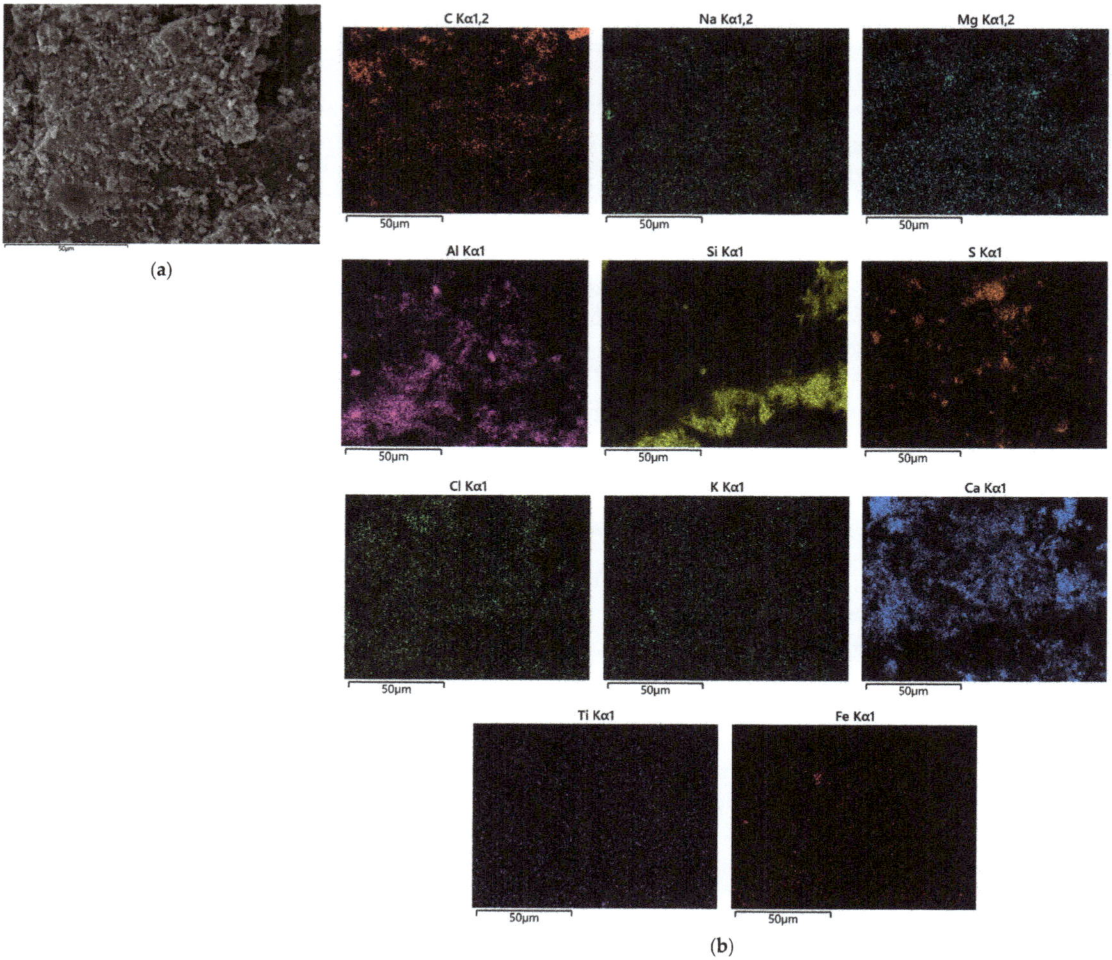

Figure 14. Element mapping results of the north wall rammed earth sample. (**a**) Investigated area; (**b**) imaging results. (Image source: compiled by the author based on SEM analysis results).

As shown in Table 5, the elemental composition of the samples from the four walls was mainly C, Si, Al, Ca, and Fe, all accounting for >93 wt% (East 97.08%, South 98.68%, West 96.90%, North 93.52%), but there were significant directional differences in the framework and cementation type. The west wall had significantly higher Si and Al content (Si 38.03 wt%, Al 18.79 wt%, Si/Al $\approx$ 2.0, Ca only 5.63 wt%, Ca/Si $\approx$ 0.15), indicating a robust framework mainly composed of quartz/aluminosilicate. The east wall had a prominent Ca content (30.36 wt%, Ca/Si $\approx$ 2.12), which better reflected the carbonate cementation characteristics. The south wall had higher Si and Al content but lower Ca content (Si 23.39 wt%, Al 13.81 wt%, Ca 14.15 wt%, Ca/Si $\approx$ 0.60), suggesting that carbonate leaching may have occurred under strong sunlight and rain. The north wall exhibits a mixed matrix (Si 20.40 wt%, Ca 20.19 wt%, Ca/Si $\approx$ 1.0). Regarding soluble salt indicators, Cl $\geq$ 1 wt% is observed in all directions except the south wall (East 1.42, West 1.33, North 1.69 wt%), with higher S on the north wall (2.77 wt%), consistent with salt accumulation and crystallization stress under high humidity/low sunlight conditions. The south wall has only 0.09 wt% Cl and 0.31 wt% S, consistent with salt erosion by rain. C content is generally high (~29.53–40.77 wt%, atomic fraction approximately 50.89–64.70%), appearing alongside higher Ca content on the east/south/north walls, possibly indicating the presence of carbonates or organic residues. In summary, the west wall is predominantly composed of a siliceous framework and exhibits low chemical solubility sensitivity. The east wall has a high proportion of carbonate cementation and is more sensitive to acid rain, dissolution, and redeposition; the north wall has the highest salt load and salt swelling risk. The south wall has the lowest salt content but may suffer from surface weakening due to carbonate loss.

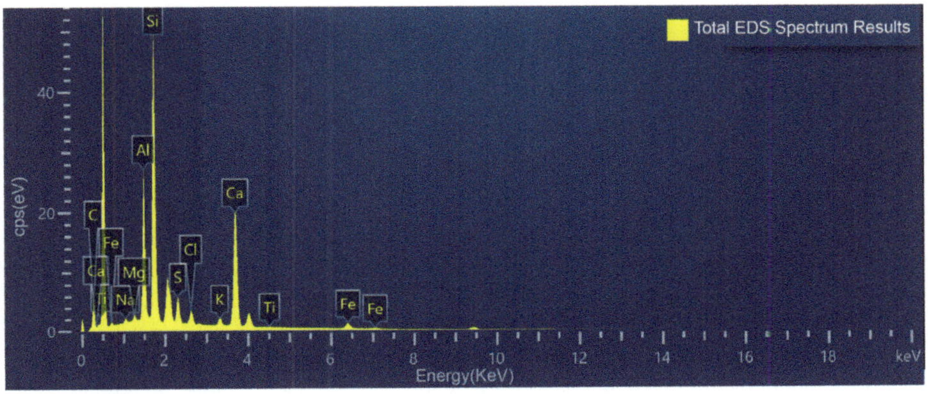

Figure 15. EDS spectrum at north wall sample. In the energy spectrum, the *X*-axis represents the energy (in keV) or wavelength of the X-rays; the *Y*-axis represents the number of X-rays detected at a specific energy/wavelength, i.e., the intensity or count. (Image source: compiled by the author based on SEM analysis results).

Table 4. The energy-dispersive spectrometer component analysis results for north wall sample.

Element	Line Type	Wt%	Wt % Sigma	At%
C	K-series	39.98	0.63	64.35
Na	K-series	0.37	0.03	0.31
Mg	K-series	0.37	0.03	0.29
Al	K-series	9.65	0.11	6.92
Si	K-series	20.4	0.22	14.04
S	K-series	2.77	0.05	1.67
Cl	K-series	1.69	0.04	0.92
K	K-series	1.13	0.04	0.56
Ca	K-series	20.19	0.23	9.74
Ti	K-series	0.14	0.04	0.06
Fe	K-series	3.30	0.1	1.15

Source: Author's statistics.

Table 5. Elemental composition from EDX analysis on the sample from the walls.

Wall	East		South		West		North	
Metric	wt%	At%	wt%	At%	wt%	At%	wt%	At%
C	35.1	60.48	40.77	64.7	29.53	50.89	39.98	64.35
Na	0.4	0.36	0.07	0.06	0.38	0.34	0.37	0.31
Mg	0.14	0.12	0.09	0.07	0.13	0.11	0.37	0.29
Al	12.76	9.78	13.81	9.75	18.79	14.42	9.65	6.92
Si	14.35	10.57	23.39	15.87	38.03	28.03	20.4	14.04
S	0.22	0.14	0.31	0.18	0.48	0.31	2.77	1.67
Cl	1.42	0.83	0.09	0.05	1.33	0.77	1.69	0.92
K	0.45	0.24	0.48	0.23	0.55	0.29	1.13	0.56
Ca	30.36	15.68	14.15	6.73	5.63	2.91	20.19	9.74
Ti	0.29	0.12	0.3	0.12	0.24	0.1	0.14	0.06
Fe	4.51	1.68	6.56	2.24	4.92	1.83	3.3	1.15

Source: Author's statistics.

4. Discussion

At this stage, this study can only provide evidence of the role of soluble salts based on the semi-quantitative and spatial distribution of S and Cl in EDS and correlate it with visible deterioration phenomena such as salt frost/dust on the wall surface so as to avoid misinterpreting the elemental signals as a definitive identification of minerals. Through on-site environmental surveys, XRD analysis, and SEM-EDS analysis, the long-term stability of the rammed earth walls of Xiaochikan Village, under the subtropical maritime monsoon climate, was revealed. The study revealed that the microstructure (particle morphology, porosity, and bond strength) and chemical composition (element distribution and mineral content) of the four walls exhibit significant directional variations due to their exposure to different environments. The east wall maintains its structural density thanks to high carbonate cementation, while the south wall is susceptible to cracking due to high temperatures and heavy rain. Salt spray deposition accelerates weathering of the west wall, while the high humidity triggers biological erosion of the north wall. Based on the "environment-microstructure-macrodegradation" linkage mechanism and combining traditional rammed earth craftsmanship with modern conservation techniques, a conservation strategy tailored to each orientation was proposed to support the living legacy of the village's rammed earth architecture.

4.1. East Wall: Potential Impacts of Carbonate Balance and Diurnal Temperature Variation

The East Wall is the most structurally stable of the four rammed earth walls in Xiaochikan Village. Its characteristics are closely related to the regional environment and material composition. Microstructurally, the rammed earth particles in the East Wall are mostly regular, spherical, or polygonal, small, and evenly distributed, with only slight agglomeration. The overall surface is relatively smooth, and the porosity is well within a reasonable range. This morphological characteristic is directly related to the local east-facing walls being more moderately exposed to the natural environment. They are shielded from the direct impact of prolonged high temperatures and heavy rains from the south and mitigate the persistent effects of long-term high humidity from the north, thus preserving the intact rammed earth particle structure. Conservation strategies for the East Wall should focus on stabilizing the carbonate balance and strengthening the cementation strength. Based on the traditional Lingnan rammed earth craft, it is recommended that a mixed coating of "oyster shell powder + glutinous rice paste" be used for protection in the future. The calcium carbonate content in oyster shell powder exceeds 90%, which can artificially supplement the carbonate in the wall and reduce natural dissolution loss. The glutinous rice paste can enhance the adhesion between the coating and the wall and prevent the coating from falling off. In addition, this formula is in line with the process logic of the traditional rammed earth "three-in-one soil + organic gelling agent" of Xiaochikan Village, ensuring that the protection measures do not damage the historical appearance of the wall [9,10].

4.2. South Wall: Synergistic Impacts of High Temperature and Heavy Rainfall

The South Wall is one of the most significantly stressed walls in Xiaochikan Village. Its characteristics are primarily due to the synergistic effects of high temperature and heavy rainfall. Xiaochikan Village has a subtropical maritime monsoon climate, with an average annual temperature of 22.6 °C (maximum 38.5 °C) and an average annual sunshine duration of 1876.3 h. Intense summer sunshine directly impacts the South Wall. Furthermore, rainfall during the rainy season (April–September) accounts for 84.7% of the annual total, with a maximum daily rainfall of 393.7 mm. Heavy rainfall exerts a significant direct erosion effect on the South Wall. Microstructurally, the rammed earth particles in the South Wall are irregular polygons with a wide range of particle sizes (3–20 μm in diameter) and uneven distribution. Significant agglomeration is evident, and the surface is rough with fine cracks, reflecting the significant environmental stresses sustained during the formation and use of this wall. The protection strategy for the south wall should focus on solving the problems of crack expansion and surface peeling caused by high temperature exposure and heavy rain while strengthening the bonding strength. In the future, it is recommended that a removable bamboo sunshade (30 cm spacing) be built along the outer side of the south wall at the physical protection level to maintain the ventilation of the wall while reducing the surface temperature by 5–8 °C and reducing the thermal expansion and contraction of particles caused by high temperature. Traditional gray tile eaves (80 cm wide) should be added to the top of the wall to replicate the "rainproof eaves" of Lingnan folk houses [33–35], guiding rainwater away from the wall and preventing heavy rain from directly eroding the surface fine particles [36].

4.3. West Wall: Weathering Effects of Salt Fog Enrichment and Insufficient Cementation

The characteristics of the West Wall are determined by salt fog erosion and insufficient cementation of the sandy skeleton, which is closely related to the location environment of Xiaochikan Village. From a macro perspective, the village is in the alluvial plain on the west bank of the Pearl River Estuary. There is no high terrain blocking the west side. The salt fog in the marine climate comes not only from the southeast monsoon but also from the local circulation of the Hutiaomen waterway on the west side (such as the southwest monsoon and westerly wind). Although Huangyang Mountain is in the southeast (the main peak is 583 m above sea level), it can weaken the offshore salt fog carried by the southeast monsoon, but it cannot block the "near-source salt fog" formed by evaporation from the nearby waterway on the west side. In addition, the West Wall is close to the village water system, and the groundwater level is high, which further aggravates the upward migration of salt, superimposed on the influence of salt fog, forming a "double salt erosion" effect [37]. Therefore, it is highly recommended that the protection strategy of the West Wall should focus on "salt blocking, salt drainage, and salt swelling prevention", considering the stability of the sandy skeleton and the prevention and control of salt fog erosion. In terms of water repellency and salt resistance, traditional tung oil can be used to enhance the water repellency of the wall surface [38] and reduce salt adhesion. This material is in line with the protection principles of Xiaochikan Village and avoids the impact of chemical reagents on the environment. In terms of drainage and moisture-proof optimization, a blue brick moisture-proof layer (50 cm high) is added at the base of the west wall to match the village's rammed earth and brick-wood mixed building structure to block salt migration caused by rising groundwater. A 10 cm wide and 15 cm deep gravel drainage ditch is excavated along the foot of the wall, and a gravel layer is laid to accelerate the drainage of rainwater and reduce the retention time of salt on the wall [39].

4.4. North Wall: Synergistic Destruction by High Humidity and Biological Activity

The synergistic effect of high humidity and biological erosion determines the characteristics of the north wall. It is the wall with the most significant loose structure in Xiaochikan Village. The north side of Xiaochikan Village has the shortest sunshine hours, and the north side accounts for the lowest proportion in the annual average sunshine hours of 1876.3 h. As a result, the north wall is in a shaded environment for a long time, and the environmental humidity often exceeds 80%, providing suitable growth conditions for organisms such as mosses and actinomycetes [40,41]. Therefore, it is recommended that the protection strategy of the north wall should focus on "humidity control-bacterial inhibition-loose structure repair", considering both biological erosion prevention and control and wall permeability. In terms of antibacterial protection, TiO_2/chitosan nanocomposite coating is used, which has an antibacterial rate of 92.19% against Staphylococcus aureus [42]. Chitosan is derived from shells (which is consistent with the oyster shell component in the rammed earth of Xiaochikan Village), which is in line with the tradition of using local materials. In terms of humidity control, moisture-resistant trees (such as local banyan trees) are planted around the north wall to reduce ground evaporation and lower the local environmental humidity through vegetation transpiration; ventilation holes are set at the bottom of the wall (with a spacing of 1 m) to promote air circulation inside and outside the wall and avoid water vapor retention. At the same time, the ventilation holes are covered with wooden grilles to coordinate with the traditional architectural style [43].

5. Conclusions

Xiaochikan Village, Doumen Town, Zhuhai City is in the subtropical maritime monsoon climate zone. Long-term exposure to high temperatures, high humidity, salt spray, and heavy rain has put the local traditional rammed earth buildings at risk of accelerated deterioration. This study aimed to reveal the compositional and microstructural differences and dominant failure mechanisms of the exterior walls of typical rammed earth dwellings in Xiaochikan Village along different orientations and to propose differentiated conservation strategies. Methodologically, targeted sampling of the exterior walls was combined with SEM-EDS and XRD to systematically characterize their morphology, elemental composition and spatial distribution, and mineralogical phases. The following conclusions were drawn:

(1) There are significant differences in orientation between the four walls. The east wall is characterized by high carbonate cementation (Ca = 30.36 wt%), with regular and dense particles. The west wall is dominated by a silicate skeleton, with the highest Si and Al contents (Si = 38.03 wt%, Al = 18.79 wt%) and the lowest Ca, indicating insufficient cementation and domination by particle friction and cohesion. The north wall shows marginal enrichment of soluble salt indicator elements (Cl = 1.69 wt%, S = 2.77 wt%), suggesting salinization risk. The south wall has irregular particles, fine cracks, and obvious agglomeration damage, reflecting the synergistic damage caused by strong sunshine, wind-driven rain, and rainstorm erosion. The above differences are consistent with the local dominant wind direction, rainfall, and sunshine exposure conditions.

(2) Mineral assembling and engineering characteristics are obvious. The four-direction samples generally contain quartz, calcite, hydrous aluminum silicates (kaolinite group), and aluminum hydroxide (gibbsite). Quartz provides framework stability; calcite is dissolved in acidic/wet environments; layered aluminum silicates have water absorption and expansion properties, and the presence of aluminum hydroxide may also enhance the volume change effect during the wet–dry cycle, jointly triggering water absorption-swelling-contraction-surface erosion.
(3) Mechanism chain and vulnerability pattern. Wind-driven rain, salt fog deposition, and high humidity in the marine climate drive carbonate dissolution, salt crystallization, and biological attachment. These effects dominate in different directions, forming a vulnerability pattern where the east wall is relatively stable, the south wall is prone to cracking, the west wall is prone to erosion, and the north wall is prone to tidal salt. This pattern provides a basis for proposing a directional protection and restoration method for rammed earth walls.

This study also has some shortcomings: (1) Due to constraints related to powder sample collection and heritage conservation, this paper does not conduct quantitative analysis of particle size and porosity based on polished cross-sections/3D stereoscopic views. Currently, assessments of homogeneity and agglomeration are only qualitative, based on images. In the future, when conditions permit, limited two-dimensional image measurements (equivalent circle diameter, aspect ratio, roundness, and pore throat dimensions, $n \geq 200$/sample) will be conducted based on acquired SEM images. The two-dimensional approximations and representative boundaries will be clarified to provide semi-quantitative support for morphological description without introducing additional damage to the material. We understand this is a certain difficulty for sampling traditional village residential heritage, and we can still work in this direction in the future. (2) To control length and minimize sampling damage, this paper did not conduct qualitative and quantitative analysis of the mineral phases of weathering products: stoichiometry of water-soluble ions (Cl^-, SO_4^{2-}), phase identification of trace secondary salts, and semi-quantitative XRD full spectrum fitting of trace phases are still lacking. Existing SEM-EDS only provides semi-quantitative evidence and pore edge enrichment patterns at the elemental level and cannot replace mineral-level confirmation. Future work will focus on: (a) Quantitative analysis of water-soluble salts and conductivity/pH measurement to calculate total salt load and ion ratio. (b) Phase confirmation of typical salt crystals using Raman/FTIR. (c) Semi-quantitative analysis of secondary salts using XRD-Rietveld. (d) Establishing a correlation model between salt content indicators and macroscopic deterioration grading and micropore structure parameters (equivalent pore throat, porosity) to improve the discriminative power and comparability of rammed earth durability assessment. (3) The time dimension is inadequate. One-time summer sampling cannot reflect the process influence of seasonal cycles (typhoon season and dry season), and there is also a lack of rapid retesting and long-term tracking after heavy rain. The next step will be to expand the grid sampling of multiple buildings and multiple depths, superimpose in situ moisture or salt monitoring and mechanical-multi-field coupling verification, and carry out field tests and evaluations of orientation-oriented materials for salt blocking, moisture control, and cementation repair to form a scalable Lingnan coastal rammed earth protection technology system.

Author Contributions: Conceptualization, L.Z. and Y.C.; methodology, L.Z. and Y.C.; software, L.Z. and Y.C.; validation, Q.D. and J.L.; formal analysis, L.Z., Q.D., J.L., Z.G., Y.Z., W.L. and Y.C.; investigation, L.Z. and Y.C.; resources, L.Z., Q.D., J.L. and Y.C.; data curation, L.Z., Q.D., J.L. and Y.C.; writing—original draft preparation, L.Z., Q.D., J.L., Z.G., Y.Z., W.L. and Y.C.; writing—review and editing, L.Z. and Y.C.; visualization, L.Z., Q.D., J.L., Z.G., Y.Z., W.L. and Y.C.; supervision, Y.C.; project administration, Y.C.; funding acquisition, Y.C. All authors have read and agreed to the published version of the manuscript.

Funding: This research was funded by the (1) Faculty Research Grants funded by Macau University of Science and Technology (FRG-MUST) (grant number: FRG-25-041-FA; FRG-25-067-FA); (2) Guangdong Provincial Department of Education's key scientific research platforms and projects for general universities in 2023: Guangdong, Hong Kong, and Macau Cultural Heritage Protection and Innovation Design Team (grant number: 2023WCXTD042); (3) Guangdong Provincial Philosophy and Social Sciences Planning 2025 Lingnan Cultural Project (grant number: GD25LN30). The funders had no role in study conceptualization, data curation, formal analysis, methodology, software, decision to publish, or preparation of the manuscript. There was no additional external funding received for this study.

Institutional Review Board Statement: Not applicable.

Informed Consent Statement: Not applicable.

Data Availability Statement: The original contributions presented in this study are included in the article. Further inquiries can be directed at the corresponding author.

Acknowledgments: We sincerely thank Junxin Song and her husband for their assistance during the field research.

Conflicts of Interest: The authors declare no conflicts of interest.

Appendix A. Other SEM Image Results

The electron microscope's magnification results (Figures A1 and A2) at 1.00 kX (for macroscopic pore observation) and 5.00 kX (for observation of particle aggregates) correspond to the content in Section 3.2. Figure A3 shows the SEM image results at 500× resolution. Figure A4 shows the SEM image results at 100× resolution.

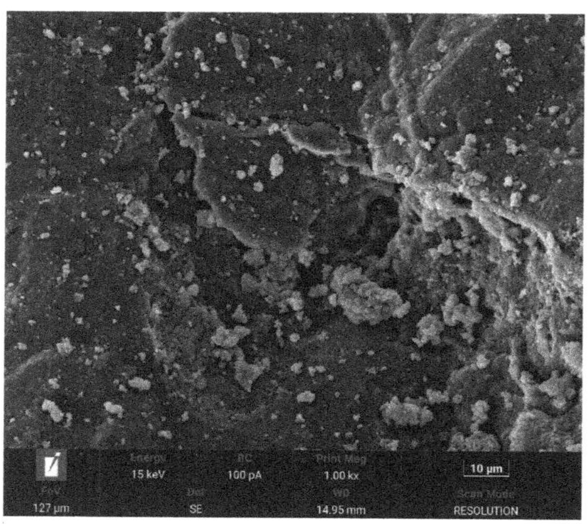

(**a**)

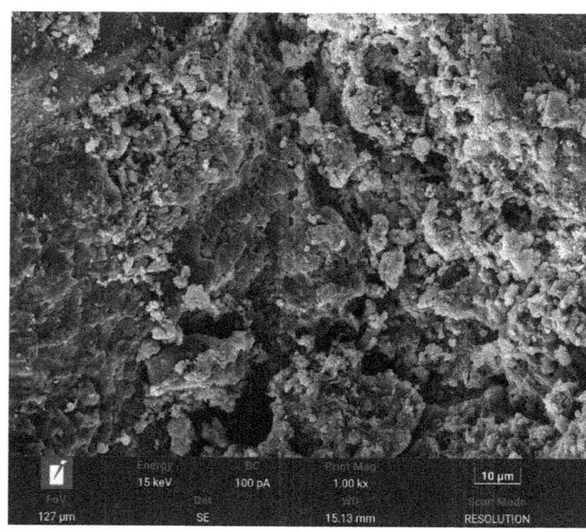

(**b**)

Figure A1. *Cont.*

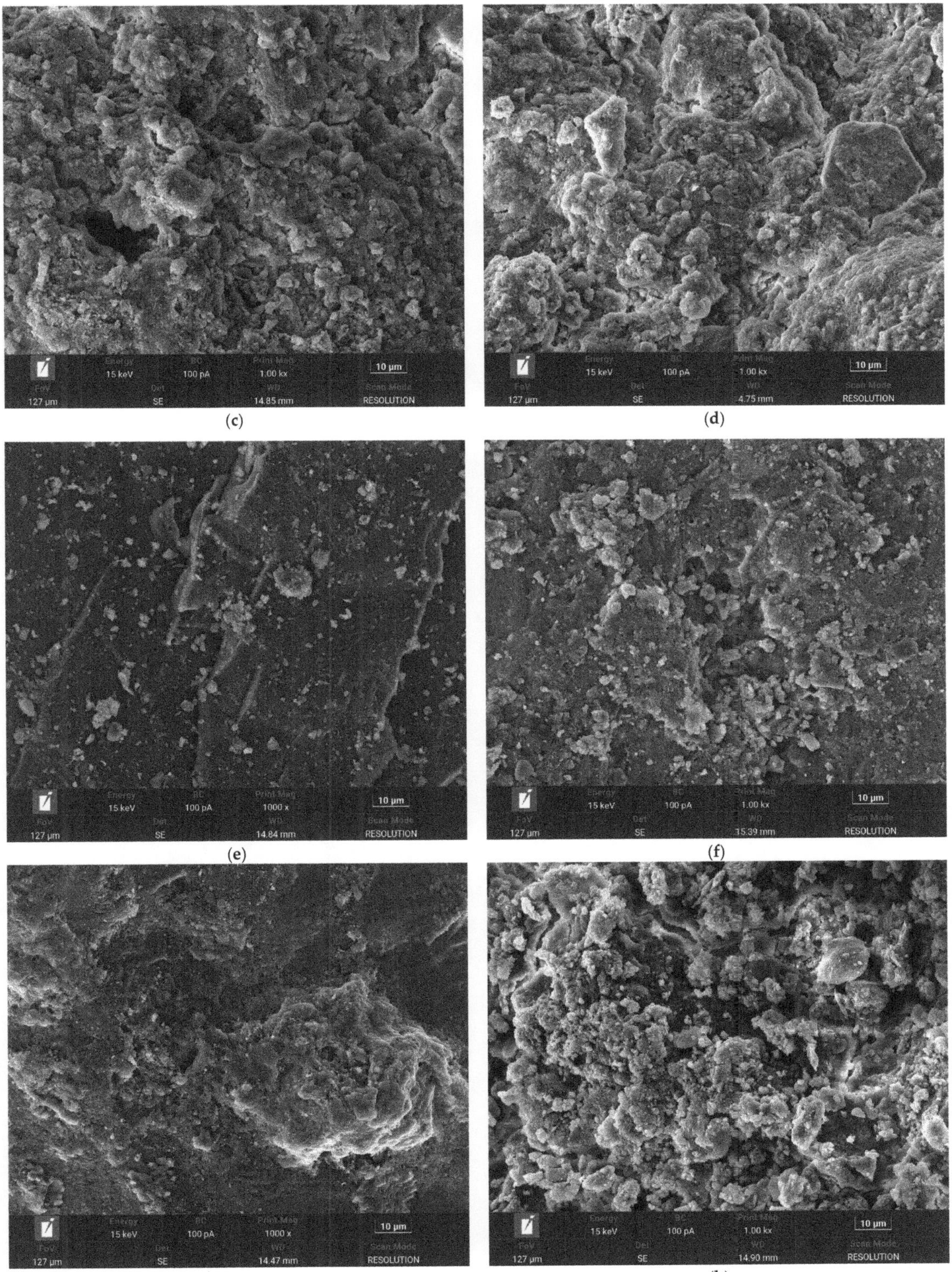

(c) (d) (e) (f) (g) (h)

Figure A1. *Cont.*

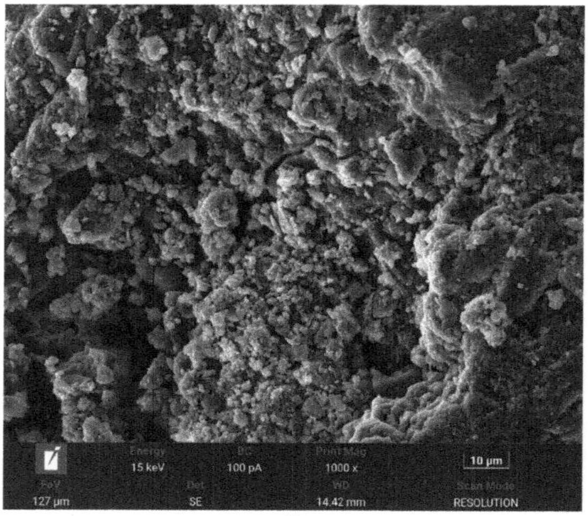

(i)

(j)

Figure A1. Comparative SEM results of four wall samples at 1.00 kX. (**a**) SEM morphology of the east wall sample particles (WD is 14.95 mm); (**b**) SEM morphology of the east wall sample particles (WD is 15.13 mm); (**c**) SEM morphology of the south wall sample particles (WD is 14.85 mm); (**d**) SEM morphology of the south wall sample particles (WD is 14.75 mm); (**e**) SEM morphology of the west wall sample particles (WD is 14.84 mm); (**f**) SEM morphology of the west wall sample particles (WD is 15.39 mm); (**g**) SEM morphology of the west wall sample particles (WD is 14.47 mm); (**h**) SEM morphology of the north wall sample particles (WD is 14.90 mm); (**i**) SEM morphology of the north wall sample particles (WD is 14.42 mm); (**j**) SEM morphology of the north wall sample particles (WD is 14.40 mm). (Image source: compiled by the author based on SEM analysis results).

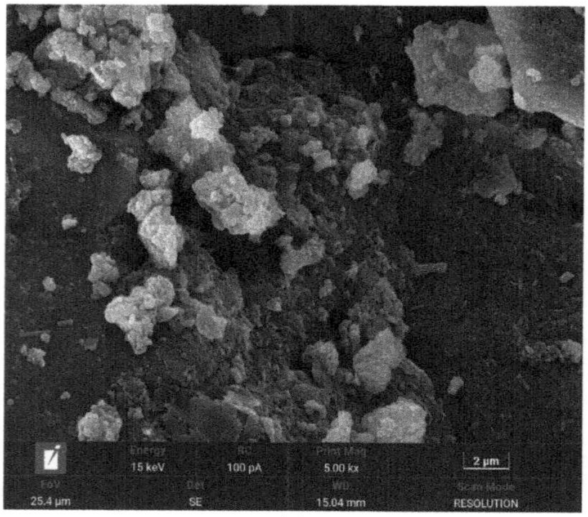

(a)

(b)

Figure A2. *Cont.*

(c)

(d)

(e)

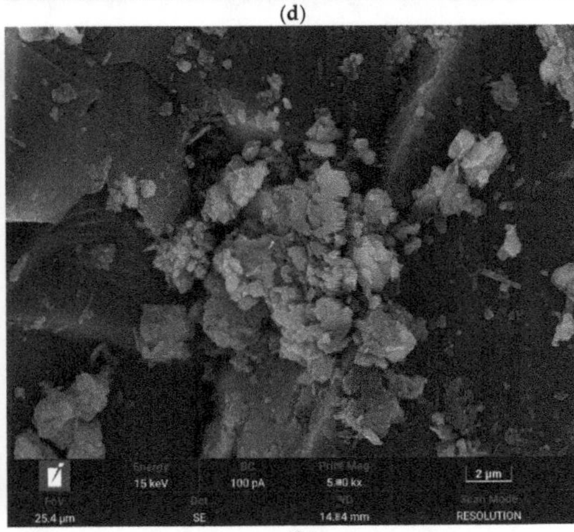

(f)

(g)

(h)

Figure A2. Comparative SEM results of four wall samples at 5.00 kX. (**a**) SEM morphology of east wall sample particles (WD is 15.04 mm); (**b**) SEM morphology of east wall sample particles (WD is

15.36 mm); (**c**) SEM morphology of the south wall sample particles (WD is 14.75 mm); (**d**) SEM morphology of east wall sample particles (WD is 14.85 mm); (**e**) SEM morphology of the west wall sample particles (WD is 14.47 mm); (**f**) SEM morphology of the west wall sample particles (WD is 14.84 mm); (**g**) SEM morphology of the north wall sample particles (WD is 14.42 mm); (**h**) SEM morphology of the north wall sample particles (WD is 14.90 mm). (Image source: compiled by the author based on SEM analysis results).

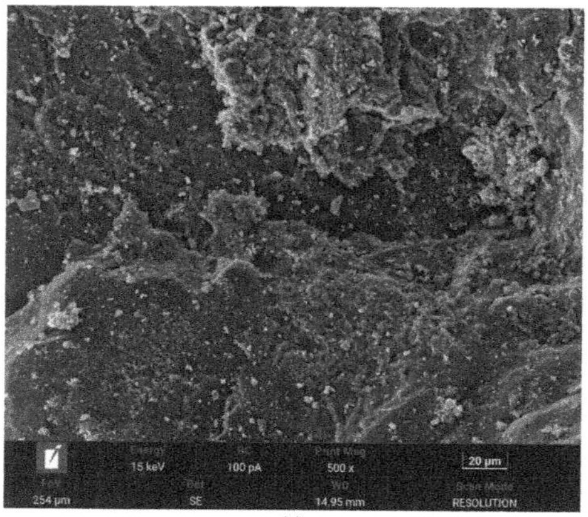

(**a**)

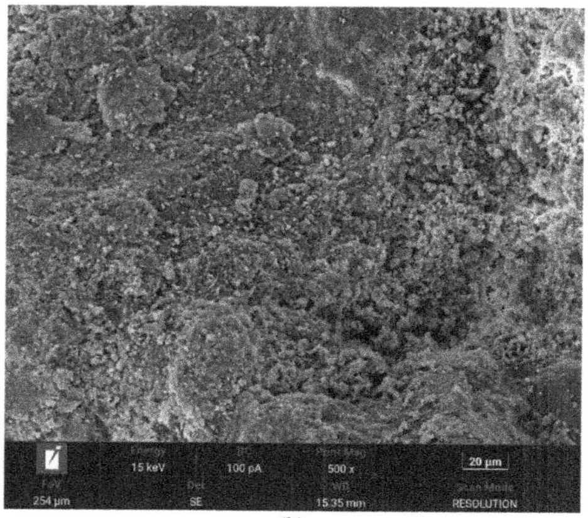

(**b**)

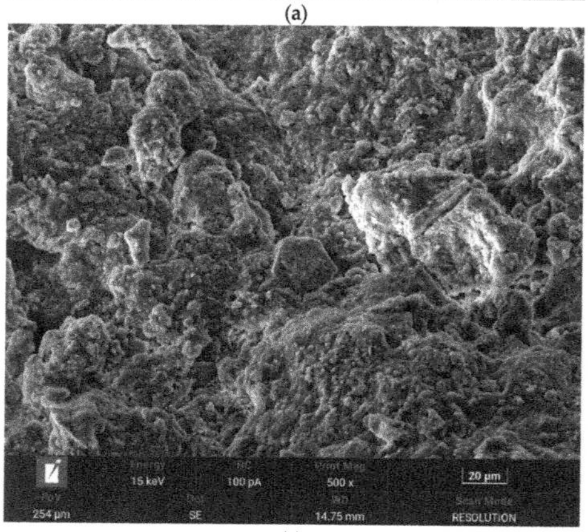

(**c**)

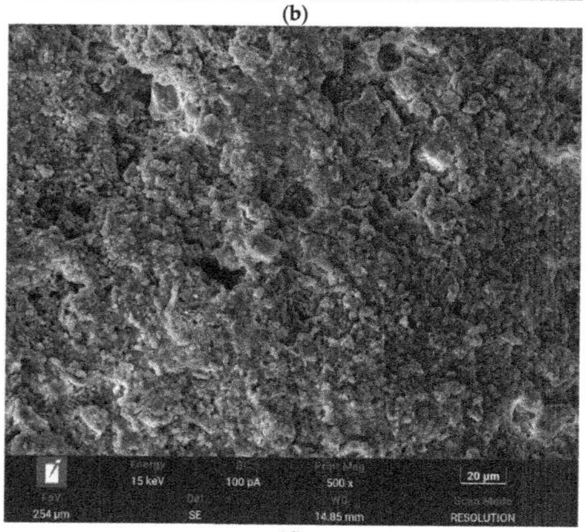

(**d**)

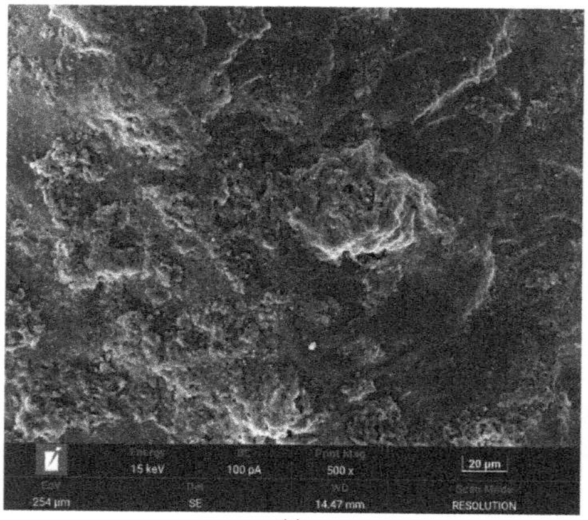

(**e**)

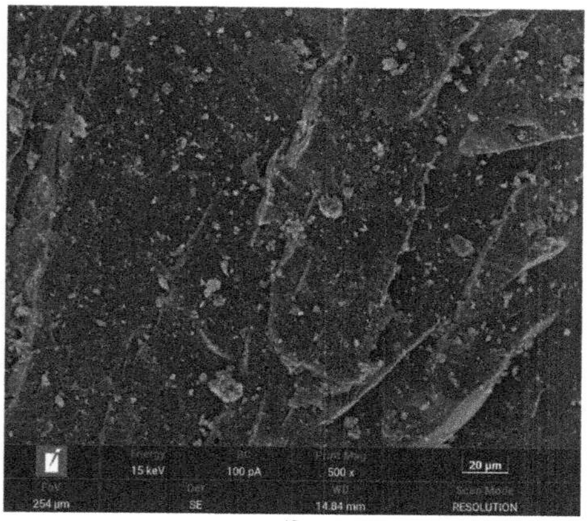

(**f**)

Figure A3. *Cont.*

(g) (h)

Figure A3. Comparative SEM results of four wall samples at 500×. (**a**) SEM morphology of east wall sample particles (WD is 14.95 mm); (**b**) SEM morphology of east wall sample particles (WD is 15.35 mm); (**c**) SEM morphology of the south wall sample particles (WD is 14.75 mm); (**d**) SEM morphology of the south wall sample particles (WD is 14.85 mm); (**e**) SEM morphology of the west wall sample particles (WD is 14.47 mm); (**f**) SEM morphology of the west wall sample particles (WD is 14.84 mm); (**g**) SEM morphology of the north wall sample particles (WD is 14.42 mm); (**h**) SEM morphology of the north wall sample particles (WD is 14.90 mm). (Image source: compiled by the author based on SEM analysis results).

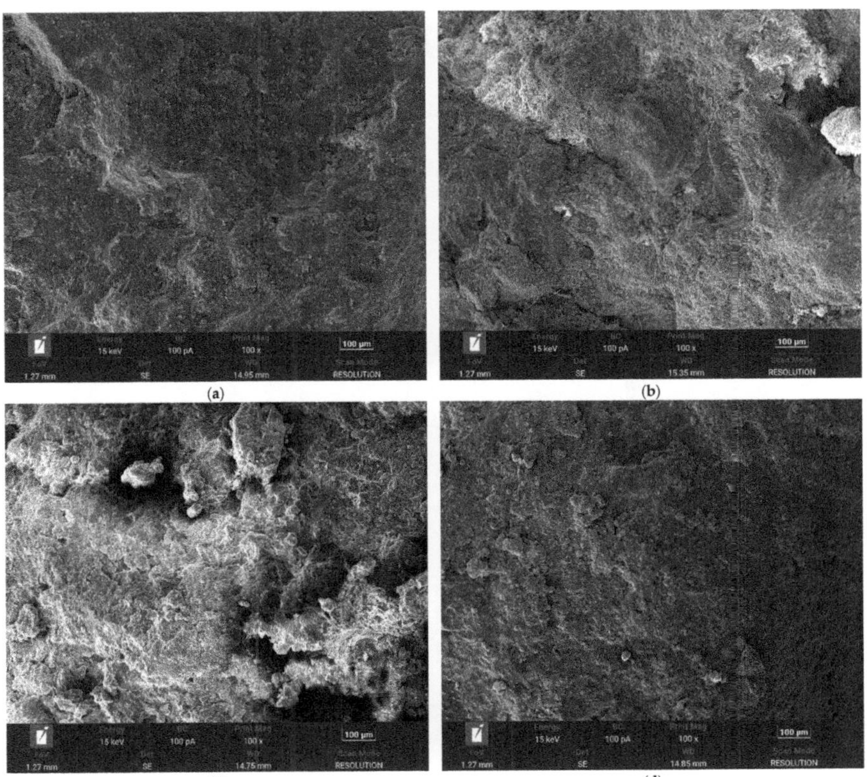

(a) (b) (c) (d)

Figure A4. *Cont.*

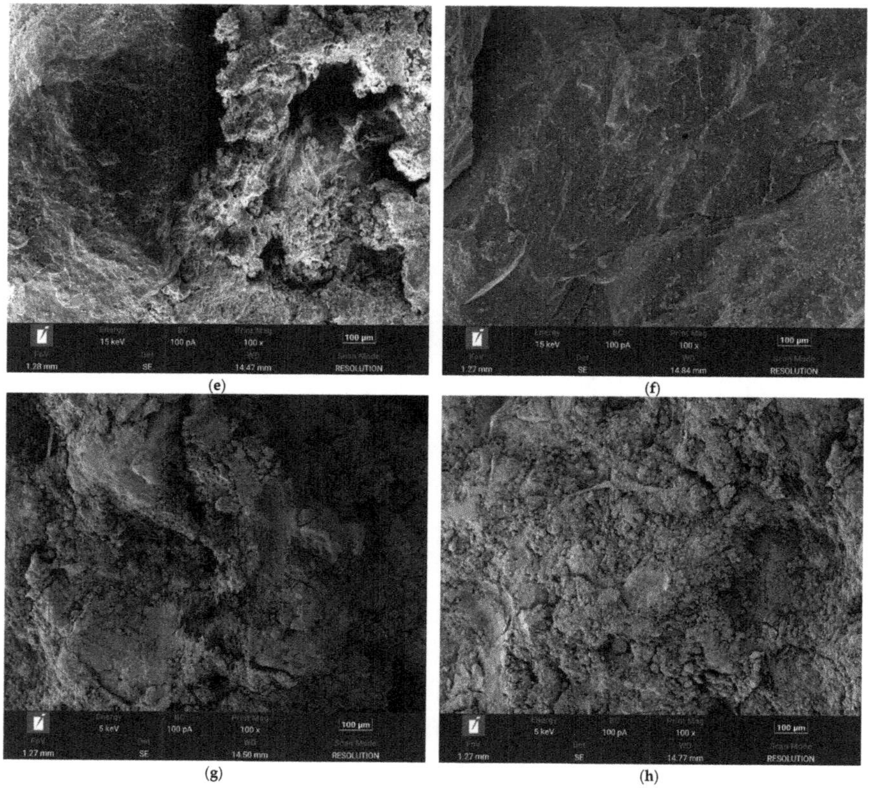

Figure A4. Comparative SEM results of four wall samples at 100×. (**a**) SEM morphology of the east wall sample particles (WD is 14.95 mm); (**b**) SEM morphology of the east wall sample particles (WD is 15.35 mm); (**c**) SEM morphology of the south wall sample particles (WD is 14.75 mm); (**d**) SEM morphology of the south wall sample particles (WD is 14.85 mm); (**e**) SEM morphology of the west wall sample particles (WD is 14.47 mm); (**f**) SEM morphology of the west wall sample particles (WD is 14.84 mm); (**g**) SEM morphology of the north wall sample particles (WD is 14.50 mm); (**h**) SEM morphology of the north wall sample particles (WD is 14.77 mm). (Image source: compiled by the author based on SEM analysis results).

References

1. Niroumand, H.; Zain, M.F.M.; Jamil, M. Various types of earth buildings. *Procedia-Soc. Behav. Sci.* **2013**, *89*, 226–230. [CrossRef]
2. Schroeder, H. *Sustainable Building with Earth*; Springer: Cham, Switzerland, 2016. [CrossRef]
3. Li, T.T.; Zhao, L.H. Investigation of Existing Situation and Study on Energy-Saving Potential of Rural Residential Buildings in Pearl River Delta. *Adv. Mater. Res.* **2012**, *374*, 199–203. [CrossRef]
4. Tao, J.; Zhuo, X.; Qin, Q.; Xiao, D.; Zhang, S.; Chen, H. Conflict, fortification, and settlement patterns: A study of the vernacular villages in Guangdong, South China. *War Hist.* **2021**, *28*, 237–262. [CrossRef]
5. Communist Youth League Committee of South China Normal University. Impressions of Xiaochikan: Explore Historical Sites and Admire Cultural Charm. Available online: https://youth.scnu.edu.cn/news/2018/0725/11209.html (accessed on 4 October 2025). (In Chinese)
6. Chen, J.; Chen, Y. Research on the Conservation and Development Paths of Traditional Rammed Earth Architecture in Zhuhai and Macau. In Proceedings of the 2022 Annual Conference and Academic Symposium of the Architectural History Branch of the Architectural Society of China: Evolving Architectural History Research and Heritage Conservation, Hangzhou, China, 11–13 November 2022; Architectural History Branch of the Architectural Society of China, School of Architecture, Huaqiao University: Xiamen, China, 2022; Volume 5. Available online: https://chn.oversea.cnki.net/KCMS/detail/detail.aspx?dbcode=CPFD&dbname=CPFDLAST2023&filename=JZSX202207001068&uniplatform=OVERSEA&v=2bCrDrC04gYfgwEr6rCrAhZnFU4MP4uiXTeEgyybWhj2_ChTLzHpsRmmi44QzRnCtDK1rPQK_pA%3d (accessed on 4 October 2025). (In Chinese)

7. Discovering Heritage and Treasures—Experiencing Traditional Rammed Earth Building Techniques, by Antiquities and Monuments Office (Hong Kong). Available online: https://www.amo.gov.hk/tc/about-us/online-articles/20240411/index.html (accessed on 4 October 2025).
8. Fu, Z.; Chen, P. Research on the ecological construction techniques of traditional Hakka residential buildings in Lingnan. *J. Shaoguan Univ.* **2024**, *45*, 1–6. (In Chinese)
9. Luo, Y.; Zhong, H.; Bao, F.; Guo, Z.; Ni, P. Insights into natural and carbonation curing of ancient Chinese rammed earth mixed with brown sugar. *Constr. Build. Mater.* **2022**, *317*, 125969. [CrossRef]
10. Liang, R.; Stanislawski, D.; Hota, G. Material characterization and structural response under earthquake loads of hakka rammed earth buildings. *Sustain. Struct.* **2021**, *1*, 3. [CrossRef]
11. Song, Y.; Liao, C. Structural Materials, Ventilation Design and Architectural Art of Traditional Buildings in Guangdong, China. *Buildings* **2022**, *12*, 900. [CrossRef]
12. Lin, X.; Zhang, Y.; Wu, Y.; Yang, Y. Assessment of Architectural Typologies and Comparative Analysis of Defensive Rammed Earth Dwellings in the Fujian Region, China. *Buildings* **2024**, *14*, 3652. [CrossRef]
13. Chen, D.; Sun, D.; Lin, Y. Preservation or revitalization? Examining the conservation status and destructive mechanisms of Tulou heritage in Raoping, Chaozhou, China. *Int. J. Archit. Herit.* **2025**, *19*, 441–456. [CrossRef]
14. Zhang, Y.; Chen, J.; Yan, S. Study on the degradation characteristics and mechanism of Kizil mudstone under periodic salt expansion. *Saf. Environ. Eng.* **2025**, *32*, 189–196. (In Chinese)
15. Liu, H.; Chen, Y.; Li, J.; Zhang, T.; Jiang, Y.; Ye, W.; He, H. Weathering disease types, mechanisms and protection and reinforcement of rammed earth sites: Status and prospects. *Earth Sci.* **2025**, *50*, 2284–2297. (In Chinese)
16. Qi, Z.; Lopez-Manzanares, F.V.; Mileto, C.; Trizio, F. Rammed earth techniques in China and the Mediterranean area: A comparative analysis. *Int. J. Archit. Herit.* **2024**, *18*, 1727–1748. [CrossRef]
17. Zhao, C.; Zhang, Y.; Wang, C.-C.; Hou, M.; Li, A. Recent progress in instrumental techniques for architectural heritage materials. *Herit. Sci.* **2019**, *7*, 36. [CrossRef]
18. Costa, T.G.; Richter, F.A.; Castro, E.T.; Gonçalves, S.; Spudeit, D.A.; Micke, G.A. Elemental identification of blue paintings traces present in historic cemeteries in the São Martinho region, southern Brazil. *J. Mol. Struct.* **2018**, *1155*, 434–442. [CrossRef]
19. Liang, R.; Hota, G.; Lei, Y.; Li, Y.; Stanislawski, D.; Jiang, Y. Nondestructive Evaluation of Historic Hakka Rammed Earth Structures. *Sustainability* **2013**, *5*, 298–315. [CrossRef]
20. Luo, Y.; Yang, M.; Ni, P.; Peng, X.; Yuan, X. Degradation of rammed earth under wind-driven rain: The case of Fujian Tulou, China. *Constr. Build. Mater.* **2020**, *261*, 119989. [CrossRef]
21. Cui, K.; Yu, L.; Zhang, X.; Dang, X.; Pan, Y.; Wang, D. Research on the Quality Evaluation System of Rammed Earth Sites in Arid Areas. *Int. J. Archit. Herit.* **2025**, 1–21. [CrossRef]
22. Zhujiang Evening News. Available online: https://pub-zhtb.hizh.cn/a/202407/27/AP66a478fae4b072e1214e34e6.html (accessed on 4 October 2025). (In Chinese)
23. Xiaochikan Village (Doumen District, Zhuhai City, Guangdong Province, China) on Google Maps. Available online: https://www.google.com/maps/place/%E4%B8%AD%E5%9B%BD%E5%B9%BF%E4%B8%9C%E7%9C%81%E7%8F%A0%E6%B5%B7%E5%B8%82%E6%96%97%E9%97%A8%E5%8C%BA%E5%B0%8F%E8%B5%A4%E5%9D%8E%E6%9D%91+%E9%82%AE%E6%94%BF%E7%BC%96%E7%A0%81:+529147/@22.2789096,113.21113,15z/data=!3m1!4b1!4m6!3m5!1s0x340185eb402252a7:0x1b0d6ce2a71bfc7e!8m2!3d22.27891!4d113.21113!16s%2Fg%2F11c61l8_88?entry=ttu&g_ep=EgoyMDI1MTAwMS4wIKXMDSoASAFQAw%3D%3D (accessed on 4 October 2025).
24. Introduction to Zhuhai City and Its Climate Characteristics, by China Weather Website. Available online: https://www.weather.com.cn/zt/qxfwzt/1135158.shtml (accessed on 4 October 2025). (In Chinese)
25. Liu, T.; Ni, X.; Sun, J.; Peng, Z. Effectiveness of Emergency Response to Typhoons: Lessons Learned from Historical Typhoon Disasters in Zhuhai, China. In Proceedings of the International ISCRAM Conference, Münster, Germany, 25–29 May 2024. [CrossRef]
26. Song, J.; Chen, Y.; Zheng, L.; Tang, Q. The evolution of traditional ancient towns in South China: Tangjiawan Ancient Town in Zhuhai city. *J. Asian Archit. Build. Eng.* **2025**, 24, 2114–2132. [CrossRef]
27. Wang, Y.; Yan, Z. *Centuries and Millennia: Tracing and Interpreting the Origins of Xiangshan Culture*; Guangdong People's Publishing: Guangzhou, China, 2006.
28. Xie, H.; Tho-ard, M. Folk culture patterns and dissemination in Xiangshan cultural area of Guangdong Province. *J. Sch. Comunity* **2025**, *3*, 198–207, (In Thai and English).
29. Doumen District Library. Xiaochikan Village, Doumen Town, by Doumen District Library Official Website. Available online: https://www.dmlib.cn/information/903 (accessed on 4 October 2025). (In Chinese)
30. *SY/T 5163-2018*; X-ray Diffraction Analysis Method for Clay Minerals and Common Non-Clay Minerals in Sedimentary Rocks. Standardization Administration of the People's Republic of China: Beijing, China, 2018.

31. Malki, A.; Mekhalif, Z.; Detriche, S.; Fonder, G.; Boumaza, A.; Djelloul, A. Calcination products of gibbsite studied by X-ray diffraction, XPS and solid-state NMR. *J. Solid State Chem.* **2014**, *215*, 8–15. [CrossRef]
32. Gu, T.; Yang, X.; Shi, D.; Yang, S.; Ma, Y.; Zheng, X.; Wu, J.; Zhang, Y.; Song, G. Scientific and technological analysis and research on rammed earth materials of Shang Dynasty buildings in Zhengzhou area: Taking Zhengzhou Shang City and Zhengzhou Xiaoshuangqiao site as examples. *Cult. Relics Prot. Archaeol. Sci.* **2024**, *36*, 89–100. (In Chinese)
33. Liang, M.; Li, L. A Study of Lingnan Garden's Adaptability to Hot and Humid Climate. In Proceedings of the International Conference on Human-Computer Interaction, Orlando, FL, USA, 26–31 July 2019; Springer International Publishing: Cham, Switzerland, 2019; pp. 61–73. [CrossRef]
34. Leung, E.Y.W. Roofs of Clans: Chinese Roofs of Vernacular Architectures of Branches of Communities in Hong Kong and Respective Restoration Techniques. In Proceedings of the International Conference on East Asian Architectural Culture, Tianjin, China, 13–17 October 2017; Springer International Publishing: Cham, Switzerland, 2017; pp. 393–406. [CrossRef]
35. Tao, J.; Xiao, D.; Qin, Q.; Zhuo, X.; Wang, J.; Chen, H.; Wang, Q. Climate-adaptive design of historic villages and dwellings in a typhoon-prone region in southernmost Mainland China. *Int. J. Archit. Herit.* **2022**, *16*, 117–135. [CrossRef]
36. Lai, D. Waterproofing Technology in Traditional Lingnan Architecture. In Proceedings of the Fifth Annual Academic Conference of China Cultural Relics Protection Technology Association, Nanjing, China, 4 September 2007; Shenzhen Cultural Relics Management Office: Shenzhen, China; China Cultural Relics Protection Technology Association, Cultural Relics Protection Technology Department of the Palace Museum: Beijing, China, 2007; pp. 91–96. Available online: https://chn.oversea.cnki.net/KCMS/detail/detail.aspx?dbcode=CPFD&dbname=CPFD9908&filename=ZGWA200709001013&uniplatform=OVERSEA&v=G3hoaHiCKsYqSvaO3HUwCz_Nx3pihet1Fyf-WJuLxpcGBdU5h6JJyVkbeWV K-0mMxnCImUtRAOM%3d (accessed on 4 October 2025). (In Chinese)
37. Xia, J.; Zhao, X.; Zhao, Z.; Chen, Y.; Liu, J. Characteristics of soil water and salt migration and their interaction effects at different groundwater depths. *Trans. Chin. Soc. Agric. Eng.* **2015**, *31*, 93–100. (In Chinese)
38. Chu, J.; Liang, H. Study on the performance of slag-tung oil-glutinous rice juice raw earth wall materials. *Sci. Technol. Inf.* **2008**, *30*, 374+385. Available online: https://chn.oversea.cnki.net/KCMS/detail/detail.aspx?dbcode=CJFD&dbname=CJFD2008&filename=KJXX200830308&uniplatform=OVERSEA&v=agGTHgkq-F-iPby70rRrfXhSbcRuxJpRMwEolpS61oiBl1M-OXujPIdt-rNThqwt (accessed on 4 October 2025). (In Chinese)
39. Zhang, Z. Research on the Creation Strategy of Contemporary Lingnan Green Buildings Based on the Inheritance of Guangfu Construction Wisdom. Master's Thesis, South China University of Technology, Guangzhou, China, 2021. Available online: https://chn.oversea.cnki.net/KCMS/detail/detail.aspx?dbcode=CMFD&dbname=CMFD202301&filename=1021894386.nh&uniplatform=OVERSEA&v=hs6HnjviKy2EFTfz2YWweTjAFS1nZL1wM3YJfsC-Rm4Wast4xCVBR6Zs7udeKYhy (accessed on 4 October 2025). (In Chinese)
40. Stanaszek-Tomal, E. Environmental Factors Causing the Development of Microorganisms on the Surfaces of National Cultural Monuments Made of Mineral Building Materials—Review. *Coatings* **2020**, *10*, 1203. [CrossRef]
41. Tiano, P. Biodegradation of Cultural Heritage: Decay Mechanisms and Control Methods. Seminar Article, New University of Lisbon, Department of Conservation and Restoration. 2002, pp. 7–12. Available online: https://www.semanticscholar.org/paper/Biodegradation-of-Cultural-Heritage-%3A-Decay-and-Tiano/ace1c3025c5a93a05e361e09efefe03609564981?utm_source=direct_link (accessed on 4 October 2025).
42. Li, H.; An, X.; Wang, Y.; Wang, W.; Zhao, Y. Research status of antibacterial coatings. *Chem. Fiber Text. Technol.* **2024**, *53*, 35–37. (In Chinese)
43. Zuo, B.; Kitagawa, K.; Chen, Y. Coincidental values of responding to climate in contemporary Chinese architectural design. *J. Asian Archit. Build. Eng.* **2025**, *24*, 154–177. [CrossRef]

coatings

MDPI

Article

Microbial Diversity Analysis on Rammed Earth Wall Surfaces in the Lingnan Region: A Case Study of Paishan Village, China

Wei Wei [1], Shuai Yang [2], Junxin Song [3] and Md Sayuti Bin Ishak [1,*]

1 School of Civil Engineering, Engineering Campus, Universiti Sains Malaysia, Transkrian, Nibong Tebal 14300, Penang, Malaysia; weiwei2022@student.usm.my
2 School of Art and Archaeology, Hangzhou City University, No.51 Huzhou Street, Gongshu District, Hangzhou 310015, China; samyang@zju.edu.cn
3 Institute for Research on Portuguese Speaking Countries, City University of Macau, Avenida Padre Tomás Pereira, Taipa, Macau 999078, China; a23092100150@cityu.edu.mo
* Correspondence: cesayuti@usm.my

Abstract: Rammed earth walls in traditional villages in the humid and hot climate of Lingnan are susceptible to microbial damage and disease. Paishan Village in Zhuhai, which is the largest extant rammed earth building complex in the Pearl River Delta with rammed earth walls dating from the Ming (1368–1644) and Qing (1644–1912) Dynasties to the Republic of China (1912–1949) period, faces weathering and hollowing issues, yet targeted microbial research is lacking. This study, with a focus on the village's rammed earth walls, aimed to reveal microbial diversity and its relationship to the environment, providing a basis for heritage conservation. We used SEM (scanning electron microscopy) to analyze the microstructure of walls facing different orientations. High-throughput sequencing (based on the 16S rRNA gene V3–V4 region) was combined with microbial community analysis. Species annotation and differential analysis were performed using QIIME2 and R. The results indicated that the west wall had the highest microbial diversity (45 at the phylum level and 2969 at the genus level), while the south wall exhibited the lowest. Different orientations shaped distinct community structures, with the north wall harboring a higher concentration of hygrophilous microorganisms, while the south wall was dominated by thermotolerant bacteria. All four walls shared only 0.29% of the core microorganisms. This study elucidates the distribution patterns of microorganisms in rammed earth walls in humid and hot areas, offering scientific support for their ecological restoration.

Keywords: rammed earth; traditional village dwellings; hot and humid climates; microbial diversity; Zhuhai City

1. Introduction

Rammed earth, a remarkably durable construction material, was widely employed in ancient times for building city walls and palaces [1]. In China, the mastery of rammed earth techniques can be traced back to the Longshan culture and later became prevalent in residential construction [2,3]. Rammed earth walls are formed by compacting moist soil—typically a mixture of clay, sand, and gravel—within wooden molds. Similarly to other unfired earthen construction methods, such as the use of adobe bricks, rammed earth exhibits a long-standing and continuous historical tradition across many regions of the world [4]. As an important component of geotechnical engineering, rammed earth construction has contributed to the development of structural systems and improvements in load-bearing capacity [5]. Paishan Village, Zhuhai, was established in 1778 during the reign of Emperor Qianlong (Qing Dynasty), and it represents a quintessential example of traditional Lingnan rural architecture. Nearly one hundred historic houses remain in the village today; they were primarily constructed with rammed earth walls and timber-framed tile roofs and adorned with intricate wooden and stone carvings typical of the Lingnan style. Among them are 70 rammed earth houses, 6 red-brick houses, and 9 blue-brick houses, all demonstrating refined craftsmanship. Over time, the combined effects of humidity and environmental exposure have fostered distinctive microbial communities on the surfaces of these rammed earth walls [6].

Rammed earth buildings remain common today, with approximately one-third of the global population living in dwellings constructed primarily from earthen materials [7,8]. The Paishan Village complex, as the largest existing rammed earth settlement in the Pearl River Delta, provides an important interdisciplinary research case for architecture, materials science, and cultural heritage conservation [9]. However, the humid subtropical climate of Zhuhai accelerates the deterioration of rammed earth structures. Compared with brick or reinforced concrete walls, rammed earth exhibits relatively low strength and durability, and insufficient maintenance and funding often result in structural safety issues [10]. Existing microbiological studies on ancient rammed earth structures have mainly focused on microbial community diversity, their effects on soil materials, and the application of detection technologies. Characterizing microbial communities provides essential insights into their distribution and ecological mechanisms within rammed earth materials [11]. Microorganisms significantly influence the physical and chemical properties of earthen structures and play crucial roles in both their preservation and degradation processes [12,13]. Some microbial biofilms can protect surfaces from erosion, while microbial-induced mineralization has been explored as a means to enhance the mechanical stability of earthen materials [14].

Recent advances in molecular biology and high-throughput sequencing technologies have greatly improved the study of microbial diversity in built heritage materials. Techniques such as 16S rRNA and ITS gene sequencing, metagenomic analysis, and fluorescence microscopy enable the comprehensive profiling of microbial community compositions and their functional potential [15]. Metagenomic approaches, in particular, facilitate the identification of biodeteriogenic species and the assessment of their metabolic pathways, providing a scientific foundation for conservation and restoration strategies [16]. Nevertheless, current research often remains confined to a single disciplinary dimension—either materials or microorganisms—without systematic integration. The lack of in-depth investigations into site-specific microbial communities in traditional villages, such as Paishan, limits our understanding of the interactions among environmental factors, microbial metabolism, and the chemical–physical evolution of rammed earth. The rammed earth materials used in Paishan Village primarily rely on locally sourced soil blended with quartz-rich aggregates and natural binders such as glutinous rice slurry and brown sugar syrup, which enhance cohesiveness and strength [17]. However, Zhuhai's high temperature and humidity accelerate weathering, microbial degradation, and wall collapse [18,19]. Consequently, many historic dwellings have been abandoned or fallen into disrepair. Addressing the challenges of durability and the biodeterioration of traditional rammed earth structures under such climatic conditions is therefore urgently needed for both cultural heritage preservation and sustainable building material research.

2. Materials and Methods

2.1. Study Area: Paishan Village in Zhuhai City

As mentioned in the Introduction, this study takes traditional rammed earth dwellings in Bajia, Paishan Village, Doumen District, Zhuhai City, as the research object (Figure 1).

There are 70 well-preserved rammed earth houses, 6 ancient red-brick houses, and 9 ancient blue-brick houses, showcasing traditional craftsmanship, delicate walls, and exquisite carvings. These rammed earth buildings have withstood the test of time, and under the long-term influence of the natural environment, their surfaces have developed unique microbial communities. Rammed earth is primarily composed of soil, water, and other materials compacted through specific processes. Its pore structure, humidity conditions, and chemical composition create an appropriate microenvironment for the survival and reproduction of microorganisms. Different environmental factors render the microbial community structure on the surface of the rammed earth complex and diverse. These microorganisms interact with the rammed earth materials, significantly affecting the durability, stability, and appearance of buildings. In-depth research on the microorganisms in the rammed earth of Paishan Village is of great significance for revealing the aging mechanisms of buildings and formulating scientific and reasonable protection strategies.

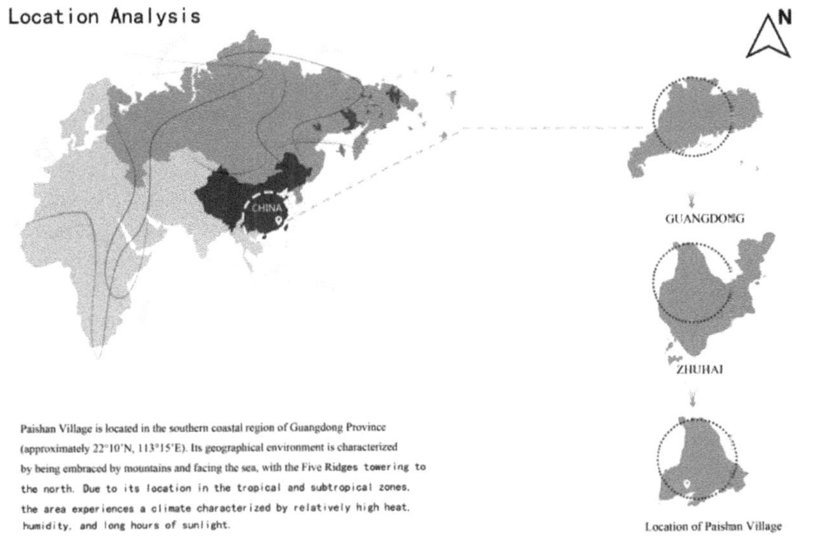

Figure 1. Location analysis of Paishan Village in Zhuhai (image source: drawn by the authors).

Through field investigation, experimental analysis, microscopic characterization, and other methods, an in-depth study is conducted on the diversity of microbial communities and protection strategies for the surface of rammed earth dwellings. In the initial stage of this study, a field survey of the rammed earth dwellings in Paishan Village was carried out, and typical damage types, such as cracks, wall discoloration, and the degree of microbial adhesion, were recorded in detail according to four orientations: east, west, north, and south. At the micro-mechanism level, the microstructure of the rammed earth wall is analyzed with the aid of a scanning electron microscope (SEM) [20,21]. The rammed earth walls of Paishan Village, mostly constructed between the Ming and Qing Dynasties and the Republic of China, are made of local clay, sand, and plant fibers (such as straw and bamboo rebar). After centuries of erosion by wind and rain, fluctuations in temperature and humidity, and human interference, they are commonly subject to surface weathering, hollowing, cracking, and mildew, directly threatening the structural safety and heritage integrity of the buildings. Microbial activity is a key "invisible factor" driving the deterioration of rammed earth walls, and analyses of its diversity provide scientific support at the microscopic level for heritage conservation. Through scientific exploration and practical operations, this study aims to protect cultural heritage, promote the sustainable development of traditional construction technology, and provide valuable references for future research and preservation efforts.

2.2. Analytical Techniques

2.2.1. Scanning Electron Microscopy (SEM)

We used a ZEISS Gemini SEM 360 (Carl Zeiss AG, Germany) scanning electron microscope (SEM) to conduct detailed microstructural characterizations of the Paishan Village rammed earth samples in four different directions: east, south, west, and north. As we were examining the surface of rammed earth walls from Paishan Village, the samples collected appeared powdery. First, we inspected the samples' surface for cracks and weathered layers. We gently brushed away loose soil and particles adhering to the samples' surface with a dust-free brush, avoiding direct rubbing, which could damage the microstructure. After preliminary cleaning, we placed the samples in a vacuum drying oven and dried them for 48 h using a low-temperature vacuum drying method (set to 35–40 °C, vacuum $\geq$ 0.1 MPa) until the sample mass remained constant. This step avoids high-temperature drying to prevent dehydration and shrinkage of clay minerals in the rammed earth, which could result in the deformation of the micropore structure. Furthermore, the vacuum environment minimizes air convection from disturbing microorganisms or fine particles on the samples' surface. We then secured the treated rammed earth powder sample to the center of the SEM stage using TedPella's Leitsilber fast-drying conductive adhesive (Ted Pella Inc., CA, USA), ensuring close contact between the sample and the stage. During the experiment, the acceleration voltage was set to 10.0 kV, and magnification was flexibly adjusted to 5.00 KX, 10.00 KX, and 20.00 KX according to actual needs. During image acquisition, the accelerating voltage (EHT) was 3.00 kV, the working distance (WD) was 8.0 mm, and the aperture size was 20.00 μm. Secondary electron (SE2) imaging was used in a high-vacuum environment to clearly present the microscopic morphology of the rammed earth sample' surface. These parameters ensure the effective observation of the sample's fine structure and provide high-quality image data for analyzing its microscopic characteristics. Through this experiment, the differences in the particle morphology and microstructure of samples were analyzed in various directions, and the possible causes of these differences were explored. At the same time, by comparing the characteristics in different directions, the anisotropic properties that may exist in the samples during synthesis or processing were preliminarily revealed. By comparing the microstructural features of samples from different orientations, we analyzed variations in particle morphology and density, revealing the anisotropic properties of the rammed earth resulting from synthesis processes and long-term environmental exposure. The obtained SEM images were analyzed qualitatively to assess particle morphology, porosity, and bonding structure using ImageJ software (https://imagej.net/ij/, accessed on 12 August 2025), and representative micrographs were selected to illustrate typical features from each orientation.

2.2.2. Biodiversity Community Identification

We analyzed microbial community diversity to characterize the composition and relative abundance of microorganisms colonizing the surface of rammed earth walls. To minimize external contamination, sampling was performed using sterile scalpels and gloves, and each sample (~5 g) was collected from undisturbed wall surfaces, with a minimum spacing of 10 cm between sampling points. Three replicates were collected for each orientation (east, south, west, and north), sealed in sterile 50 mL tubes, placed into cryogenic tubes, frozen quickly in liquid nitrogen, transported with dry ice, and stored at −20 °C prior to analysis. A total of four samples were used for DNA metabarcoding analysis (one from each of the four orientations). Then, the sample was removed from the refrigerator to further partition an appropriate 0.2 g sample. This partition was then added to a centrifuge tube containing an extraction and lysis buffer for grinding. The grinding instrument used is the Tissuelyser-48 (Shanghai Jingxin®, Shanghai, China) multi-sample tissue grinder from Shanghai Jingxin Company, with a grinding frequency of 60 Hz.

(1) Extraction of Total DNA from the Microbiome: We used the MagBeads FastDNA Kit for Soil (MP Biomedicals, CA, USA) to extract total genomic DNA from the rammed earth wall samples. This choice was intentional, as soil samples (including the rammed earth wall matrix) often contain impurities such as humus and heavy metals that inhibit DNA extraction. The kit's magnetic bead method effectively removes impurities, ensuring DNA purity. DNA quality was subsequently verified using a NanoDrop NC2000 spectrophotometer (Wilmington, DE 19810 USA) (to measure nucleic acid concentration and purity, such as using the A260/A280 ratio to assess protein contamination) and 0.8% agarose gel electrophoresis (to visually assess DNA integrity and prevent degradation fragments from interfering with subsequent experiments), providing a reliable foundation for subsequent amplification and sequencing. All DNA extraction procedures were performed in a sterile laminar flow hood to prevent airborne contamination.

(2) PCR Amplification of Target Fragments: We amplified the bacterial 16S rRNA gene targeting the V3–V4 hypervariable region (~470 bp) using the universal primers 338F (5′-ACTCCTACGGGAGGCAGCAG-3′) and 806R (5′-GGACTACHVGGGTWTCTAAT-3′) [11]. PCR amplification was performed in 25 μL reaction volumes containing 12.5 μL of 2× Pfu Master Mix (TransGen Biotech, Beijing, China), 0.5 μL of each primer (10 μM), and 1 μL of DNA template. The thermocycling conditions were as follows: initial denaturation at 95 °C for 3 min; 27 cycles of 95 °C for 30 s, 55 °C for 30 s, and 72 °C for 45 s; followed by a final extension at 72 °C for 10 min. Negative controls were included to monitor potential environmental or reagent contamination, and any contaminated samples were excluded from downstream analysis. All amplification steps were performed in a sterile laminar flow hood using aerosol-resistant filter tips to prevent cross-contamination. Each reaction was run in triplicate, and the products were pooled before purification to reduce amplification bias and ensure data reliability.

(3) PCR products were purified using Vazyme VAHTS™ DNA Clean Beads (Vazyme, China) and quantified with the Quant-iT™ PicoGreen™ dsDNA Assay Kit (Life Technologies Corporation, Eugene, OR, USA) via a BioTek FLx800 microplate reader (BioTek Instruments, Winooski, VT, USA).

(4) Fluorescence Quantification of Amplified Products: We then performed fluorescence quantification on PCR amplification products. The fluorescent reagent used was from the Quant-iT™ PicoGreen™ dsDNA Assay Kit (Life Technologies Corporation, Eugene, OR, USA), and the quantitative instrument was a microplate reader (BioTek, FLx800, BioTek Instruments, Winooski, VT, USA).

(5) Sequencing Library Preparation: We prepared sequencing libraries using Illumina's TruSeq Nano DNA LT Library Prep Kit (https://sapac.illumina.com/products/by-type/sequencing-kits/library-prep-kits/truseq-nano-dna.html, accessed on 10 February 2025) according to the manufacturer's protocol. Briefly, PCR amplicons were end-repaired, A-tailed, and ligated to indexed adapters for library construction. Adapter-ligated fragments were purified using AMPure XP magnetic beads (Beckman Coulter Inc., CA, USA) and subsequently amplified to enrich sequencing templates. The final libraries were size-selected (~200–450 bp) via 2% agarose gel electrophoresis, and their quality was verified using an Agilent Bioanalyzer (Agilent Technologies, Santa Clara, CA, USA) before sequencing.

(6) High-Throughput Sequencing: Before high-throughput sequencing, the library needs to be strictly tested for quality and quantitatively analyzed. First, the library quality inspection was completed on the Agilent Bioanalyzer (Agilent Technologies, Inc., Santa Clara, CA, USA) using the Agilent High Sensitivity DNA Kit (Agilent Technologies, Inc., Santa Clara, CA, USA). The qualified standard for the library is the presence of a single peak without adapter contamination. Then, the library is quantified using the Quant-iT™ PicoGreen™ dsDNA Assay Kits (Life Technologies Corporation, Eugene, OR, USA) on the QuantiFluor® dsDNA System (Promega Corporation, Madison, WI, USA) to ensure that the concentration of the qualified library reaches 2 nM or above. Libraries that passed quality inspection and quantification (with each library having a unique index sequence) were mixed in appropriate proportions based on the required sequencing depth. Considering the short read length of MiSeq sequencing and the need for sequencing quality assurance, the optimal sequencing length range of the target fragment is set to 200~450 bp; the determination of this range is related to the structural characteristics of ribosomal RNA (ribosomal RNA contains multiple conserved regions and highly variable regions). We built the primers based on the parts of the 16S rRNA gene that are the same in all organisms. We utilized the universal primer pairs 338F (5′-ACTCCTACGGGAGGCAGCAG-3′) and 806R (5′-GGACTACHVGGGTWTCTAAT-3′). Moreover, the V3-V4 hypervariable sections (~470 bp) might be preferentially amplified based on the conserved parts of the 16S rRNA gene. In this study, however, an amplicon sequence variant (ASV) was utilized to examine the diversity of the microbial population.

Raw sequencing data were processed using QIIME2 (version 2019.4) for quality control, denoising, sequence merging, and chimera removal. Each sequencing library was analyzed independently, and the resulting amplicon sequence variant (ASV) feature tables were merged to obtain the final dataset. Singletons were removed prior to downstream analyses to minimize background noise and improve data reliability. Taxonomic classification was performed using a naïve Bayes classifier trained on the SILVA 138 database with a confidence threshold of 0.8. Alpha diversity (Shannon and Simpson) and beta diversity (Bray–Curtis dissimilarity) indices were calculated, and microbial community structures were visualized using principal coordinate analysis (PCoA). To comprehensively evaluate the alpha diversity of microbial communities, this study used the Chao1 [22] and observed species indices to represent richness; the Shannon [23,24] and Simpson [25] indices to represent diversity; Faith's PD [26] index to represent evolutionary diversity; Pielou's evenness [27] index to represent evenness; and Good's coverage [28] index to represent coverage. The beta diversity index focuses on the comparison of diversity between different habitats: that is, the differences between samples. We used the most classic unconstrained sorting (classical multidimensional scaling, cMDScale) analysis method [29]: principal coordinate analysis (PCoA). This method projects the sample distance matrix and expands it in a low-dimensional space, while retaining the distance relationship of the original sample to the maximum extent. We use the flattened ASV/OTU table, called the "qiime diversity core-metrics-phylogenetic" command, to calculate four different distance matrices, and PCoA analysis was performed on these distance matrices to output QZV files. For visualization, the QZV file was then inserted into the corresponding areas found on its website (https://view.qiime2.org/, accessed on 12 July 2025).

In addition, R software (version 4.3.2) was used to generate bar plots of taxonomic composition at each classification level (phylum, class, order, family, and genus) to visually compare the relative abundance differences among samples.

3. Results

3.1. Analysis of SEM Images

Through SEM imaging analysis of samples from the east, south, west, and north walls of Paishan Village, it was observed that there were obvious differences in their micromorphology and surface characteristics. From the SEM results, the rammed earth particles of the east wall are randomly distributed in the form of agglomerates (Figure 2). Blocky and flaky particles of different sizes are stacked and bonded to each other, forming a relatively loose aggregate structure; the surface is rough and concave, with widespread pores and depressions. Some pores are "nested" due to particle accumulation, reflecting the heterogeneity of the microstructure and the combined influence of raw material grading, preparation technology, and later environmental erosion. As shown in Figure 3, the rammed earth particles of the south wall exist in the form of irregular agglomerates with significant size differences. It can be observed that the long strips and blocky particles are stacked and bonded to each other, forming a relatively loose structure; the surface is rough and uneven, with pores and depressions widely distributed, and some pores are "connected" due to particle accumulation, reflecting the heterogeneity of the microstructure—this feature is a result of both the gradation of the raw materials (such as the ratio of coarse and fine particles and composition differences) and subsequent environmental erosion and transformation. The rammed earth particles of the west wall exhibit an irregular agglomeration state (Figure 4). Particles of different sizes adhere to and stack upon each other, forming a relatively loose aggregate structure. The surface exhibits a pronounced rough texture with noticeable bumps, depressions, and numerous pores. If the raw materials are not mixed evenly (for example, with too many large or small particles), the small particles cannot fill the spaces between the larger ones, resulting in gaps; moreover, if pressure, layering, and moisture are not managed well during the ramming process, the particles will not be packed tightly enough, resulting in tiny holes. The west wall is affected by sunlight, temperature, dry–wet cycles, and other environmental effects, which may aggravate the change in the bonding force between particles, causing the surface particles to fall off and the pores to expand, as observed via the concave–convex and pore characteristics in the image. Under a high magnification of 30.00 KX, the rammed earth particles of the north wall present an irregular agglomeration morphology, with obvious differences in particle size (Figure 5). In addition, the particle sizes of the four rammed earth walls were statistically analyzed (Table 1). Small particles adhere to each other and pile up to form a relatively loose agglomerate structure. The surface is rough and uneven, with many pores and depressions. Zhuhai has a subtropical marine climate with high air humidity and frequent rainfall. The north-facing wall remains shaded for long periods, receiving little sunlight and maintaining a low temperature. When interacting with hot and humid air, a "cold bridge effect" easily forms (the wall's temperature is lower than the dew point temperature of the hot and humid indoor and outdoor air), causing moisture to condense and penetrate the wall's surface. Repeated freezing and thawing of moisture (under low temperatures) or dry–wet cycles weaken the bonding forces between rammed earth particles, causing surface particles to detach and form depressions, while moisture erosion gradually enlarges the pores. This morphology reflects the heterogeneity of rammed earth at the microscopic level, resulting from the combined effects of its composition and formation process.

Figure 2. SEM morphology of the particles of the east wall sample (image source: photographed by the authors).

Figure 3. SEM morphology of the particles of the south wall sample (image source: photographed by the authors).

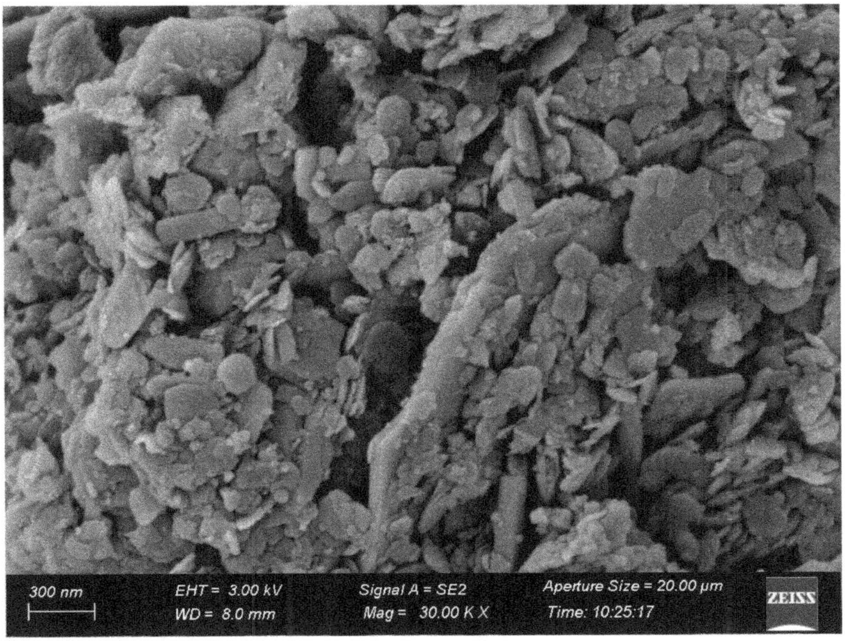

Figure 4. SEM morphology of the particles of the west wall sample (image source: photographed by the authors).

Figure 5. SEM morphology of the particles of the north wall sample (image source: photographed by the authors).

Table 1. Particle size statistics.

Wall Orientation	Average Particle Size (nm)	Particle Size Standard Deviation (nm)	P10 (nm)	P50 (nm)	P90 (nm)
East	185.841	74.159	118.378	276.216	434.054
South	173.797	74.290	122.653	323.896	525.14
West	216.280	117.208	146.329	389.663	632.997
North	191.117	91.097	129.888	333.762	537.636

Source: Statistics conducted by the authors.

3.2. *Species Taxonomic Annotation Results*

To ensure the correspondence between microstructure and community data, we selected one representative sample from each of the four orientations for SEM imaging. Additionally, we conducted 16S rRNA metabarcoding sequencing and bioinformatics analysis on three replicates per orientation, totaling 12 samples. The sequencing statistics and data quality summaries for each sample are presented in Table 2 and Figure 6.

Table 2. Statistics of the sequencing volume per sample.

Wall Location	Sample ID	Input	Filtered	Denoised	Merged	Non-Chimeric	Non-Singleton
North Wall	N_bei	127,181	120,550	115,401	98,694	93,708	93,444
South Wall	S_nan	137,950	131,690	129,114	116,730	103,735	103,576
West Wall	W_xi	136,255	130,191	123,274	99,464	93,017	92,428
East Wall	E_dong	115,859	110,653	107,973	94,577	75,148	74,915

Note: The second column in this table indicates the sample's name; the third column presents the original data size; the fourth column presents the data size after removing low-quality sequences; the fifth column presents the sequence data size after denoising, that is, the effective sequence size; the sixth column indicates the sequence size after splicing; the seventh column presents the sequence size after removing chimeras, that is, the high-quality sequence size; the eighth column indicates the sequence size after removing singletons.

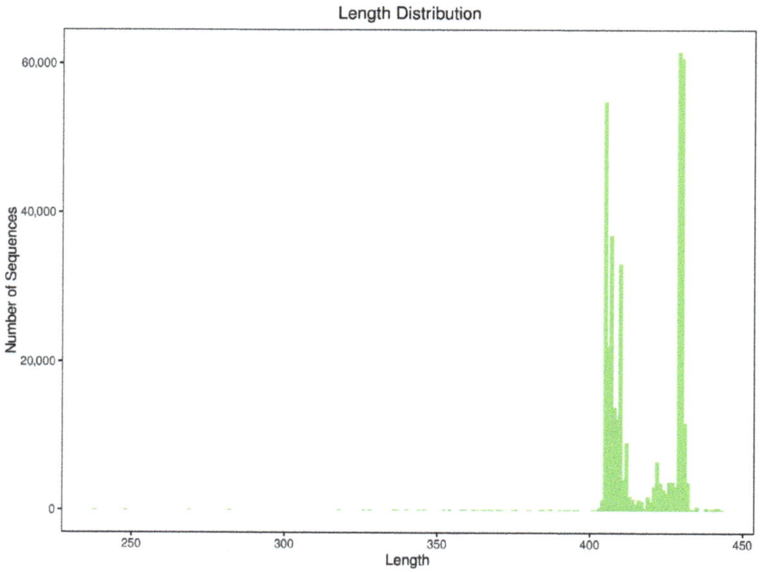

Figure 6. Sequence length distribution (image source: author-produced graph based on experimental results).

The statistical analysis of species annotation results is summarized in Table 3. Figure 7 illustrates the taxonomic composition of the microbial communities at different classification levels (from phylum to genus) for samples collected from the east (E_dong), south (S_nan), west (W_xi), and north (N_bei) walls. Significant differences were observed in both the number and relative abundance of microbial taxa among the four orientations. The south-facing wall exhibited a higher proportion of Firmicutes and Actinobacteria, while the north-facing wall showed an increased abundance of Proteobacteria and Bacteroidetes. These findings indicate that environmental conditions, such as sunlight exposure and moisture content, strongly influence microbial diversity and community composition on rammed earth wall surfaces.

Table 3. Statistics of species taxonomic annotation results.

Wall Location	ID	Domain	Phylum	Class	Order	Family	Genus	Species	Unclassified
South Wall	S_nan	5	6	35	85	186	933	44	0
North Wall	N_bei	23	43	72	198	575	2058	129	0
East Wall	E_dong	26	21	21	105	171	1012	45	0
West Wall	W_xi	42	45	105	283	814	2969	149	0

Note: "Unclassified" refers to the number of ASV/OTUs that cannot be assigned to any known taxonomic unit.

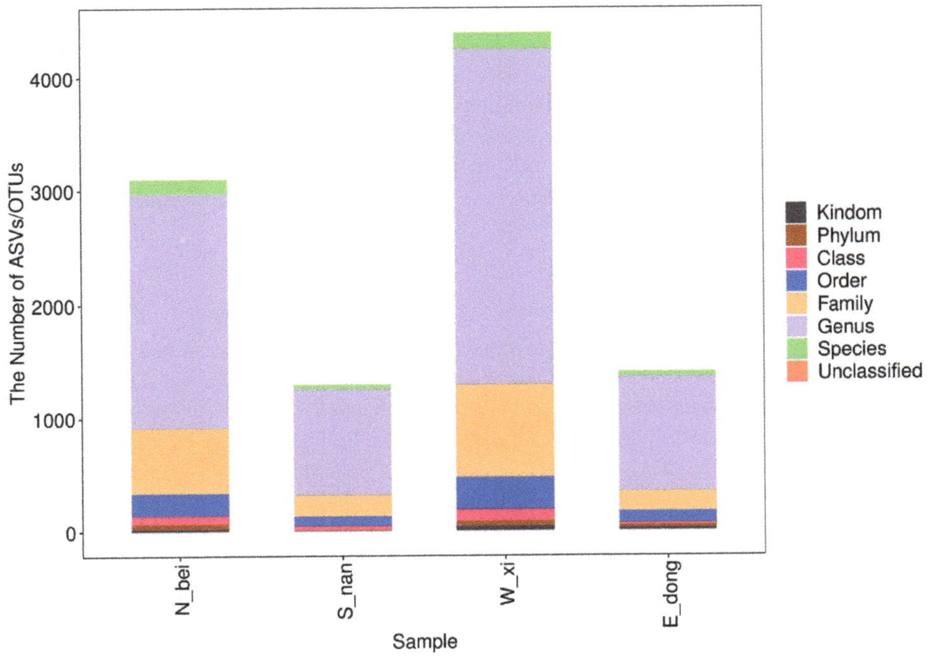

Figure 7. Statistics of species taxonomic annotation results. The horizontal axis is arranged according to the sample name, and the vertical axis is the number of ASV/OTUs in each sample that can only be classified with respect to phylum, class, order, family, genus, and species levels. This visualization can be used to understand the resolution of species annotations. The longer the columns at the genus and species levels and the shorter the columns at the phylum level, the higher the resolution of the annotation (image source: author-produced graph based on experimental results).

(1) The west wall (sample W_xi) had the greatest variety of species at every taxonomic level, with no unclassified microorganisms: 42 at the domain level, 45 at the phylum

level, 105 at the class level, 283 at the order level, 814 at the family level, 2969 at the genus level, and 149 at the species level. This shows that the microbial community on this wall was very diverse and that the taxonomic classification was largely complete.

(2) The microbial richness of the north wall (sample N_bei) ranked second. With the exception of the domain (23) and phylum (43) levels, which were slightly lower than those of the west wall, the number of classes (72), orders (198), families (575), genera (2058), and species (129) was significantly higher than those of the east and south walls. Notably, the genus level was close to 70% of that of the west wall, indicating a more complex community structure.

(3) The microbial richness of the east wall (sample E_dong) exhibited a "polarized" characteristic: the number at the domain level was the highest (26), but the number at the class level was the lowest (21). The numbers of phyla (21), orders (105), families (171), genera (1012), and species (45) were all at a medium level, indicating that the microorganisms on the wall exhibited some diversity in high-order taxonomic units (such as domains), but the degree of differentiation in medium-order and low-order units (such as classes and species) was low.

(4) The south wall (sample S_nan) had the lowest microbial richness. The number of taxa at each level (domain: 5; phylum: 6; class: 35; order: 85; family: 186; genus: 933; species: 44) was significantly lower than those of other walls, with the genus count only slightly higher than that of the east wall, indicating that the microbial community structure of this wall was relatively simple and the species diversity was low.

It is worth noting that the number of unclassified microorganisms on all walls is 0, indicating that the coverage and accuracy of the taxonomic annotation are high, and this can better reflect the taxonomic characteristics of the microbial communities on walls facing different directions. This difference in microbial richness may be closely related to the microenvironment of the wall (such as light, humidity, temperature, etc.), providing basic data for further exploring the interaction mechanism between the building environment and the microbial community.

3.3. *Species Composition Analysis*

3.3.1. Statistics of the Number of Taxonomic Units

By analyzing the ASV/OTU tables obtained after sequencing and quality control, the taxonomic composition of the microbial communities in each wall sample was determined across multiple classification levels (phylum, class, order, family, and genus). The number of taxonomic units identified in each sample is summarized in Table 4, while their distribution patterns are visualized in Figure 8. Rather than simply presenting raw counts, this section highlights the comparative differences among samples and their ecological implications. The south-facing wall (S_nan) exhibited the highest number of taxonomic units across most classification levels, suggesting greater microbial richness, likely influenced by higher temperature, humidity, and sunlight exposure. In contrast, the north-facing wall (N_bei) contained fewer identified taxa, indicating lower microbial diversity; this is possibly a result of reduced light and cooler surface conditions. The east (E_dong) and west (W_xi) walls displayed intermediate values, reflecting moderate environmental variability. These quantitative differences in taxonomic richness correspond well with the compositional trends shown in Figure 8, providing a clear statistical basis for understanding spatial heterogeneity in microbial diversity across the four wall orientations.

Table 4. Statistics of the number of microbial taxa at each level.

Wall Location	Sample	Phylum	Class	Order	Family	Genus	Species
North Wall	N_bei	26	74	162	226	315	91
South Wall	S_nan	19	48	98	131	152	36
West Wall	W_xi	33	83	172	244	337	79
East Wall	E_dong	16	36	69	83	92	30

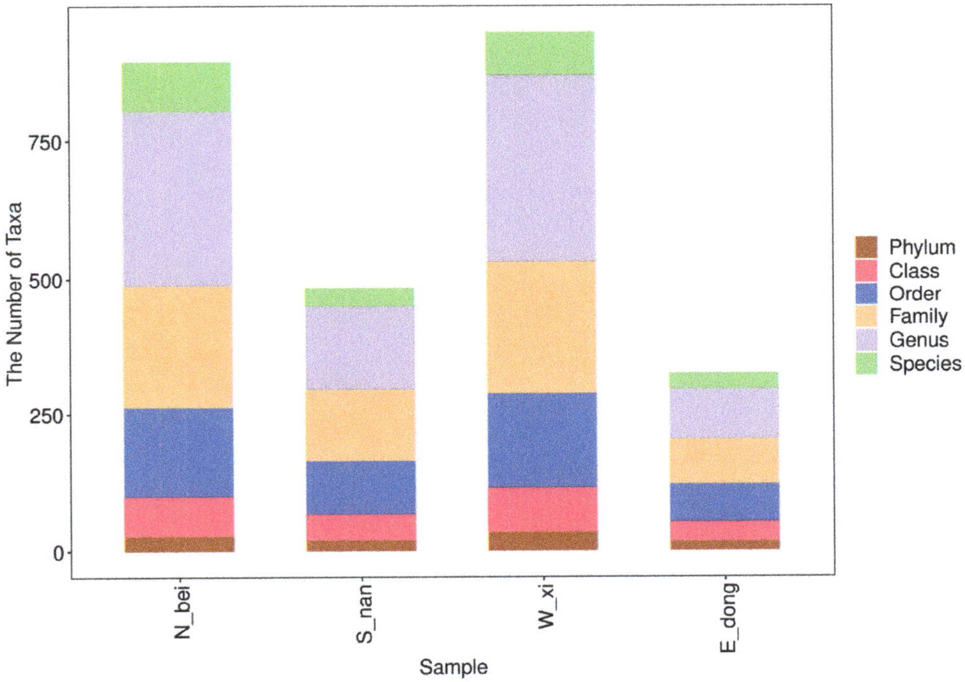

Figure 8. Statistics of the number of microbial taxa at each level (image source: author-produced graph based on experimental results).

3.3.2. Taxonomic Composition Analysis

Figure 9 shows the relative abundance of microbial taxa at different classification levels across samples from the four wall orientations. Distinct compositional patterns were observed among the samples. The south-facing wall (S_nan) was dominated by Firmicutes and Actinobacteria, reflecting higher tolerance to heat and desiccation, while the north-facing wall (N_bei) had higher proportions of Proteobacteria and Bacteroidetes, indicating adaptation to cooler and more humid microenvironments. The east (E_dong) and west (W_xi) walls displayed intermediate compositions with a balanced distribution of these phyla. These findings demonstrate that wall orientation and associated environmental factors significantly shape the microbial community structure on rammed earth surfaces.

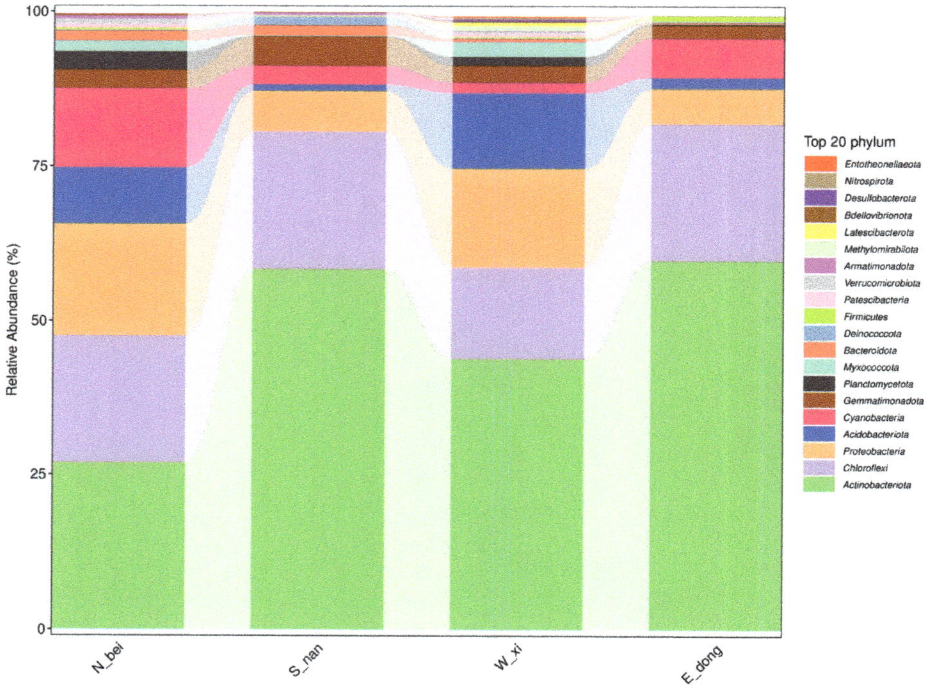

(**a**) Top 20 phyla

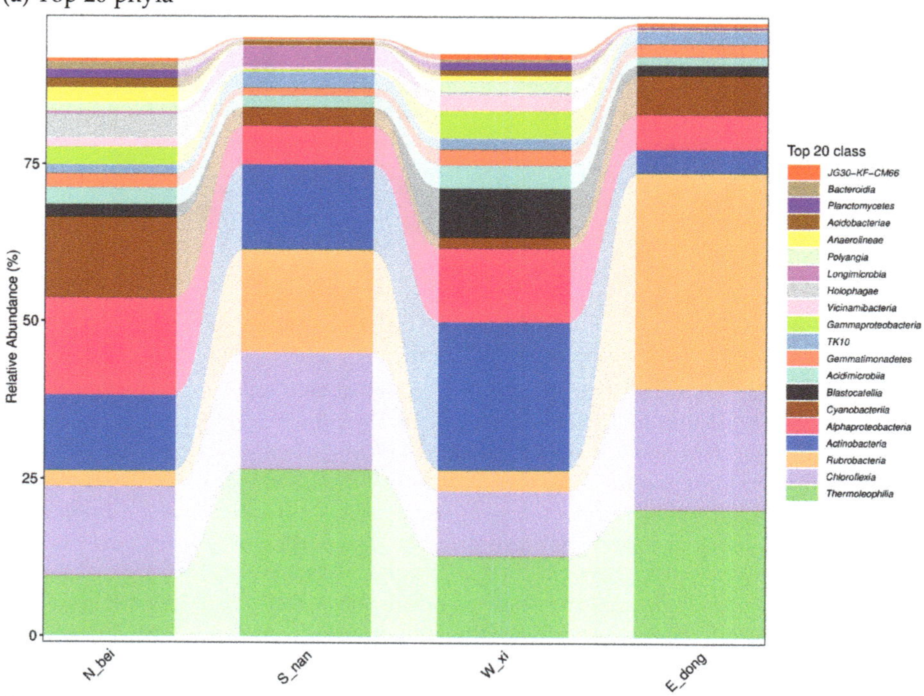

(**b**) Top 20 classes

Figure 9. *Cont.*

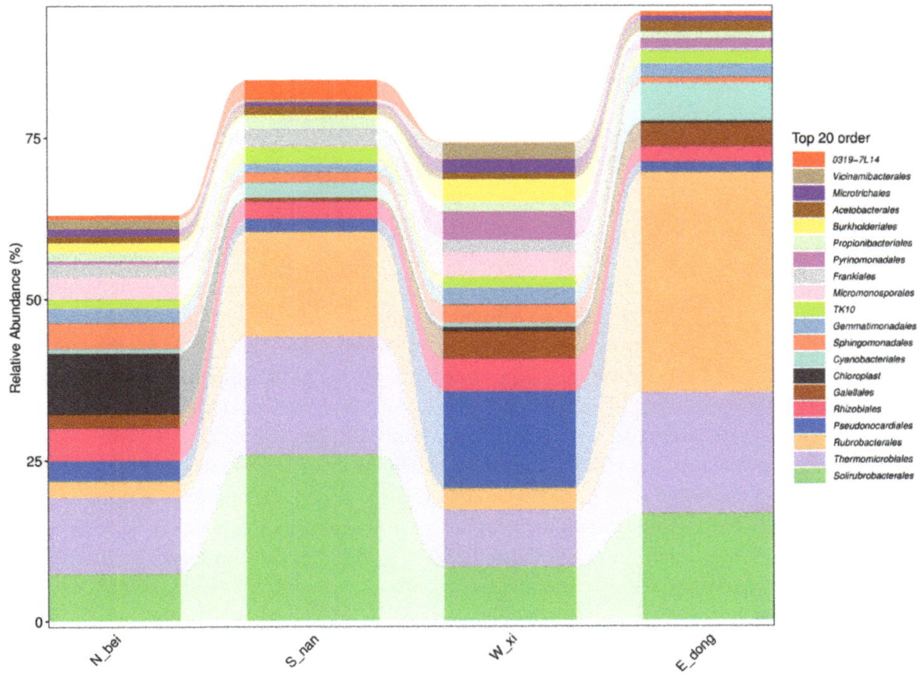

(**c**) Top 20 orders

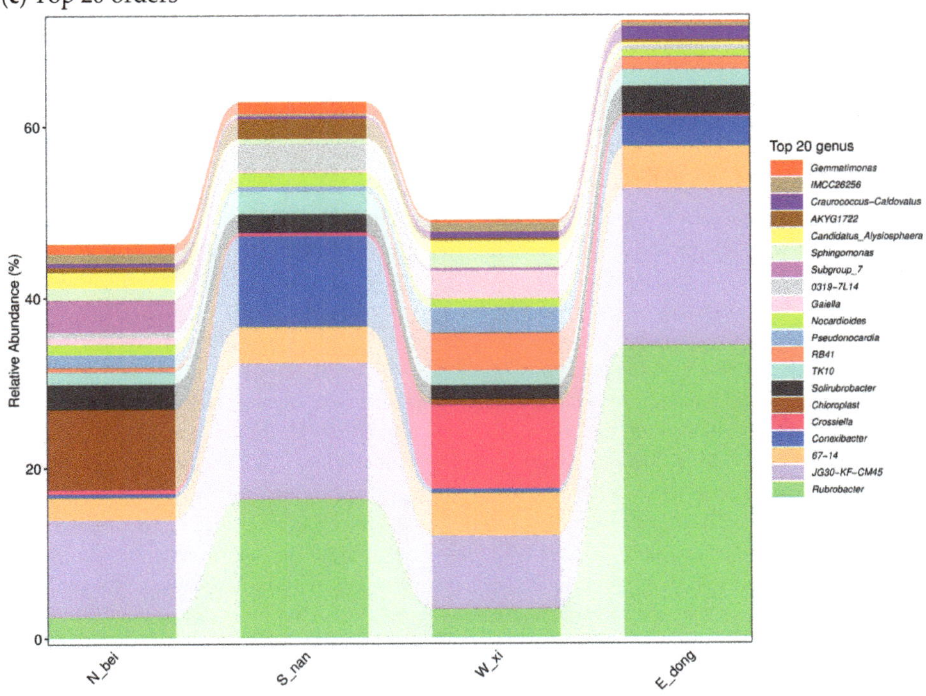

(**d**) Top 20 genera

Figure 9. Species composition at each level (image source: author-produced graphs based on experimental results).

Figure 9a shows the analysis results of the relative abundance of the top 20 microbial phylum levels in the four rammed earth samples obtained from the north (sample N_bei), south (sample S_nan), west (sample W_xi), and east walls (sample E_dong). (1) In the north wall (sample N_bei), microbial phyla richness is high, and the dominant phyla are scattered. *Firmicutes* (pink in Figure 9a), *Bacteroidota* (red in Figure 9a), and *Proteobacteria* (light orange in Figure 9a) all have a certain proportion, reflecting the community characteristics of multiple microbial phyla coexisting under the low-light and high-humidity environment of the north wall. (2) In the south wall (sample S_nan), the community structure is "dominated by a few phyla". *Actinobacteriota* (green in Figure 9a), *Proteobacteria* (light orange in Figure 9a), and *Bacteroidota* (red in Figure 9a) comprise a high proportion, suggesting that the strong sunlight and high-temperature environment of the south wall selectively favors these adaptive phyla, which have become dominant. (3) As for the west wall (sample W_xi), the abundance of dominant phyla fluctuates violently; for example, the proportions of *Proteobacteria* (light orange in Figure 9a), *Actinobacteriota* (green in Figure 9a), and *Bacteroidota* (red in Figure 9a) change dramatically, indicating that the west wall's microbial community may be strongly influenced by environmental pressures such as strong winds and salt deposition. (4) In the east wall (sample E_dong), the microbial community structure is relatively concentrated, with *Actinobacteriota* (green in Figure 9a), *Proteobacteria* (light orange in Figure 9a), and *Bacteroidota* (red in Figure 9a) as the dominant phyla. This reflects the adaptive succession of the microbial community under the moderate light and humidity gradient conditions of the east wall.

Figure 9b presents the relative abundance of the top 20 classes. (1) North Wall (Sample N_bei): The microbial community is rich, and the dominant communities are scattered. The *JG30-KF-CM66* (red in Figure 9b), *Bacteroidia* (purple in Figure 9b), *Planctomycetes* (brown in Figure 9b), and other communities all have certain proportions, reflecting the community characteristics of multiple microorganisms coexisting under the low-light and high-humidity environment of the north wall. Although *Planctomycetes* and other taxa are low in abundance but widespread, they may participate in nutrient cycling (such as nitrogen and carbon transformation) and support the basic ecological processes of the community. (2) South Wall (Sample S_nan): The community structure presents a "dominant minority" model. *Actinobacteria* (dark blue in Figure 9b), *Chloroflexia* (light orange in Figure 9b), and *Thermoleophilia* (green in Figure 9b) comprise high proportions, suggesting that the strong sunlight and high-temperature environment of the south wall have a screening effect on microorganisms, and adaptive communities have become dominant. These microorganisms may be involved in the metabolism of soil cementing substances and potentially contribute to the structural stability of rammed earth structures; their high abundance may be related to the demand for organic matter decomposition in the high-temperature environment of the south wall. (3) West Wall (Sample W_xi): The abundance of dominant communities fluctuates dramatically. The proportions of *Vicinamibacteria* (pink in Figure 9b) and *Gammaproteobacteria* (light green in Figure 9b) change dramatically, reflecting the strong impact of environmental pressures, such as strong winds and salt deposition, on the microbial community. (4) East Wall (Sample E_dong): The community structure is relatively concentrated. *Longimicrobia* (light purple in Figure 9b), *Holophagae* (light pink in Figure 9b), and *Cyanobacteria* (dark pink in Figure 9b) are the main dominant groups, reflecting the adaptive succession of the microbial community under the moderate light and humidity gradient environment of the east wall.

Figure 9c shows the relative abundance distribution of the top 20 microbial groups (order level). Overall, the four sample groups have certain commonalities; for example, some dominant orders (such as *Solirubrobacterales*) are distributed in multiple groups. Figure 9d shows the relative abundance distribution of the top 20 microbial groups (genus level) in the rammed earth samples at four locations. It can be further observed that all four groups of samples show the diversity of microbial genera, and there is a certain overlap in the dominant genera (*Rubrobacter*, *Subgroup_7*, etc., are dominant in multiple groups), reflecting the common selection of some adaptive microbial genera in different rammed earth microenvironments. However, there are also certain differences between them, such as the fluctuation in distribution patterns and the abundance of dominant genera: The north wall (sample N_bei) exhibits high genus richness and dispersed dominance, reflecting a multi-coexistence feature; the south wall (sample S_nan) is dominated by a few genera (*Rubrobacter*, *Pseudonocardia*, etc.) and has concentrated dominance; the abundance of dominant genera (*Crossiella*, *RB41*, etc.) on the west wall (sample W_xi) fluctuates dramatically and is significantly affected by environmental disturbances; the dominant genera (*Rubrobacter*, *Gaiella*, etc.) on the east wall (sample E_dong) comprise a high proportion and exhibit stable structures, reflecting the results of adaptive succession. These similarities and differences are due to variations in directional microenvironments (light, humidity, salinity, etc.): low light and high humidity on the north wall promote dispersed coexistence; strong sunlight and high temperatures on the south wall select dominant genera; strong winds and salt disturbances on the west wall cause fluctuations in microbial abundance; and the mild, gradual environmental conditions on the east wall drive stable succession. Together, these shape the spatial differentiation patterns observed at the microbial genus level.

Figure 9d displays the top 20 microbial genera in the four wall samples. All samples exhibited high genus-level diversity, with partial overlap in dominant taxa (e.g., Rubrobacter, Subgroup_7). Labels such as Subgroup_7 and TK10 represent placeholder names automatically assigned by the SILVA 138 database to sequences that could not be confidently classified at the genus level (classification confidence $\geq$ 0.8). These names were retained to preserve community structure visualization but should be interpreted as "unclassified or provisional groups" and not as established genera. The north wall (N_bei) exhibited high genus richness and dispersed dominance, indicating the coexistence of multiple microbial lineages. The south wall (S_nan) was dominated by a few genera (e.g., *Rubrobacter*, *Pseudonocardia*), displaying the strong selective effects of heat and light. The west wall (W_xi) exhibited sharp fluctuations in dominant genera (e.g., Crossiella, RB41), reflecting environmental disturbance, while the east wall (E_dong) was dominated by *Rubrobacter* and *Gaiella*, with a stable structure suggesting adaptive succession. Together, these spatial differences demonstrate that microenvironmental factors—light exposure, humidity, and salinity—strongly influence microbial community compositions and stability at the genus level.

3.3.3. Species Classification Tree

The species classification hierarchy tree presents the phylogenetic relationship and inter-group distribution characteristics of the microbial groups of the four rammed earth samples. The pie charts (threshold 0.5%) at each branch node of the taxonomic hierarchy tree show the composition ratio of the taxonomic unit in each sample/group. Figure 10 shows the phylogenetic tree of the dominant microbial genera detected in rammed earth samples from the four wall orientations. The sector size represents the relative abundance of each taxonomic unit in the corresponding sample or group, and the sector colors indicate the sampling sites (north, south, west, and east walls). As shown in the figure, several genera, such as *Crossiella* and *Rubrobacter*, appear on multiple colored branches, indicating cross-clustering among different wall orientations and suggesting shared community components. In contrast, branches dominated by a single color correspond to site-specific taxa that are unique to particular environments—such as the north or east wall—and these are likely influenced by local microclimatic factors (light, humidity, and salinity). Overall, the phylogenetic distribution illustrates both the overlap and differentiation of microbial communities among wall orientations, highlighting the spatial heterogeneity of the microbial assemblages on the rammed earth surfaces.

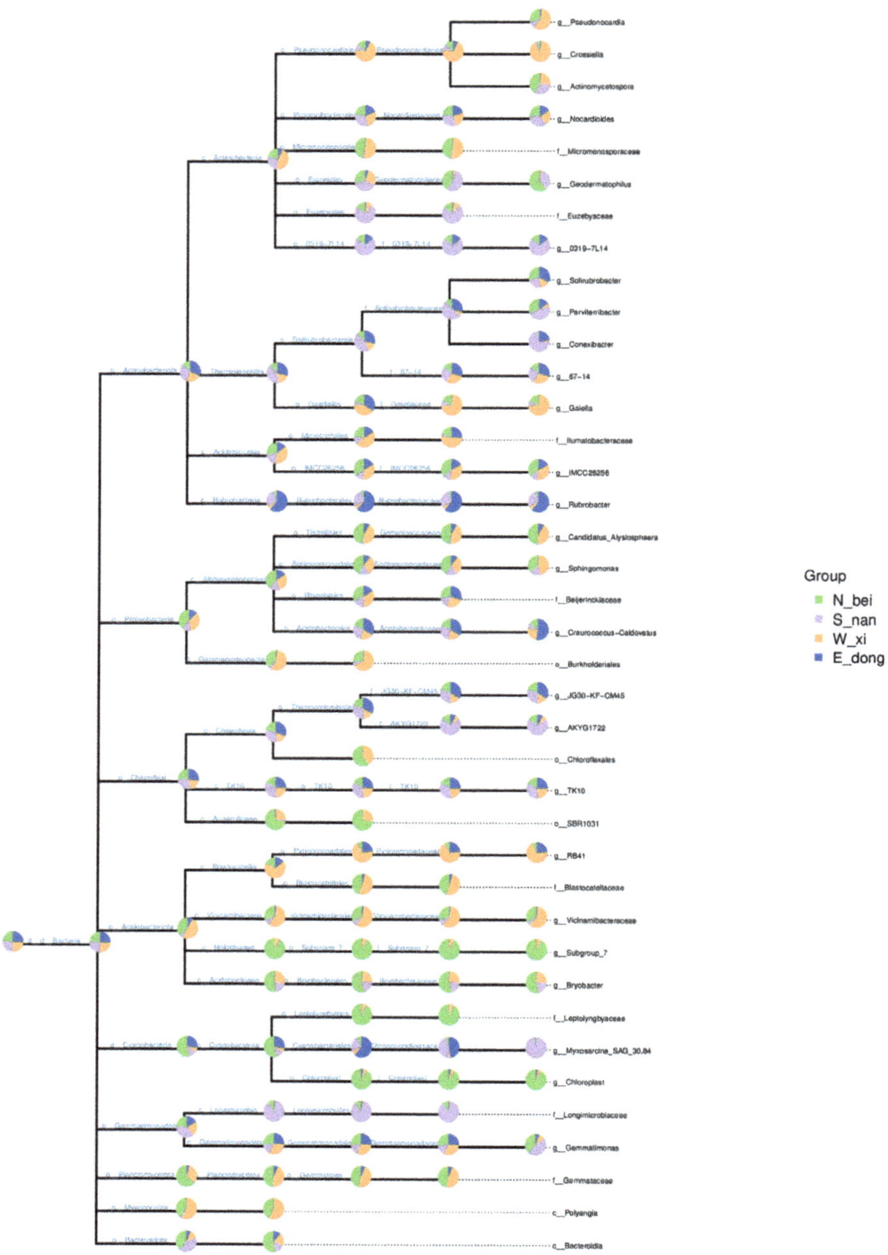

Figure 10. Species taxonomic hierarchy tree with abundance information (image source: author-produced tree based on experimental results).

3.3.4. GraPhlAn Evolutionary Tree

Figure 11 displays the circular GraPhlAn evolutionary tree, which integrates both the phylogenetic relationships and relative abundance of microbial taxa within the rammed earth community. Unlike Figure 10—which emphasizes the spatial clustering and overlap of microbial genera among different wall orientations—Figure 11 focuses on the hierarchical organization of taxa within the overall microbial community. In the GraPhlAn tree, the outer colored rings correspond to genus-level groups (e.g., *Rubrobacter*, JG30-KF-CM45), and the color gradient indicates relative abundance (blue for higher abundance and red for lower abundance). The inner rings represent higher taxonomic ranks (e.g., *Actinobacteriota*, *Chloroflexi*), illustrating their phylogenetic positions and clustering patterns. The figure shows that dominant phyla such as *Actinobacteriota* contain highly abundant genera such as *Rubrobacter*, while certain taxa (e.g., *Solirubrobacter*) display abundance differentiation along evolutionary branches, likely influenced by local environmental conditions such as light and humidity.

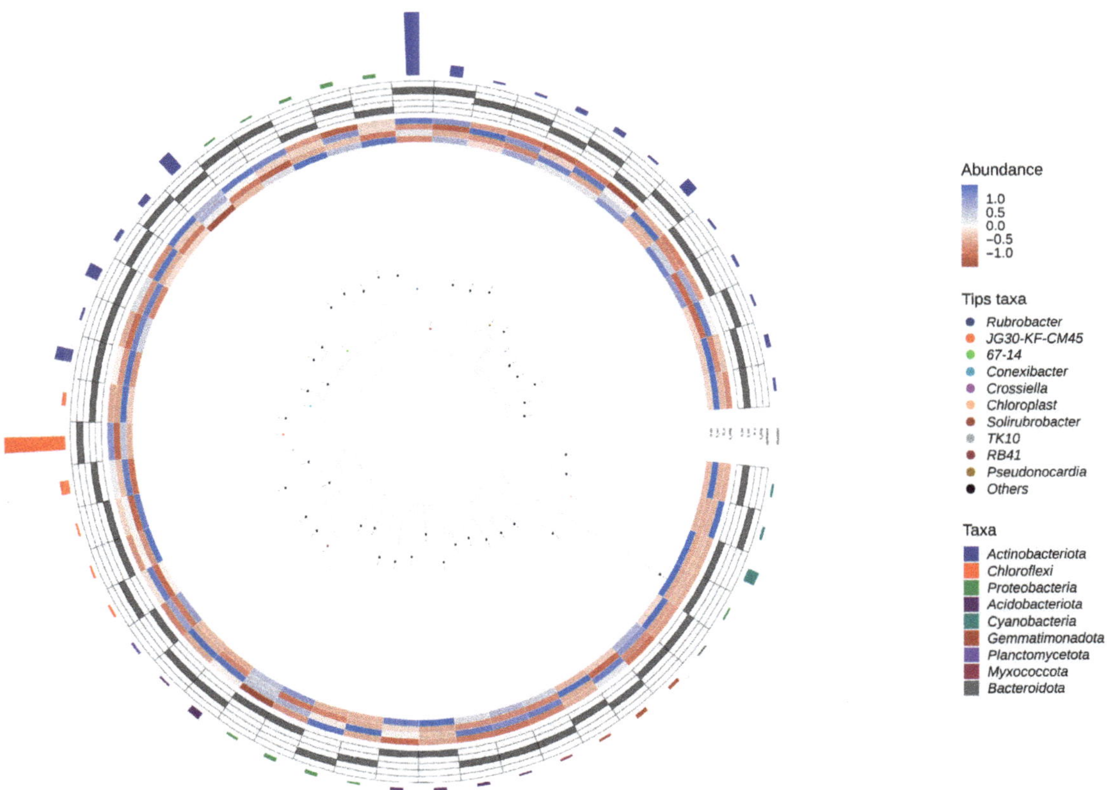

Figure 11. GraPhlAn evolutionary tree (image source: author-produced image based on experimental results).

3.4. Alpha Diversity Analysis

Alpha diversity refers to the richness, diversity, and evenness of species in a locally uniform habitat, and it is also known as within-habitat diversity [30]. Table 5 shows the microbial diversity index of the bacterial community.

Table 5. The microbial diversity index of the bacterial community.

Sample	Chao1	Simpson	Shannon	Pielou_e	Observed_species	Faith_pd	Goods_coverage
E_dong	1404.39	0.958144	7.39641	0.707485	1403.2	66.0369	0.999681
N_bei	3139.7	0.989159	9.50049	0.819165	3099.4	171.202	0.997469
S_nan	1315.24	0.981772	7.73356	0.748609	1287.7	80.7529	0.99865
W_xi	4469.27	0.994199	10.3129	0.851907	4407.1	189.049	0.995869

Source: Statistics based on experimental results.

The analysis results show that the species richness-related indicators have the following characteristics: (1) Chao1 index: This index is used to estimate the total number of species in the community. The Chao1 index is highest in the west (W_xi) (4469.27), followed by the north (N_bei) (3139.7), the east (E_dong) (1404.39), and the south (S_nan) (1315.24). This indicates that the species richness of the west rammed earth wall's microbial community is the highest, and the north wall also exhibits relatively high richness, while the east and south walls demonstrate relatively low richness. (2) Observed_species: The number of observed species in the west (4407.1) and north (3099.4) walls is substantially higher than that in the east (1403.2) and south (1287.7) walls, which is consistent with the Chao1 index trend, further indicating that the species richness of the west and north walls' microbial community is higher.

Secondly, from the perspective of indicators related to diversity and evenness, there are also three significant characteristics: (1) The Shannon index reflects the diversity of the community, showing the following trend: west (10.3129) > north (9.50049) > south (7.73356) > east (7.39641). The highest diversity was observed in the west wall, and the lowest was observed in the east. (2) For the Simpson index, a value closer to 1 indicates higher community diversity; the analysis indicated west (0.994199) > north (0.989159) > south (0.981772) > east (0.958144). This trend is consistent with the Shannon index. (3) The Pielou_e index reflects the evenness of species in the community: west (0.851907) > north (0.819165) > south (0.748609) > east (0.707485). The most even distribution of bacterial species was observed in the west wall, and the lowest evenness was observed in the east.

The results of phylogenetic diversity and sequencing coverage analysis also revealed two significant characteristics. (1) Faith_pd represents phylogenetic diversity. The west (189.049) and north (171.202) values are much higher than the east (66.0369) and south (80.7529), indicating that the microbial communities on the west and north rammed earth walls are more diverse in terms of evolutionary relationships. (2) Goods_coverage: This index is close to 1 for all samples (all above 0.995), indicating that the sequencing depth is sufficient, which can more comprehensively reflect the diversity of microbial communities, and data reliability is high.

In summary, the microbial communities of the west and north rammed earth walls of Paishan Village's residential buildings are generally superior to those of the east and south walls in terms of species richness, diversity, uniformity, and phylogenetic diversity; the microbial diversity of the microbial communities of the east and south rammed earth walls is relatively low.

From the rank abundance curve in Figure 12, the bacterial community structure of the rammed earth walls in different directions of Zhuhai Paishan Village shows the following characteristics: (1) The broken lines corresponding to the west (W_xi) and north (N_bei) extend longer on the horizontal axis (the broken line in the west extends to an OTU rank of about 4000, and the broken line in the north extends to an OTU rank of about 3000), indicating that the bacterial communities of the walls in these two directions contain more ASVs/OTUs and higher species richness. (2) The broken lines corresponding to the east

(E_dong) and south (S_nan) walls extend less along the horizontal axis (with the east ending at an OTU rank of more than 1000 and the south even shorter), indicating that the bacterial communities have fewer ASVs/OTUs and relatively lower species richness. This is consistent with the conclusion that species are richer in the west and north walls (based on Chao1, Observed_species, and other indices). However, in terms of community evenness, the curve for the west wall (W_xi) is the flattest, indicating minimal abundance differences among ASVs/OTUs in the rammed earth wall microbiome there and further reflecting the most even community composition.

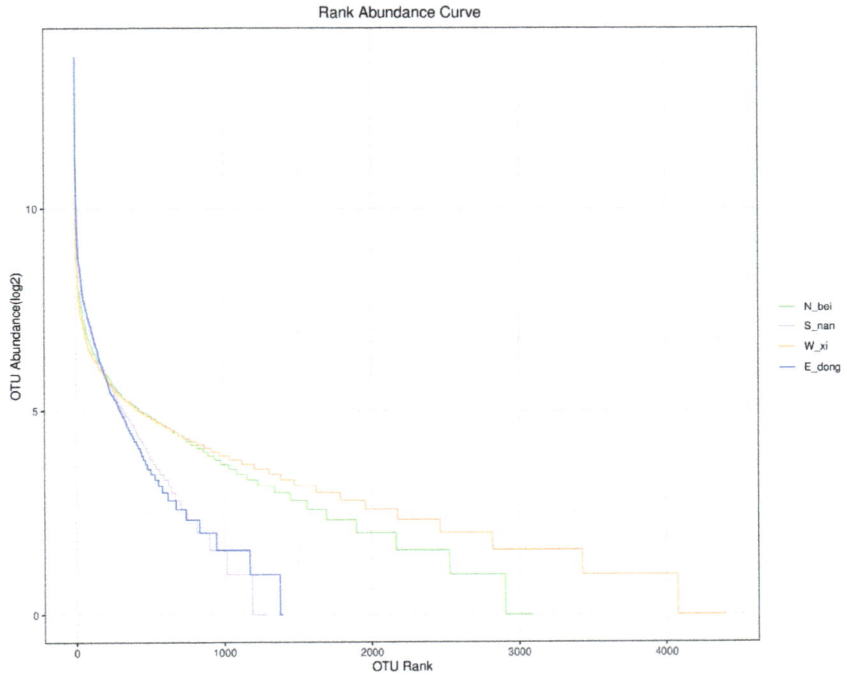

Figure 12. Rank Abundance Curve (image source: author-produced image based on experimental results).

The curve for the north (N_bei) wall is the second flattest, also indicating high evenness, although it is slightly lower than that of the west wall. The curves for the east (E_dong) and south (S_nan) walls are relatively steep, with the east wall being the steepest. This indicates that the abundance differences among ASVs/OTUs in the rammed earth wall microbiome in these two walls are large, and the community composition is less even, with the east wall exhibiting lower evenness than the south wall.

Overall, the microbial community on the west (W_xi) rammed earth wall exhibited high species richness, low variability in abundance among species, and high uniformity. The north (N_bei) microbial community also exhibited high species richness and good uniformity but was slightly lower than the west. The east (E_dong) and south (S_nan) microbial communities exhibited low species richness, large variabilities in species abundance, and low uniformity, with the east exhibiting the worst uniformity. These differences may be related to the microenvironmental conditions of the rammed earth walls (such as light, humidity, and material exchange with the surrounding environment), which in turn influenced the microbial community's structure and diversity.

3.5. Beta Diversity Analysis

The results of the beta diversity index revealed differences among the four rammed earth wall samples. Figure 13 shows a two-dimensional PCoA (principal coordinate analysis) ordination diagram based on Jaccard distance, illustrating the similarities and differences in the composition of bacterial communities.

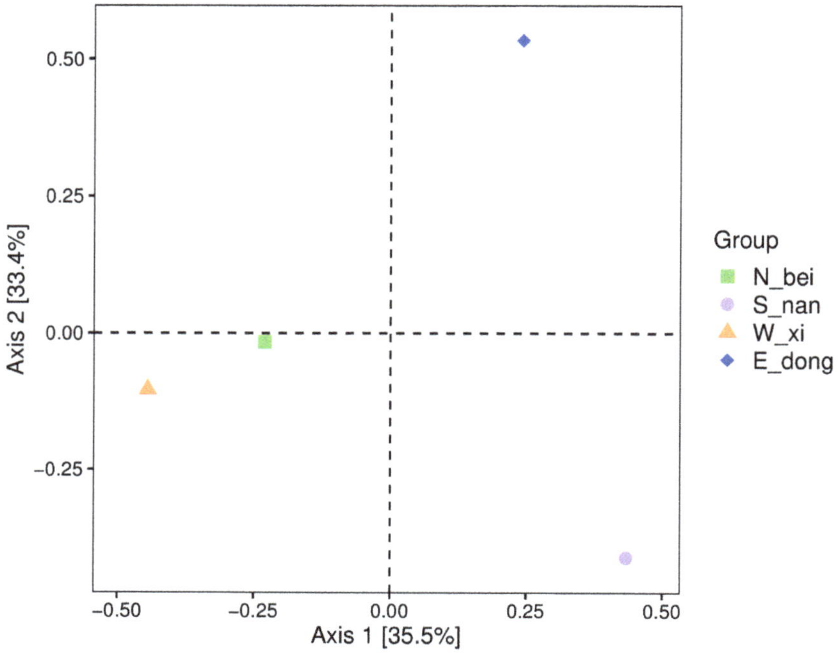

Figure 13. Abundance level curve (image source: author-produced image based on experimental results).

In the figure, Axis 1 (horizontal axis) explains 35.5% of the variation in community composition, and Axis 2 (vertical axis) explains 33.4% of the variation. The two axes together explain approximately 68.9% of the differences in community composition, reflecting the changes in the composition of the bacterial community to a large extent. Three notable features also emerge in terms of sample distribution and community similarity.

East (E_dong): The sample point (blue diamond) is in the upper right corner of the image, relatively far from the sample points in the other three directions. This indicates that the microbial community composition of the east rammed earth wall is substantially different from that of the other directions, and it is highly unique.

South (S_nan): The sample point (purple dot) is in the lower right corner of the image, and it is also somewhat distant from the sample points in the other directions. This indicates that the microbial community composition of the south rammed earth wall is also unique, differing significantly from the microbial communities in the east, west, and north walls.

West (W_xi): The sample point (orange triangle) is located in the lower left corner of the image.

North (N_bei): The sample point (green square) is located slightly at the center of the left side of the image. The relatively close proximity of the west and north sample points indicates that the microbial community composition of the west and north rammed earth walls is highly similar, and their microbial structures are more similar in terms of species composition.

Overall, the microbial community compositions of the east and south rammed earth walls in Paishan Village, Zhuhai, are highly unique and differ significantly from those in other locations. In contrast, the microbial community compositions of the west and north rammed earth walls are highly similar. This difference in community composition may be related to the varying microenvironments in different locations (such as light, humidity, surrounding vegetation, or human influence), which in turn shape the unique microbial community structures in each location.

Based on the UPGMA (unweighted group average method) clustering tree of the sample distance matrix (Figure 14), the microbial communities of the rammed earth walls in different directions in Paishan Village, Zhuhai, exhibited two characteristics with respect to sample clustering and similarity. (1) West (W_xi) and North (N_bei): The branch length of the two was 0.44, which was relatively short among all sample pairs, indicating that the microbial community composition of the rammed earth walls on the west and north sides was highly similar, and they were closer in species composition. (2) East (E_dong) and South (S_nan): The branch length of the two was 0.451, which is classified as a relatively short branch length, indicating that the microbial community composition of the rammed earth walls on the east and south sides also had certain similarities, and the differences in community structure between them were relatively small.

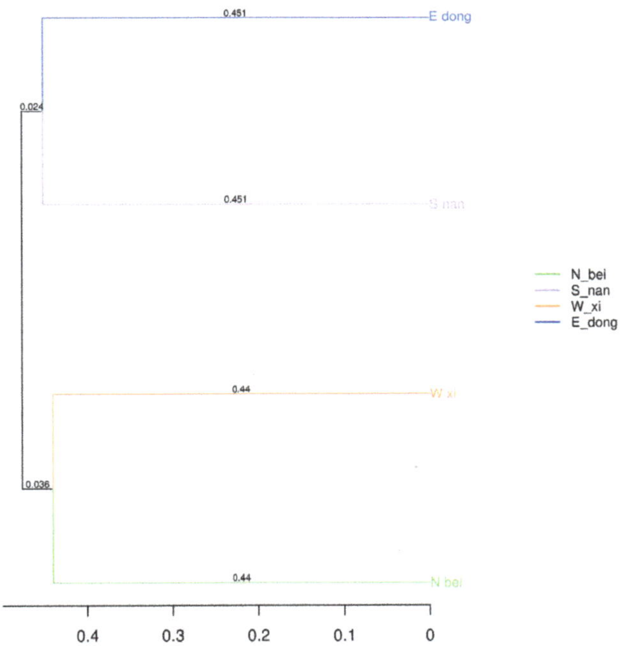

Figure 14. UPGMA cluster tree based on sample distance matrix (image source: author-produced image based on experimental results).

However, there were also differences between groups on different wall surfaces. For example, branch lengths differed between the west (W_xi) and north (N_bei) directions and between the east (E_dong) and south (S_nan) directions (0.44 for the west–north direction and 0.451 for the east–south direction). Additional branch lengths were observed between other pairs, such as between west and east and between north and south. This suggests that the bacterial communities in the west and north directions differed significantly from those in the east and south directions, forming relatively independent clusters.

This clustering result further verified the similarities and differences in the composition of bacterial communities in different directions obtained through PCoA analysis, reflecting that the rammed earth walls in different directions in Paishan Village have shaped the bacterial community structure with certain grouping characteristics due to differences in microenvironments and other factors.

3.6. Species Difference Analysis and Endemic Species

The Venn diagram (Figure 15) illustrates the distribution and overlap of amplicon sequence variants (ASVs) among rammed earth samples collected from the north (N_bei), south (S_nan), west (W_xi), and east (E_dong) walls. It should be noted that the diagram represents ASV-level sequence features rather than species-level classifications; therefore, it reflects compositional similarity and differentiation among microbial communities rather than direct taxonomic differences. The number of unique ASVs in each group varied considerably: The west wall contained the highest number of unique ASVs (3906, 41.73%), followed by the north wall (2499, 26.7%). The south and east walls exhibited fewer unique ASVs—1115 (11.91%) and 1157 (12.36%), respectively—indicating relatively similar community compositions between these two orientations. Only 27 ASVs (0.29%) were shared across all four groups, representing a small core community with broad environmental tolerance. Limited overlaps between two or three wall orientations (e.g., 344 shared ASVs between north and west walls, 22 between south and east walls) further demonstrate the spatial heterogeneity of microbial assemblages associated with microclimatic variations in light exposure, humidity, and salinity.

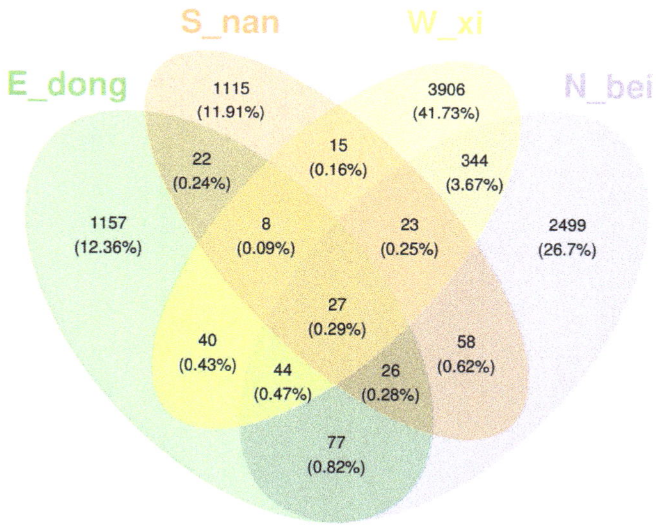

Figure 15. Venn diagram of samples (groups) ASV/OTU. Each ellipse represents a sample (group), the overlapping area between the ellipses indicates the common ASV/OTU between the samples (groups), and the number in each block indicates the number of ASV/OTU contained in the block. (image source: author-produced image based on experimental results).

Figure 16 presents the frequency distribution of operational taxonomic units (OTUs) at the genus and phylum levels in the rammed earth samples. This figure complements the previous compositional analysis (Figure 9) by illustrating the relative frequencies of dominant taxa rather than their proportional abundances. The bars do not sum to 100% because only the most representative genera and phyla are shown, while low-abundance groups were excluded for clarity. At the genus level, variations in the frequency of taxa, such as *Conexibacter* and *Geodermatophilus*, across orientations indicate orientation-related differences in microbial enrichment. At the phylum level, the frequencies of *Verrucomicrobiota* and *Deinococcota* fluctuate among samples, suggesting subtle shifts in phylum-level distributions.

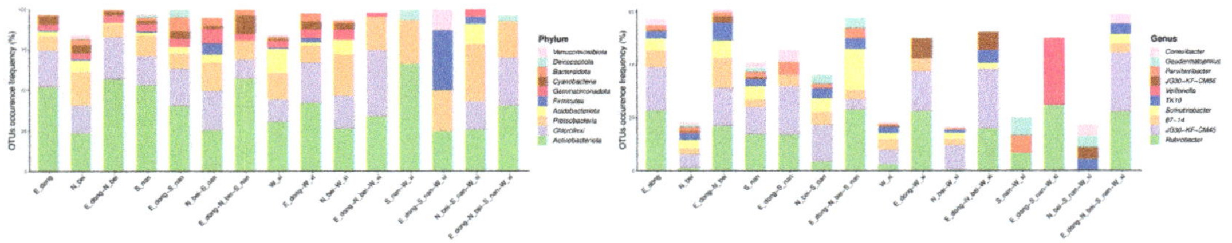

Figure 16. ASV/OTU statistics in different regions of the Venn diagram (image source: author-produced image based on experimental results).

Figure 17 shows the relative abundance distribution of microbial groups in rammed earth samples from different combinations at the phylum and genus levels. At the phylum level, the proportions of phyla, such as *Actinobacteriota*, in each combination are different, reflecting the selection of phylum-level communities by the microenvironment; at the genus level, the abundance of genera such as *Rubrobacter* fluctuates between combinations, reflecting the environmental adaptability of genus-level groups.

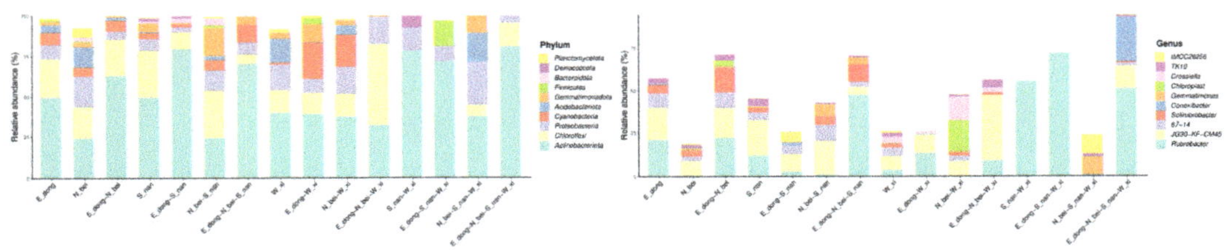

Figure 17. ASV/OTU abundance columns in different regions of the Venn diagram (image source: author-produced image based on experimental results).

To explore the ecological differences among walls, we performed clustering analysis based on Bray–Curtis dissimilarity and visualized the results as heatmaps (Figure 18). Red blocks indicate higher abundance, and blue blocks indicate the lower abundance of each genus, revealing distinct clustering patterns among wall orientations. This form of clustering corresponds well with environmental gradients such as humidity and sunlight exposure, confirming spatial heterogeneity in microbial communities.

In the heat map, the red–blue gradient is used to intuitively map the abundance differences at the genus level. Red blocks (such as *Solirubrobacter* and *67-14* in the N_bei group) indicate high-abundance groups, reflecting the low-light and high-humidity environment of the north wall and its effect on hygrophilous and shade-tolerant microorganisms, while blue blocks (such as some actinomycetes in the E_dong group) correspond to low-abundance groups, suggesting the adaptive succession of the community under the mild microenvironment of the east wall. Among the specific groups, *JG30-KF-CM45* is highlighted via a darker shade of red (high abundance) in the S_nan group, which is highly consistent with the selection pressure inflicted by the strong sunlight and high-temperature environment of the south wall. *TK10* is highlighted in blue (low abundance) in the W_xi group, reflecting the inhibitory effect of salt stress and wind disturbance on the genus in the west wall. This dual "cluster association + abundance gradient" visualization accurately analyzes the ecological differentiation of microbial genera driven by the orientation microenvironment and provides a quantitative basis at the genus level for exploring the microecological construction mechanism and environmental response laws of historical rammed earth walls.

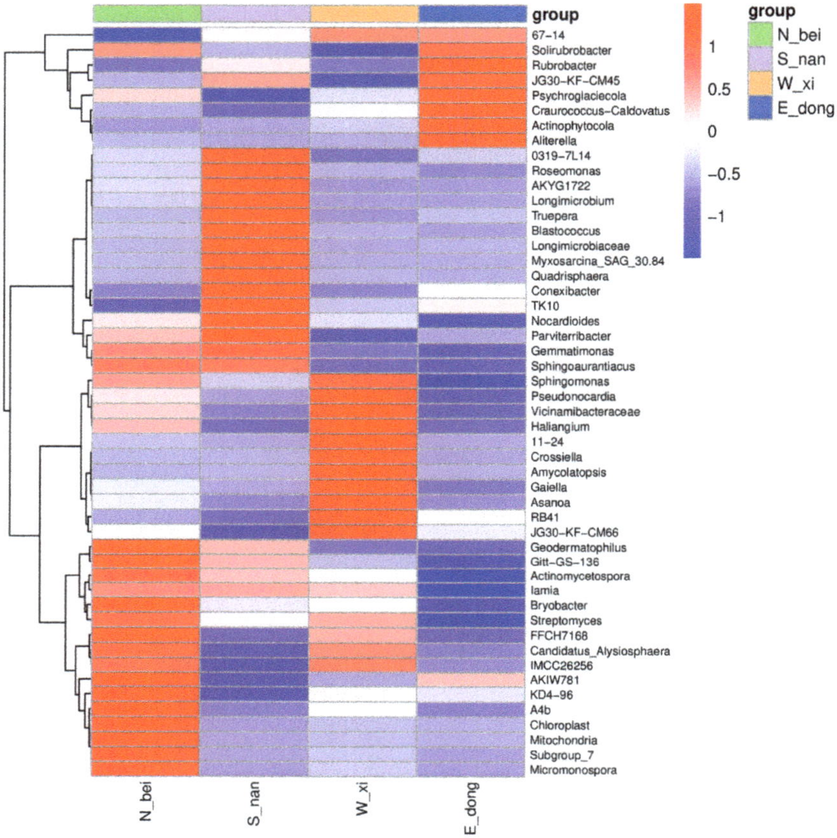

Figure 18. Heat map of species composition at the genus level with species clustering (image source: author-produced image based on experimental results).

3.7. Predictive Calculation of Metabolic Pathways

Figure 19 presents the predicted KEGG secondary functional pathway distribution of the microbial community, inferred from 16S rRNA gene sequencing data using the Tax4Fun2 functional annotation tool [31]. The analysis workflow followed three key steps: first, the high-quality OTU (Operational Taxonomic Unit) table was normalized by 16S rRNA gene copy number to reduce taxonomic bias; second, OTUs were matched to the SILVA database (v138) for taxonomic annotation; third, annotated taxa were mapped to the KEGG Orthology (KO) database to calculate the relative abundance of secondary metabolic pathways. It should be noted that this analysis reflects potential rather than directly measured metabolic functions. 16S-based functional prediction relies on the correlation between taxonomic composition and known functional profiles of reference microbes, so the results are approximate estimations [32]. This method has been validated in studies of historical building-associated microbes [33], showing consistent trends with metagenomic data for core metabolic pathways (e.g., amino acid metabolism), though it may underestimate rare functional modules.

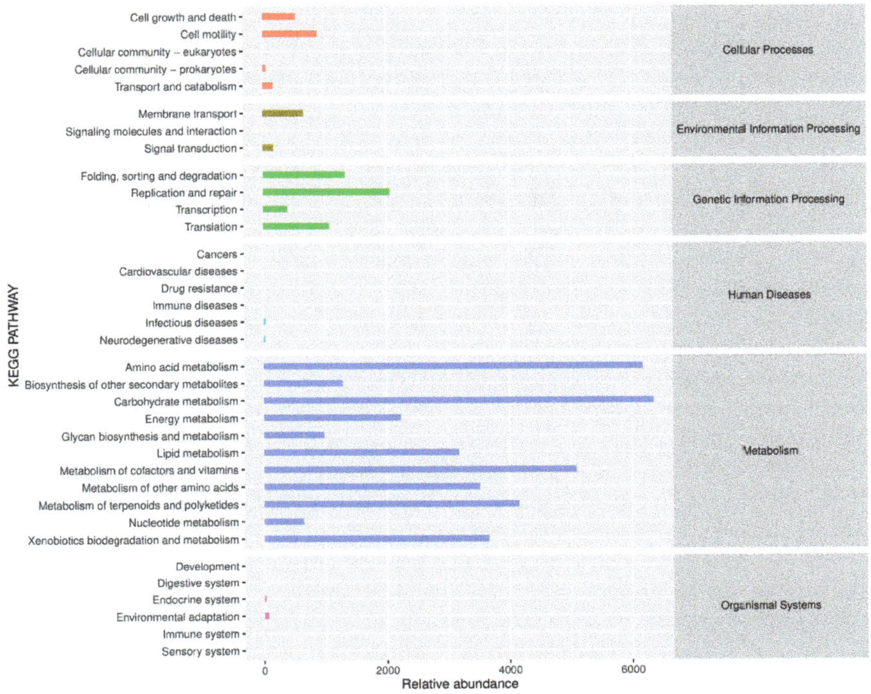

Figure 19. Predicted KEGG secondary functional pathway abundance. The horizontal axis represents the abundance of functional pathways (per million KO), the vertical axis represents the functional pathways at the second level of KEGG classification, and the rightmost axis represents the first-level pathway to which the pathway belongs. The average abundance of all samples is shown here.

The predicted functions are distributed across six major KEGG categories, including Metabolism, Environmental Information Processing, and Cellular Processes. Among these, pathways related to amino acid and carbohydrate metabolism show relatively high predicted abundance. This is particularly relevant to the study of historic rammed earth materials: these pathways suggest the microbial community may participate in nutrient cycling (e.g., decomposition of organic matter in rammed earth) and energy metabolism, which are key to understanding microbial impacts on the material's physical and chemical stability (e.g., preventing organic acid accumulation that causes structural erosion).

The Environmental Information Processing module, particularly membrane transport pathways, may indicate adaptive responses to environmental changes (e.g., fluctuations in moisture and temperature within historic rammed earth structures). In contrast, pathways associated with Genetic Information Processing and Human Diseases display lower predicted abundance, reflecting limited direct host-related activity and supporting the hypothesis that the microbial community is primarily adapted to the abiotic historic rammed earth environment.

The occurrence of unclassified taxa at different taxonomic levels reflects the limitations of current reference databases and the complexity of microbial communities inhabiting historic rammed earth materials. We therefore only interpreted confidently annotated taxa, avoiding overgeneralization of uncertain classifications. This conservative approach provides a more robust and scientifically sound basis for discussing microbial community patterns.

Figure 20 presents the Predicted MetaCyc secondary functional pathway abundance map. The MetaCyc database covers a large amount of microbial metabolic pathway information. Through the predicted abundance map, we can understand the potential activity of various metabolic pathways (such as substance synthesis, decomposition, energy metabolism, etc.) in the microbial community and thus infer the metabolic functions that the community may have. Among biosynthetic pathways, Amino Acid Biosynthesis showed an extremely high abundance, approaching 20,000, suggesting that the microbial community may be actively synthesizing amino acids. Amino acids are the building blocks of proteins, and this high abundance may indicate a high demand for protein for the microbial growth and reproduction within the community, or that the microbial community is providing amino acids to the surrounding environment. Furthermore, pathways such as Carrier and Vitamin Biosynthesis also showed a certain abundance, suggesting that microbes have the potential to synthesize substances such as cofactors and vitamins that are crucial for life, helping to maintain metabolic balance within the community itself and potentially that of surrounding organisms. Under the Degradation/Utilization/Assimilation category, the abundance of degradation pathways for various substances is relatively moderate. Pathways such as amino acid degradation and aromatic compound degradation show some activity, indicating that the microbial community has the potential to degrade these organic substances. Rammed earth walls may contain amino acids and aromatic compounds from the environment. Microorganisms degrade these substances to obtain energy and nutrients, participating in the material cycle surrounding the rammed earth wall and contributing to the transformation of substances in the wall's microenvironment.

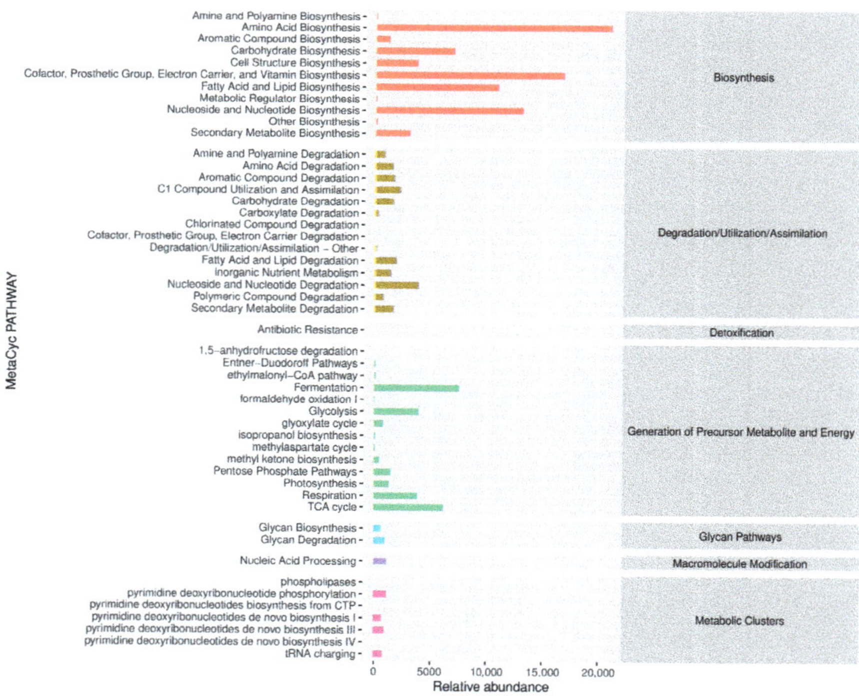

Figure 20. Predicted MetaCyc secondary functional pathway abundance map. The horizontal axis is the abundance of functional pathways (unit: per million PWY), the vertical axis is the functional pathway of the second classification level of MetaCyc, and the rightmost is the first-level pathway to which the pathway belongs. The average abundance of all samples is shown here.

It should be noted that these results represent predicted metabolic potential derived from 16S rRNA data rather than experimentally verified metagenomic or metatranscriptomic evidence. Therefore, the functional interpretations remain hypothetical. Future studies should validate them using whole-genome sequencing (to confirm functional gene presence) or transcriptome sequencing (to verify gene expression), which will further clarify the actual metabolic activities of the microbial community in historic rammed earth.

4. Discussion

4.1. Benefits of Microbial Diversity in Rammed Earth Walls

This study provides the first systematic characterization of microbial community compositions and potential ecological functions on historic rammed earth walls in the Lingnan region. The SEM microstructural differences are closely related to the microbial diversity patterns observed in the sequencing results. The north and west walls, characterized by looser particle structures and higher porosity, provide favorable conditions for microbial attachment and nutrient retention, explaining their higher microbial richness. Conversely, the south and east walls exhibit denser structures with lower pore connectivity, which limits microbial colonization and corresponds to lower diversity levels dominated by heat-tolerant and light-tolerant taxa. This correlation between SEM observations and microbial community composition highlights the influence of microstructural features on microbial distribution across different wall orientations. Specifically, 16S rRNA metabarcoding and SEM analyses revealed the clear spatial differentiation of microbial diversity across the four wall orientations (north, south, east, and west). The north and west walls exhibited higher ASV richness and more complex community structures, likely due to higher humidity and lower light exposure, while the south and east walls were dominated by *Actinobacteriota* and *Proteobacteria* under stronger sunlight. These results indicate that microclimatic factors such as light, moisture, and salinity strongly influence microbial community assembly and may serve as bioindicators of wall preservation states. The predicted metabolic pathways related to amino acid and carbohydrate metabolism further suggest that microorganisms play potential roles in nutrient cycling, soil stabilization, and biogeochemical transformation processes within the earthen matrix.

Our findings generally align with prior research on earthen heritage microbiology, which indicates that microorganisms can both damage and protect such structures. For instance, surface biofilms have been reported to reduce erosion on historic rammed earth walls. Similarly, the dominance of Actinobacteria on our sun-exposed south wall echoes observations from other hot-climate sites, where thermophilic Actinobacteria are predominant. Similarly, the moisture-loving microbes enriched on the north wall reflect findings from humid-climate studies, in which shaded, damp sections of old walls exhibited increased algal and mold growth that accelerated biodeterioration.

The novelty of this study lies in bridging microbial ecology and architectural conservation through a spatially resolved analysis of ancient rammed earth materials. Unlike previous studies that examined either material properties or microbial taxonomy alone, our research integrates microstructural observation, molecular diversity analysis, and functional prediction to uncover how microorganisms adapt to orientation-specific microenvironments. Our findings contribute to understanding the interaction between environmental factors and microbial colonization mechanisms, revealing how microbial diversity patterns can inform preventive conservation strategies. These interdisciplinary insights expand current knowledge on the ecological processes that underpin the degradation and stabilization of traditional earthen heritage. By distinguishing between wall orientations, our study offers a more granular view than studies that sample whole structures without microhabitat differentiation. For example, a recent metagenomic survey of a

Song Dynasty city wall found that well-preserved sections had much richer and distinct microbiomes than heavily weathered sections, suggesting that microbial activity can both mitigate and exacerbate decay. This supports our conclusion that fine-scale environmental factors drive community structure and function, directly impacting conservation outcomes. Additionally, by integrating SEM-observed microstructure with microbial sequencing, we answered calls for a multi-dimensional approach linking biology to material changes—a strategy increasingly advocated in heritage science.

Our study also highlights certain knowledge gaps. Although some researchers have improved earthen material durability using specific microbial inoculants or bio-mineralization techniques, the in situ cause–effect relationships remain complex. We found clear correlations between environment and community traits, but pinpointing which microbes cause material decay versus those that offer protection will require targeted experiments. Moreover, as with most current studies (including ours), our analysis is essentially a snapshot; the seasonal dynamics of these wall microbiomes remain largely uncharted and merit future investigation. Overall, our results reinforce the view that microbes are key agents in earthen wall weathering. We advance this understanding by providing fine-scale evidence of niche partitioning on one structure and by demonstrating the value of an integrated, interdisciplinary analysis.

4.2. Multi-Dimensional Conservation Strategy Recommendations

Building on these insights, we propose a multi-dimensional conservation strategy that combines scientific, ecological, and community-based measures. Preservation efforts should balance material protection with public engagement, effectively turning the microbial "problem" into an educational and protective resource. By linking material heritage with its micro-ecology, this approach also provides a novel pathway for cultural empowerment as the village pursues rural revitalization:

(1) Ecological Restoration Technologies: Bio-based consolidation methods should be explored to naturally strengthen rammed earth walls. For example, introducing beneficial native microbes or applying microbial-induced calcite precipitation (MICP) can bind soil particles and increase cohesion. These nature-inspired techniques treat biofilms as living protective agents, offering a symbiotic restoration approach. They can complement traditional chemical consolidants to improve structural resilience while preserving material authenticity.

(2) Heritage Science Education and Tourism: Educational tourism activities centered on the walls' microbiology should be developed. Rather than excessive commercialization, Paishan Village can host "microbial exploration" tours where visitors (especially students) observe wall microbes—for example, via portable microscopes—and learn how different areas harbor different micro-ecologies. Framing preservation as a scientific adventure not only raises public awareness of the value of rammed earth sites but also fosters a sense of wonder, reinforcing that these historic walls are dynamic "living" heritage sites rather than inert relics.

(3) Community Engagement and Co-Governance: Local residents should be encouraged to participate in protecting their heritage. Scientific findings should be shared in accessible formats (e.g., workshops or illustrated posters) to help residents understand that "moldy" wall spots are part of a natural ecosystem and that proper maintenance can extend the life of their ancestral homes. This understanding fosters pride and a sense of ownership, motivating residents to help monitor and maintain the walls. Shifting from a top–down approach to a co-governance model—with residents collaborating alongside experts—ensures that interventions are culturally appropriate and sustainable. For instance, communities could form volunteer maintenance teams trained in

basic microbial monitoring and preventive care, guided by conservation specialists. Such community-driven stewardship, supported by scientific guidance, establishes a sustainable framework that keeps rammed earth heritage alive and collectively cared for.

Furthermore, our survey revealed that some residents in Paishan Village lack awareness of rammed earth wall preservation, arbitrarily modifying and discarding traditional building materials. By translating the findings of microbial diversity research into accessible language, residents can understand the vitality of rammed earth walls from a microecological perspective. For example, we can produce posters that explain the following: "mold spots on walls are not merely a disease but part of the village's ecology, and that scientific restoration can extend the life of old houses." This helps residents understand that these are not merely "old houses" but "living heritage" embodying the village's ecological memory and cultural identity. This cultural identity, grounded in scientific understanding, can effectively inspire residents to proactively participate in heritage protection, promote the transition from "government-led conservation" to "government + villager co-governance", and provide a foundation for the sustainable development of Paishan Village.

In combination, these strategies reinforce one another and exemplify the symbiosis between ecological protection and cultural heritage. Recognizing the microbiological dimension of conservation not only aids in safeguarding the physical integrity of the walls but also enriches their cultural significance, ensuring that preservation is both scientifically informed and socially resilient.

5. Conclusions

This study combined scanning electron microscopy (SEM) and high-throughput 16S rRNA sequencing to investigate the microbial diversity and environmental adaptation mechanisms of rammed earth walls in Paishan Village, a representative site of Lingnan's hot and humid climate. The results revealed clear spatial differentiation in both microstructure and microbial compositions across different wall orientations. The north and west walls exhibited looser textures and higher microbial richness due to greater humidity and wind exposure, while the south and east walls exhibited more compact structures and specialized microbial communities dominated by heat-tolerant and light-tolerant taxa. Only a small fraction of core microorganisms was shared among all orientations, confirming that light, temperature, humidity, and salinity are major factors shaping microbial community differentiation. The predicted metabolic functions suggested that microorganisms contribute to nutrient cycling and material stabilization within rammed earth, with low representations of human-related pathogenic pathways.

The findings highlight the complex interactions between environmental conditions, material properties, and microbial ecology in traditional earthen heritage, offering new perspectives for conservation science. By linking microstructural analysis with microbial functional profiling, this study contributes to a better understanding of how microorganisms participate in both the degradation and preservation of rammed earth materials, providing a scientific foundation for developing ecological restoration strategies.

Nonetheless, this study remains a preliminary exploration. Future studies should conduct long-term monitoring to capture seasonal changes and environmental fluctuations and perform controlled laboratory experiments to verify the effects of key microbial groups on material performance. Moreover, the potential application of beneficial indigenous microorganisms or microbial-induced calcite precipitation (MICP) could be explored to enhance the durability and ecological resilience of rammed earth structures. These efforts will promote an integrated approach combining microbial ecology and cultural heritage conservation for the sustainable protection of historic villages such as Paishan.

Author Contributions: Conceptualization: W.W., S.Y. and J.S.; methodology: S.Y. and J.S.; software: S.Y. and J.S.; validation: W.W., S.Y. and J.S.; formal analysis: W.W., S.Y. and J.S.; investigation: J.S.; resources: J.S.; data curation: S.Y. and J.S.; writing—original draft preparation: W.W., S.Y. and J.S.; writing—review and editing: W.W., S.Y. and J.S.; visualization: W.W., S.Y. and J.S.; supervision: M.S.B.I.; project administration: M.S.B.I.; funding acquisition: M.S.B.I. All authors have read and agreed to the published version of the manuscript.

Funding: This research received no external funding.

Institutional Review Board Statement: Not applicable.

Informed Consent Statement: Not applicable.

Data Availability Statement: The datasets used and analyzed during the current study are available from Shuai Yang (samyang@zju.edu.cn) upon reasonable request. Shareable data associated with this article are available in the following: https://doi.org/10.6084/m9.figshare.30121138.v1.

Acknowledgments: We are grateful to all authors who contributed to the writing and data compilation of this study. We also sincerely thank the alumni of the City University of Macau, where we completed our master's degree, for their valuable assistance during field research.

Conflicts of Interest: The authors declare no conflicts of interest.

References

1. Jaquin, P.A. Analysis of Historic Rammed Earth Construction. Ph.D. Thesis, Durham University, Durham, UK, 2008.
2. Tang, X.; Shen, S.; Su, X. From rammed earth to stone wall: Chronological insight into the settlement change of the Lower Xiajiadian culture. *PLoS ONE* **2022**, *17*, e0273161. [CrossRef]
3. Guo, Z.; Qi, Q.; Zhang, S.; Chen, W.; Yang, L.; Zhang, Y. Characterization of spatial and chrono-cultural evolution of earthen sites construction technology in China based on GIS. *Archaeol. Anthropol. Sci.* **2025**, *17*, 142. [CrossRef]
4. Maniatidis, V.; Walker, P.; A Review of Rammed Earth Construction. Innovation Project "Developing Rammed Earth for UK Housing", Natural Building Technology Group, Department of Architecture & Civil Engineering, University of Bath, 12. 2003. Available online: https://people.bath.ac.uk/abspw/rammedearth/review.pdf (accessed on 12 July 2025).
5. Pei, Q.; Guo, Q.; Wang, X.; Zhao, J.; Zhao, G.; Luo, J. Evolution and Characteristics of Traditional Construction Techniques in Earthen Building Sites. *Cult. Relics Conserv. Archaeol. Sci.* **2019**, *31*, 1–13. [CrossRef]
6. Huang, X. *A Brief History of Zhuhai*; Social Sciences Academic Press: Beijing, China, 2011; ISBN 9787509725603.
7. Meng, X.; Zhang, L.; Wang, J.; Jin, Y. Research on the Zoning of Traditional Residential Buildings in Multicultural Areas. *Archit. J.* **2020**, 1–7.
8. Qian, C.; Yin, P. Research on the construction technology of Fujian Tulou based on the principle of protection. *Cent. China Archit.* **2014**, *32*, 7–10. Available online: https://chn.oversea.cnki.net/KCMS/detail/detail.aspx?dbcode=CJFD&dbname=CJFD2014&filename=HZJZ201404002&uniplatform=OVERSEA&v=wFwJk5CsL2cudUkp5pPPaw-kdhjmQjcpUGIhtIIEqwzfJW__WNX358bdPw4xsBuW (accessed on 12 July 2025).
9. Wan, L.; Wu, E.; Chi, X.; Bai, W.; Zhou, L. A new rammed earth construction strategy based on the endogenous development model of rural areas: A case study of the "One Specialty One Village" project in rural areas of Southwest China. *Archit. J.* **2024**, 105–111. [CrossRef]
10. Bai, W.; Chi, X.; Zhou, L. Research on Structural Technology for the Conservation and Utilization of Rammed Earth Diaolou in Kaiping. *New Archit.* **2021**, 144–147. [CrossRef]
11. Yang, J.; Wang, N.; Huang, Z.; Huang, Y.; Lv, W.; Yang, S. Anti-Erosion Mechanism of Biological Crusts and Eco-Protection Technology Using Composite Biofilms for Traditional Rammed Earth Dwellings in Songyang County. *Coatings* **2025**, *15*, 608. [CrossRef]
12. Zhao, M.; Li, Y.; Chen, H.; Chen, Y.; Zheng, L.; Wu, Y.; Wang, T. Metagenomic study of the microbiome and key geochemical potentials associated with architectural heritage sites: A case study of the Song Dynasty city wall in Shou County, China. *Front. Microbiol.* **2024**, *15*, 1453430. [CrossRef]
13. Liu, L.; Zhang, Y.; Guo, L.; Cao, H.; Dai, Z.; Zhao, Z.; Li, T. Experimental test and mechanism analysis of soil crust erosion resistance of rammed earth Great Wall in rainy season. *Sci. Rep.* **2024**, *14*, 9069. [CrossRef]
14. Kraus, C.; Hirmas, D.; Roberts, J. Microbially indurated rammed earth: A long awaited next phase of earthen architecture. *ARCC Conf. Repos.* **2013**, 58–65. [CrossRef]

15. Schopf, J.W. Fossils and Pseudofossils: Lessons from the Hunt for Early Life on Earth. Paper Presented at the Size Limits of Very Small Microorganisms: Proceedings of a Workshop. 1999. Available online: https://books.google.com/books?hl=zh-CN&lr=&id=cAmdAgAAQBAJ&oi=fnd&pg=PA88&dq=Schopf,+J.+W.+(1999).+Fossils+and+pseudofossils:+lessons+from+the+hunt+for+early+life+on+Earth.+Paper+presented+at+the+Size+Limits+of+Very+Small+Microorgan-isms:+Proceedings+of+a+Workshop.&ots=ouZrxpp1op&sig=hjECNCr_N4mmgZqs62M0pB2S4KQ#v=onepage&q&f=false (accessed on 24 March 2025).
16. Chen, W.; Yan, B.; Guo, S.; Liu, Y.; Yang, F.; Zhang, K.; Mao, W. An in-situ conservation method of the rammed earth sites using a new silica protective agent. *Constr. Build. Mater.* **2024**, *452*, 138960. [CrossRef]
17. He, Y.; Zhang, M.; Ren, W.; Wu, Z. Improvement strategies and applications of traditional earth construction technology: Reflections on architectural practice in rural revitalization. *New Archit.* **2021**, 38–43. [CrossRef]
18. Liu, Y.; Zhou, R.; Zhao, Q.; Zhai, J.; Tian, Y.; Yu, H. Effect of moisture on weathering of ancient, rammed earth walls. *J. Cap. Norm. Univ. (Nat. Sci. Ed.)* **2025**, *46*, 12–17. [CrossRef]
19. Xiong, F.; Lan, L.; Li, S. Exploring the Development and Changes of Traditional Villages Under the Background of New Urbanization: A Case Study of Bajia Paishan Village, Doumen Town, Zhuhai City. Sharing and Quality: Proceedings of the 2018 China Urban Planning Annual Conference (18 Rural Planning). 2018. Available online: https://chn.oversea.cnki.net/KCMS/detail/detail.aspx?dbcode=CPFD&dbname=CPFDLAST2019&filename=ZHCG201811018075&uniplatform=OVERSEA&v=9CPJG6c4waNeKEtc9HBirNwqP75X1_NHBC5qH0UqYScuT_kh6TKan1hPpRVv7QVaHfsDj-sU1gE%3d (accessed on 24 March 2025).
20. Li, D.; Bai, X.; Zuo, J.; Wang, Y.; Wang, X.; Han, P. Preventive Protection of Earth Heritage: Influence of Silane Application Methods on Surface Effectiveness of Consolidation–Pingyao Rammed Earth Wall, a World Heritage Site. *Int. J. Archit. Herit.* **2025**, 1–15. [CrossRef]
21. Bui, Q.B.; Morel, J.C.; Reddy, B.V.; Ghayad, W. Durability of rammed earth walls exposed for 20 years to natural weathering. *Build. Environ.* **2009**, *44*, 912–919. [CrossRef]
22. Chao, A. Nonparametric estimation of the number of classes in a population. *Scand. J. Stat.* **1984**, 265–270. Available online: https://www.jstor.org/stable/4615964 (accessed on 24 March 2025).
23. Shannon, C.E. A mathematical theory of communication. *Bell Syst. Tech. J.* **1948**, *27*, 379–423. [CrossRef]
24. Shannon, C.E. A mathematical theory of communication. *ACM SIGMOBILE Mob. Comput. Commun. Rev.* **2001**, *5*, 3–55. [CrossRef]
25. Simpson, E.H. Measurement of diversity. *Nature* **1949**, *163*, 688. [CrossRef]
26. Faith, D.P. Conservation evaluation and phylogenetic diversity. *Biol. Conserv.* **1992**, *61*, 1–10. [CrossRef]
27. Pielou, E.C. The measurement of diversity in different types of biological collections. *J. Theor. Biol.* **1966**, *13*, 131–144. [CrossRef]
28. Good, I.J. The population frequencies of species and the estimation of population parameters. *Biometrika* **1953**, *40*, 237–264. [CrossRef]
29. Ramette, A. Multivariate analyses in microbial ecology. *FEMS Microbiol. Ecol.* **2007**, *62*, 142–160. [CrossRef]
30. Tomašových, A.; Kidwell, S.M. Predicting the effects of increasing temporal scale on species composition, diversity, and rank-abundance distributions. *Paleobiology* **2010**, *36*, 672–695. [CrossRef]
31. Wemheuer, F.; Taylor, J.A.; Daniel, R.; Johnston, E.; Meinicke, P.; Thomas, T.; Wemheuer, B. Tax4Fun2: Prediction of habitat-specific functional profiles and functional redundancy based on 16S rRNA gene sequences. *Environ. Microbiome* **2020**, *15*, 11. [CrossRef]
32. Langille, M.G.; Zaneveld, J.; Caporaso, J.G.; McDonald, D.; Knights, D.; Reyes, J.A.; Clemente, J.C.; Burkepile, D.E.; Vega Thurber, R.L.; Knight, R.; et al. Predictive functional profiling of microbial communities using 16S rRNA marker gene sequences. *Nat. Biotechnol.* **2013**, *31*, 814–821. [CrossRef] [PubMed]
33. Otlewska, A.; Adamiak, J.; Gutarowska, B. Clone-based comparative sequence analysis of 16S rRNA genes retrieved from biodeteriorating brick buildings of the former Auschwitz II–Birkenau concentration and extermination camp. *Syst. Appl. Microbiol.* **2015**, *38*, 48–55. [CrossRef] [PubMed]

coatings

MDPI

Article

Analysis of Rammed Earth Wall Erosion in Traditional Village Dwellings in Zhuhai City

Yanjun Wang [1], Junxin Song [2,*], Jiahao Zhang [1], Yue Huang [3] and Shuai Yang [1,*]

[1] School of Art and Archaeology, Hangzhou City University, No.51 Huzhou Street, Gongshu District, Hangzhou 310015, China; 32310061@stu.zucc.edu.cn (Y.W.); 32210129@stu.hzcu.edu.cn (J.Z.)
[2] Institute for Research on Portuguese Speaking Countries, City University of Macau, Avenida Padre Tomás Pereira, Taipa, Macau 999078, China
[3] Hangzhou Animation & Game College, Hangzhou Vocational & Technical College, No.68 Xueyuan Street, 17 Xiasha Higher Education Park, Qiantang District, Hangzhou 314423, China; vivianwhang@zju.edu.cn
* Correspondence: a23092100150@cityu.edu.mo (J.S.); samyang@zju.edu.cn (S.Y.)

Abstract: (1) Background: this article focuses on the durability decline problem of rammed earth buildings in Paishan Village, Zhuhai City under the influence of complex environments. It aims to reveal the erosion mechanisms of rammed earth walls caused by different environmental factors (acid rain, salt spray, humidity, biological activities, etc.), and provide a scientific basis for formulating targeted remediation strategies. (2) Methods: a technical framework combining macroscopic investigation and microscopic analysis was adopted. Ion chromatography, scanning electron microscopy (SEM), and characterization (XRD) were used to study the damage to buildings in Paishan Village under the influence of different environmental factors. (3) Results: The acid rain and sulfate buildup could cause cracks and peeling on the south wall of the rammed earth wall. Salt spray and high humidity conditions exacerbated surface weathering on the west wall. Vibrant biological activity and high humidity made the north wall's minerals easily dissolve, leaving the structure loose. The east wall was affected by the changing dynamics of carbonate rocks, which made it more vulnerable to weathering. (4) Conclusion: according to the analysis of different walls, specific steps should be taken during future restoration to improve the durability of rammed-earth buildings.

Keywords: rammed earth; traditional village dwellings; wall erosion; ion chromatography; scanning electron microscopy (SEM); Zhuhai City

1. Introduction

1.1. Background and Purpose

As one of the most typical dwellings in China, rammed earth buildings have a long history and are the crystallization of the wisdom of the ancient Chinese working people. Rammed earth construction is a type of wall or structure that is built by filling moist soil (usually a mixture of clay, sand, and gravel) into formwork in layers and compacting it layer by layer [1]. Its core principle is to use mechanical force to tightly combine soil particles to form strong and durable building material. It has the advantages of a simple process, low cost, energy saving, and being warm in winter and cool in summer [2]. Paishan Village in Zhuhai City was first built during the Qianlong period of the Qing Dynasty (1778) and has a history of nearly 250 years. As one of the best-preserved ancient villages in Zhuhai, it was listed in the third batch of China's traditional villages list in 2014 and was named "Guangdong Provincial Ancient Village" in 2015. It is an important sample for studying

Lingnan farming culture and immigration history. There are more than 40 rammed earth buildings in the village dating from the Qing Dynasty to the Republic of China period, with the largest thickness reaching 50 cm [3].

The "earth" of Paishan Village comes from the local sand and soil. They use traditional three-in-one earth (soil, sand, and ash) as the base and combine organic glues like glutinous rice paste and brown sugar paste with Lingnan characteristics to make a building that is both authentic [4]. This ratio not only enhances the density of the wall but also improves the crack resistance and weather resistance through the bonding effect of organic matter so that the building remains structurally stable after more than 200 years. However, Zhuhai's high temperature and high humidity climate accelerates the weathering of rammed earth walls, and the frequent occurrence of extreme weather also causes the surface erosion of the wall to far exceed the traditional maintenance cycle [5]. The government-led restoration project in 2021 discovered a 30% decrease in the core structural strength of some century-old walls. As time goes by, some old houses are currently facing the risk of collapse due to disrepair. We urgently need to explore and solve problems such as the traditional rammed earth's resistance to rainwater erosion, earthquake resistance, and the durability of organic materials.

This study takes the traditional rammed earth residential buildings in Paishan Village, Bajia Village Committee, Doumen Town, Doumen District, and Zhuhai City as the research objects. We systematically studied the erosion mechanism and protection mechanism of rammed earth buildings through field investigation, experimental analysis, ion chromatography, and microscopic characterization research. First, a general survey of the rammed earth residential buildings in Paishan Village was conducted, and the types, distribution characteristics, and severity of typical damage such as cracks were recorded in detail. For research into microscopic mechanisms, scanning electron microscopy (SEM) was used to describe the microstructure of rammed earth walls. The main goal was to examine how shifts in elemental composition (e.g., key cations/anions) correlate with degradation mechanisms and erosion resistance over time. Ion chromatography was also used to look at the different anions and cations that were present in the top layer of rammed earth. This showed how the change of law changed at the microscopic level during erosion and how different erosion methods affected the wall [6,7]. We paid particular attention to the impact of acidic substances on the wall. The research results will provide a scientific basis and technical support for the protection and restoration of traditional rammed earth buildings of Paishan Village. The research will analyze the status and causes of damage to the rammed earth buildings in Paishan Village, explore the scientific understanding of the erosion resistance of rammed earth building materials in Paishan Village, and provide innovative solutions for the protection and restoration of traditional rammed earth buildings. Through scientific research and practice, the sustainable development of traditional building technology can be promoted while protecting cultural heritage, and it can provide valuable references and references for future research and protection work.

1.2. Literature Review

Residential villages are important material carriers of regional civilization, and the protection of ancient buildings is also a way to inherit local traditional culture [8,9]. Traditional Chinese villages widely use traditional rammed earth houses, one of the most typical residential buildings in China [10]. As a typical representative of Lingnan traditional villages and the largest existing rammed earth building complex in the Pearl River Delta, the rammed earth building complex in Paishan Village provides a cross-border research sample for architecture and material science and has a high research and protection value. In the process of historical evolution and inheritance, it has continuously adapted to the environment and formed a unique rammed earth culture [11]. However, as time passed and modern humanistic activities intensified in the village, the humidity of the coastal terrain caused inevitable natural and cultural damage to many ancient, rammed-earth houses. In recent years, the academic community has conducted in-depth research on the rammed earth buildings in Paishan Village. Some studies have compared the traditional rammed earth buildings in Zhuhai and Macau and analyzed the similarities and differences between the two places in terms of construction technology, material selection, and cultural background [12]. Some scholars have analyzed the causes of erosion of rammed earth materials and are committed to modifying and upgrading them by exploring the microstructure of rammed earth to protect the ancient houses in traditional villages. Some scholars have analyzed the causes of eroding rammed earth materials from a biological perspective [13]. At the same time, some researchers have noticed that the soluble salts in rammed earth go through repeated deliquescence and crystallization when water meets them. This causes salt damage, which eventually causes the wall to become unstable and fall [14]. In addition, the different chemical and mineral compositions of the soil affect its mechanical properties; the different particle compositions of the soil affect its durability. From the plastic forming of the soil to the hardening of the soil, the process from the combination of soil particles to the evaporation of water molecules will inevitably cause the volume of the material to change to varying degrees. This change may cause cracks, which will seriously affect the bearing capacity of the rammed earth wall [14,15]. In the process of the transformation and upgrading of rammed earth materials, domestic and foreign scholars have begun to improve the erosion resistance of rammed earth materials from different perspectives [16,17]. Some scholars choose traditional cement and lime to reinforce buildings in a modern way [18]. Some researchers have used natural fibers, such as straw and date palm polypropylene, to stabilize the mechanical properties of rammed earth. Due to the large carbon emissions associated with cement production, some researchers have chosen to evaluate Xanthan gum and animal glue as bio-binders [19]. All in all, further research is necessary to analyze the causes of erosion in rammed earth materials and to test material modification [20].

2. Materials and Methods

2.1. Study Area: Rammed Earth Dwellings in Paishan Village, Zhuhai City

The research area is located in Zhuhai City, Guangdong Province, China. It is situated on the west bank of the Pearl River Estuary and has a South subtropical Marine monsoon climate, featuring both tropical and subtropical climate characteristics. The seasonal fluctuations of the climate and extreme weather jointly shape its highly variable characteristics. The region has an average annual temperature of 22.6 °C, 2080.6 mm of rainfall, 80% humidity, 1876.3 h of sunshine, and an accumulated temperature of 8166.6 °C. The accumulated temperature above 10 °C exceeds 8000 °C. The rainy season is from April to September, and the rainfall accounts for 84.7% of the annual total. The average annual number of rainstorm days is 11, with 3 heavy rainstorms, and the maximum daily rainfall is 393.7 mm. Each year, it is directly affected by a tropical cyclone. The peak period of typhoons is from July to September, accounting for 75% of the annual total.

Paishan Village is in the southern part of Doumen Town, Zhuhai City, Guangdong Province, China (Figure 1), about 2 km away from the town government, and is in the hilly area of Huangyang Mountain [21,22]. The village was established in the eighth year of Qianlong (1743). Because the village field follows the slope and goes from high to low, it was named "Xie Pai" and later named "Paishan". Its geographical environment is characterized by mountains and seas, and the Five Ridges stand majestically to the north. Due to its location in the tropics and subtropics, it has a relatively hot, humid, and sunshine climate. In this natural environment, the ancient village of Paishan was able to emerge and develop and was also deeply influenced and restricted by it, thus forming a unique architectural culture (Figure 2). The yellow silty soil rammed earth houses in Doumen Town, Zhuhai City, are typical representatives of traditional residential buildings in Lingnan [23]. Among them, the most historically and culturally valuable is Paishan Village, which was built during the Qianlong period of the Qing Dynasty. This village is not only the best-preserved traditional rammed earth building complex in Zhuhai, but also the largest existing rammed earth residential settlement in Guangdong Province. In 2012, Paishan Village was selected as one of the "Third Batch of Ancient Villages in Guangdong Province", further highlighting its important historical and cultural status.

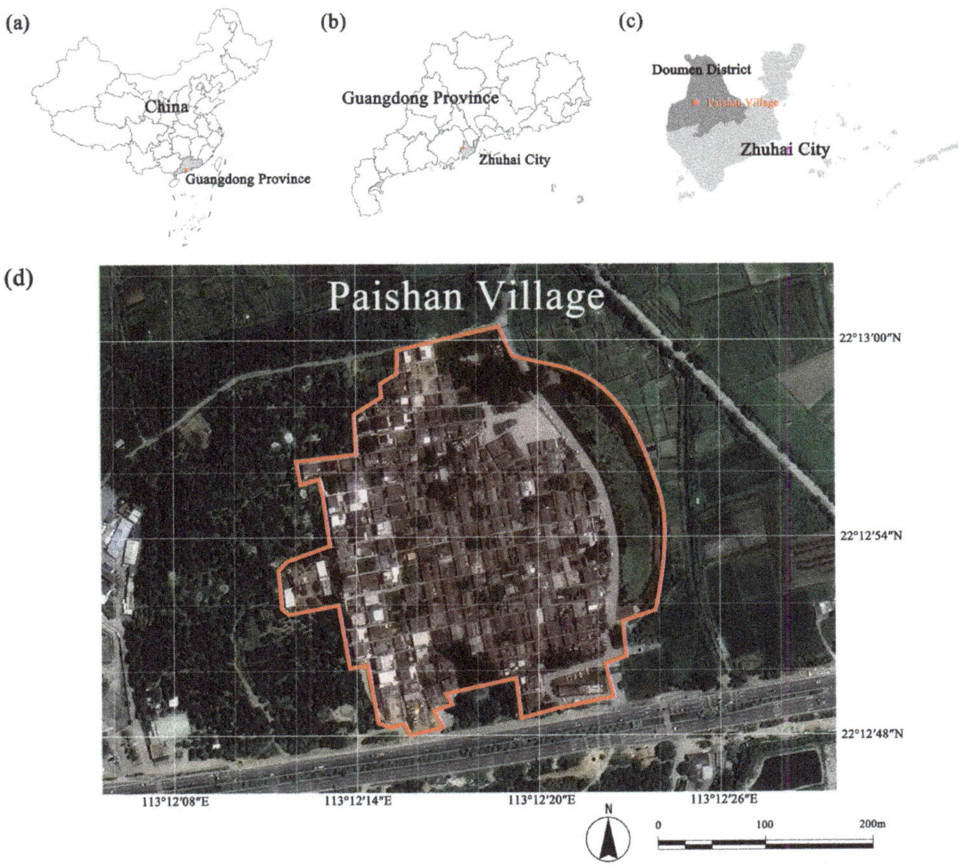

Figure 1. Location analysis of Paishan Village in Zhuhai. (Image source: drawn by the author). (**a**) The location of Guangdong Province in China; (**b**) The location of Zhuhai City in China; (**c**) The location of Paishan Village in Zhuhai City; (**d**) The geographical scope of Paishan Village.

Paishan Village is built on the mountain. The layout of the village is scientific and reasonable. The buildings are arranged in lines horizontally and vertically, showing a high degree of regularity and order. Its drainage system is exquisitely designed, reflecting the wisdom of ancient craftsmen in adapting to local conditions. More than 70 yellow silty soil rammed earth houses are completely preserved in the village. These buildings are mainly made of rammed earth walls and tile and wood structures, with an area of about 70 square meters each. Mud carvings, brick carvings, and wood carvings are dotted in between, with dragons and phoenixes carved, vivid and unique, showing superb traditional craftsmanship and artistic standards. Although most of the ancient residential complexes are vacant, the overall original architectural layout and traditional historical style are still maintained. With their unique building materials and construction technology, these buildings have become important samples for studying the architectural technology, cultural heritage, and geographical adaptability of traditional residential buildings in Lingnan [24–27]. At the same time, they also provide a valuable physical basis for exploring the formation and development of Paishan Village's architectural culture.

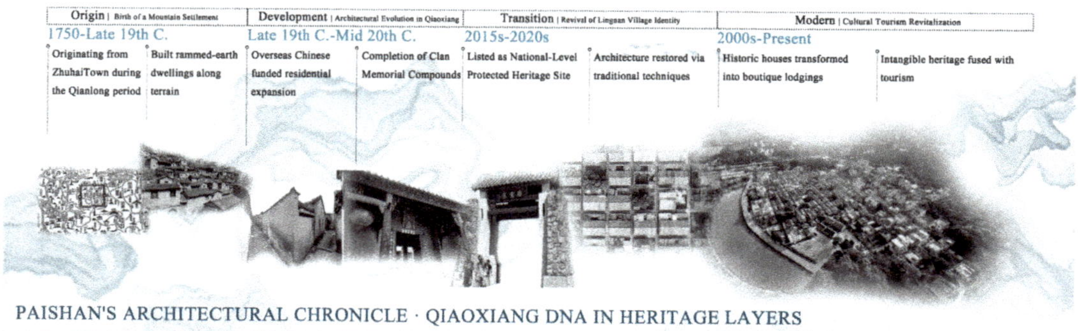

Figure 2. Analysis of architectural characteristics of Paishan Village. (Image source: drawn by the author).

2.2. Sample Collection

The walls of the yellow rammed earth house are built using traditional rammed earth technology. The materials are mainly a mixture of yellow mud (yellow silty soil), lime, glutinous rice, brown sugar, oyster shells, and other natural substances [28,29]. This unique rammed earth formula not only gives the wall excellent adhesion and structural strength but also shows the extraordinary creativity of ancient craftsmen in the science of building materials. The rammed earth wall is evenly made, delicate, strong in compaction, uniform in color, and comfortable. As the core architectural heritage of Paishan Village, the yellow mud (yellow silty soil) rammed earth house is not only a precious material for studying the traditional architectural technology of Lingnan, but also a window to understand the social environment, economic form, and cultural characteristics of Zhuhai since the Qing Dynasty.

To fully understand the weathering characteristics of the rammed earth walls of residential buildings and their environmental impact, the researchers selected one of the representative and well-preserved traditional residential buildings as the research object (Figure 3) and systematically collected rammed earth samples in the four directions of south, east, north, and west. For studies of cooling system size/cost, a sky matrix derived from the STAT file's clear sky radiation should be used (Figure 4). They are also useful for evaluating the impact of a building's Damage to the surface of rammed earth buildings caused by the expansion and contraction due to the heating and cooling from solar radiation. It can be seen from the three pictures that the solar radiation of the building changes significantly in winter and summer. The drastic temperature fluctuations compromise the structural integrity of the building, inducing repeated expansion and contraction in the rammed earth materials. This thermal stress leads to structural cracking or deformation over time.

Axonometric drawing of rammed earth building in Doumen District

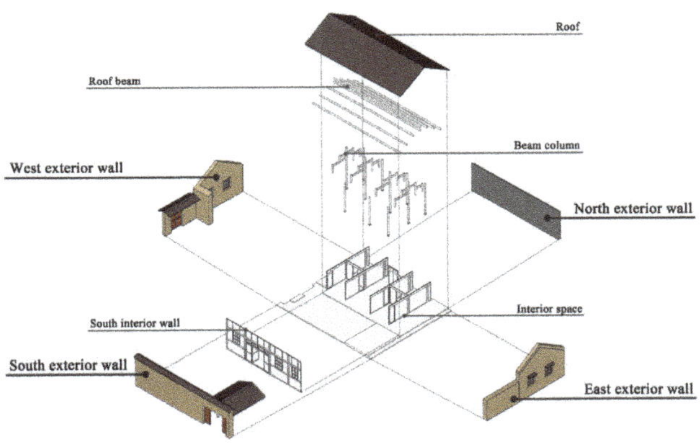

Explosive drawing of rammed earth building in Doumen District

Figure 3. Rammed earth building structure drawing. (Image source: drawn by the author).

The selection of sampling points fully considered the differences in factors such as orientation (Figure 5), light, and wind erosion to ensure that the samples can objectively reflect the environmental characteristics of the building in different directions. The south wall of the residential building is exposed to direct sunlight all year round; the wall is dry, and there are some signs of weathering; the east environment is relatively humid, and some walls have turned black due to humidity; the west wall is dark and humid, and most of the walls have turned gray due to humidity. The north also has sufficient sunlight, less wall shade, and is relatively dry. By selecting these sampling points with obvious environmental differences, the researchers hope to fully explore the potential impact of the natural environment on the aging and damage of rammed earth.

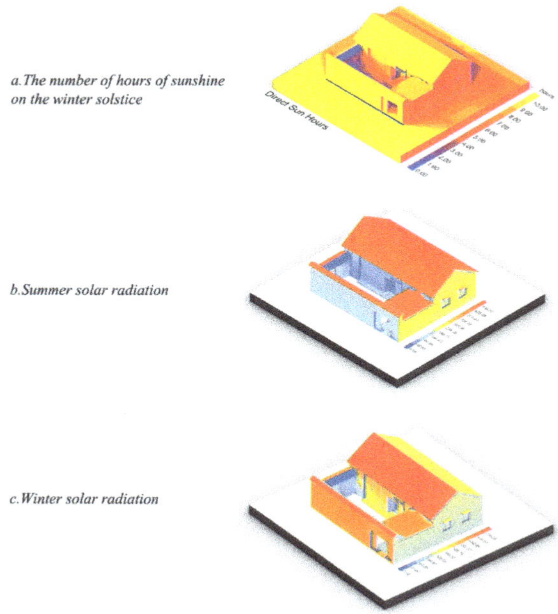

Figure 4. Sky radiation drawing based on Rhino mode. (Image source: drawn by the author).

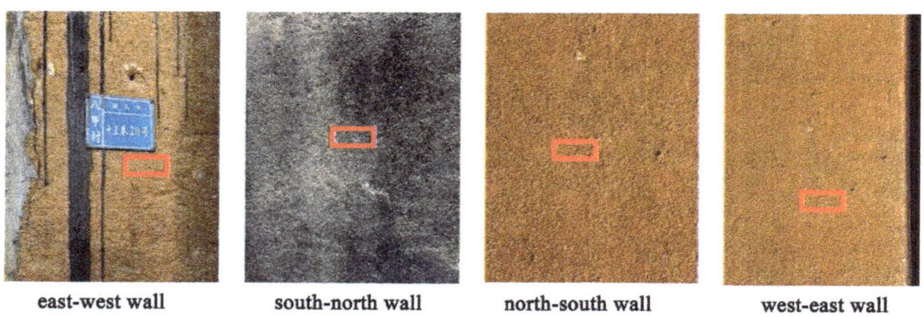

Figure 5. Sampling wall west, north, south, and east conditions and sampling locations. (Image source: photographed by the author).

We set the specific sampling location at a height of about 1.5 m from the ground. This height can effectively avoid the interference of ground moisture and plant roots on the samples, and it can fully preserve the true state of the rammed earth wall after long-term exposure to the natural environment [29]. The sampling time is 3 pm on 12 January 2025. The temperature during this period is relatively stable, which can effectively reduce the impact of environmental temperature and humidity fluctuations on the samples. During the sampling process, the researchers used a strictly disinfected medical scalpel as the main tool and slowly scraped the samples longitudinally along the surface of the rammed earth wall at a 30-degree angle. The researchers controlled the sampling depth within a range of 3–5 mm. The surface samples were chosen because they are the most exposed to environmental factors like weathering, biological erosion, and rain. This is a good starting point for studying how the chemicals in rammed earth walls change over time. This ensures that the samples can fully reflect the weathering characteristics of the rammed earth surface and minimize the damage to the wall structure. At each sampling point, we collected about 10 g of samples and stored them in 20 mL sterilized centrifuge tubes. We meticulously marked the sampling time, direction, height, and other pertinent information on the centrifuge tubes to guarantee the data's integrity and traceability. A high-definition digital camera was used to take panoramic and detailed photos of the sampling points to fully preserve the environmental characteristics of the sampling area and the surface state of the wall. After the sampling was completed, the researchers immediately put the samples into a pre-cooled sterile cold box and evenly placed several ice packs in the box for low-temperature storage to ensure that the samples were always maintained at a constant temperature of 2–8 °C during transportation to maximize the original state and chemical stability of the samples. These samples have been quickly sent to professional laboratories equipped with advanced testing equipment.

2.3. *Analytical Techniques*

This study used three experimental methods—ion chromatography, scanning electron microscopy (SEM), and characterization (XRD)—to look into how ancient, rammed earth homes in Paishan Village keep their structures from eroding. In a macro sense, these methods can look at the structure and make-up of rammed earth walls. In a micro sense, they can show how certain parameters affect the wall's ability to stop erosion.

2.3.1. Mineralogical Characterization (XRD) Analysis

Systematic sampling was conducted from the rammed earth site in the east, west, north, and south directions, and mineral composition analysis was carried out using the SmartLab high-resolution the X-ray diffractometer produced by Rikaku Co., Ltd of Japan. According to the determination procedure of *X-ray Diffraction Analysis Method for Clay Minerals and Common Non-clay Minerals in Sedimentary Rocks (SY/T 5163-2018)*, non-destructive testing was carried out on each group of block samples with a mass of approximately 10 g and a thickness precisely controlled at 10 mm. Through the fine analysis of XRD characteristic diffraction patterns, the phase composition of clay and non-clay minerals such as quartz, plagioclase, potassium feldspar, calcite, and montmorillonite is identified with focus, and their crystallinity parameters and semi-quantitative relative contents are obtained simultaneously. Mineralogical characteristic analysis indicates that the differences in quartz content in samples from different orientations directly affect the pH value distribution of the soil. The expansibility characteristics of layered silicate minerals (such as montmorillonite) show a significant correlation with the degree of pore development, while the dissolution characteristics of carbonate minerals control the ion exchange capacity of the soil site. This kind of refined research based on the theory of mineral crystal fields provides crucial microstructure data support for revealing the weathering mechanism of rammed earth sites and formulating targeted protection plans.

2.3.2. Ion Chromatography Detection

An ion chromatograph from ZEISS Gemini SEM 360 (Carl Zeiss AG of Germany) and a potentiometric titration robot from Wantong 855 were used to find the main anions (SO_4^{2-} and HCO_3^-) in the rammed earth samples used in this study. The experimental method was based on the relevant standards of the *Chinese Geotechnical Testing Methods (2023 edition)*. In the sample pretreatment stage, the ultrasonic dispersion water dissolution method was used to treat the rammed earth samples, which aims to improve the extraction efficiency of soluble ions. At the same time, the samples were stirred with the help of high-speed shaking equipment to promote the full release of ions, thereby improving the measurement accuracy. Additionally, we used the solid-phase extraction column to filter impurities, thereby reducing the interference of non-target ions and enhancing the test stability. After following the steps in the *Chinese national standard DZ/T 0064.49-2021* [30], the Wantong 855 potentiometric titration robot was used with a pH electrode and a 0.1 M hydrochloric acid titration solution to find SO_4^{2-} and HCO_3^- anions in the ion chromatography detection link [31,32]. We weighed 0.3 g–0.4 g of solid sample to the nearest 0.01 mg during specific operations. To improve data reliability, each sample was measured three times, and the average value was calculated. The concentration of each ion (mg/g or %) was then calculated and correlated with the environmental exposure characteristics.

2.3.3. Scanning Electron Microscope (SEM)

In addition, the researchers used scanning electron microscopy (SEM) to conduct a detailed microstructural characterization of the samples in four different directions (east, south, west, and north), and analyzed the differences in particle morphology, particle size distribution (this refers to a qualitative morphological assessment, such as using SEM), and microstructure, and explored the possible causes of these differences [33]. By comparing different directions, the anisotropic characteristics of the samples during synthesis or processing were revealed. In terms of experimental methods, the Zeiss SEM scanning electron microscope was selected as the core equipment [33,34]. During the experiment, the acceleration voltage was set to 10.0 kV, and the magnification was flexibly adjusted according to actual needs, with values of 5.00 KX, 10.00 KX, and 20.00 KX. Imaging was performed in a high vacuum environment. The specific operation steps were as follows: first, mark the east, south, west, and north directions of the sample and firmly fix them on the SEM sample holder; second, place the sample in the SEM sample chamber and perform the vacuum operation; then, reasonably set the acceleration voltage and working distance, and finely adjust the focus and astigmatism; then, scan different areas of the four faces of east, south, west, and north in turn, and simultaneously record the morphology images at different magnifications; finally, use image analysis software to conduct statistical analysis on the morphology and particle size of the particles, to compare the differences between the four directions, respectively.

3. Results

3.1. Ion Detection Results

3.1.1. Anion Content (SO_4^{2-}, HCO_3^-)

In this experiment, the anion content (SO_4^{2-}, HCO_3^-) of the four walls was measured, and the data are as follows (Figure 6):

(1) Sulfate (SO_4^{2-})

The content of the south wall is the highest (1.5266 mg/g), which is related to industrial acid rain deposition or the enrichment of sulfur-oxidizing bacteria metabolites. Its reaction with wall minerals may cause microcracks to expand and reduce structural strength.

The west wall has the lowest content (0.2525 mg/g), which is mainly caused by the dominant wind direction weakening the direct erosion of acid rain.

(2) Bicarbonate (HCO_3^-)

The highest content is in the south wall (9.717%), aligning with observations that rainfall-driven accumulation of alkaline ions (e.g., Ca^{2+}) enhances calcium carbonate precipitation, improving particle cementation [35]. The lowest content is in the west wall (2.192%), which reflects that the limited dissolution of carbonates has led to a decrease in the wall's cementation ability and poor erosion resistance.

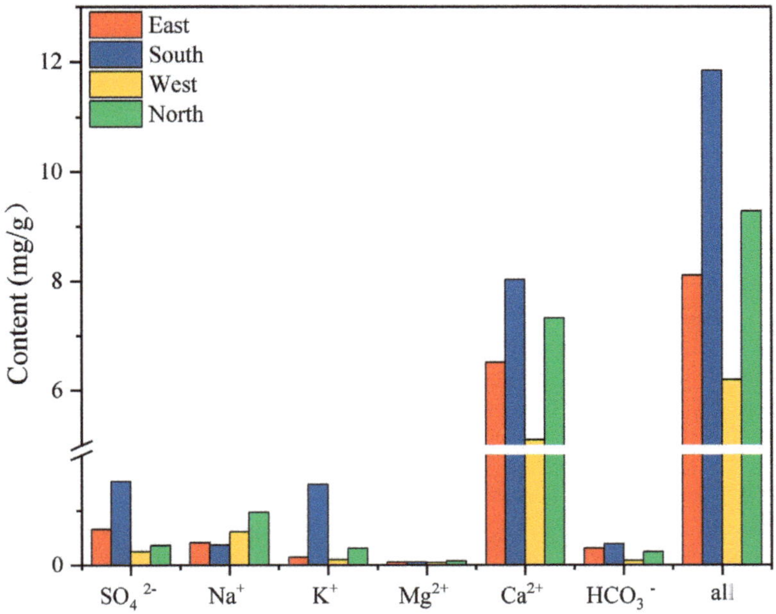

Figure 6. Ion detection results. (Image source: drawn by the author).

3.1.2. Effect of Ion Distribution on Wall Corrosion Resistance

By analyzing the distribution of anions in the surface soil of the four walls of the ancient rammed earth village in Zhuhai, it was found that after long-term exposure to the natural environment, rammed earth buildings will be affected by various factors, such as weathering, humidity changes, acid rain, salt spray, and biological activities, which will cause the wall structure to change to varying degrees. Since the sampling points are all from the outermost layer of the wall, this part is directly exposed to the environment, so its chemical composition gradually differs over time. These differences determine that some walls can remain relatively stable, while others are more susceptible to erosion and damage.

(1) South wall (higher SO_4^{2-}): sulfation may cause microcracks on the wall surface to expand, and peeling may occur in the long term. High HCO_3^- content helps to enhance structural stability in the short term but may cause loose structure due to humidity changes in the long term.
(2) West wall (lower SO_4^{2-}): The sulfate concentration is low, and the wall is relatively less affected by acid rain. However, due to the low HCO_3^- content, the cementing capacity of calcium carbonate is insufficient, and the durability of the wall is poor.
(3) North wall (higher HCO_3^-): humidity has a greater impact, and HCO_3^- enrichment may be due to the dissolution of minerals by microbial activities, resulting in the wall being in a dynamic equilibrium of carbonate chemical reactions for a long time.
(4) East wall (HCO_3^--SO_4^{2-} balance): short-term enhancement of structural density, and long-term changes in environmental factors may lead to calcium carbonate dissolution-precipitation cycles, making the wall structure tends to be loose.

Furthermore, the influencing factors can be summarized into three aspects. (1) Influence of physical factors: The south wall is exposed to the sun for a long time, and the temperature and humidity change greatly, which may accelerate the evaporation of water, cause the soluble salts inside the wall to migrate, and cause the deposition of gypsum ($CaSO_4 \cdot 2H_2O$). This situation can make the wall more compact in the short term, but as the gypsum continues to accumulate and absorb water and expand, the wall is prone to fine cracks, which in turn affects the overall strength. Because the west wall faces a distant river, it is affected by wind and salt spray all year round. The salt in the air adheres to the surface of the wall, making it more susceptible to erosion, causing the surface to powder and reduce durability.

(2) Influence of chemical factors: Acid rain is an important factor affecting the chemical changes of the wall, and the south and east walls are most affected. The sulfate (SO_4^{2-}) content of the south wall is the highest, indicating that the industrial acid rain settles more significantly in this direction. The acidic substances will react with the wall minerals, resulting in material dissolution and loss, and promote the formation of gypsum crystals. While this process may initially increase wall density, progressive gypsum accumulates and its subsequent expansion effect will ultimately compromise structural integrity. The higher content of bicarbonate (HCO_3^-) in the east wall indicates that it may be affected by a strong carbonic acid balance, causing carbonates to dissolve and precipitate continuously. In the short term, this helps the cementation of the wall, but in the long term, it may increase porosity and make the wall more susceptible to weathering and damage.

(3) Influence of biological factors: The north wall is suitable for the growth of mosses and microorganisms due to its shady location and high humidity. These microorganisms produce organic acids (such as oxalic acid) during metabolism, which accelerates the dissolution of wall minerals and increases the concentration of bicarbonate (HCO_3^-). This biological weathering will gradually weaken the structure of the rammed earth wall, making the wall loose, making it more susceptible to erosion and damage than in other directions.

In general, the four walls of rammed earth buildings evolve differently under the influence of a long-term environment. The south wall may become denser first due to acid rain and changes in temperature and humidity, but it is prone to cracking over time; the west wall is severely eroded by salt spray, and the surface powdering problem is prominent; the north wall has high humidity and vigorous microbial activity, which leads to mineral dissolution, loose walls, and reduced stability; the east wall is affected by bicarbonate, which may strengthen the structure in the short term, but in the long run, it may increase the porosity of the wall and make it more susceptible to weathering. These findings not only reveal the different deterioration mechanisms of rammed earth walls but also provide important references for future building protection and restoration.

3.2. Comparison Results of Scanning Electron Microscope (SEM)

3.2.1. Comparative Analysis of SEM Images

Through SEM imaging analysis of samples from the east, south, west, and north, it was found that there were significant differences in their microscopic morphology, particle distribution, and surface characteristics.

The particles of the east samples are regular spherical or polygonal (Figure 7), small in size and evenly distributed, with slight agglomeration, and a relatively smooth overall surface. In contrast, the particles in the south are more irregular in shape, mostly polygonal in structure, with large size variations, uneven distribution, and obvious agglomeration (Figure 8). Its surface is rough and accompanied by cracks, indicating that this orientation may have been affected by greater stress during the formation process.

Figure 7. SEM morphology of particles of the east sample. (Image source: photographed by the author).

Figure 8. SEM morphology of particles of the south sample. (Image source: photographed by the author).

The particle morphology of the west samples is between the east and south, mainly spherical and ellipsoidal, with relatively consistent size, more concentrated distribution, less agglomeration, and a relatively flat surface, showing good particle uniformity (Figure 9). The particles of the north samples show fragmented or flaky characteristics, significant size differences, a wide distribution range, and more obvious agglomeration, a rough surface, cracks, and pores (Figure 10).

Figure 9. SEM morphology of particles of the west sample. (Image source: photographed by the author).

Figure 10. SEM morphology of particles of the north sample. (Image source: photographed by the author).

In general, the particles of the east and west samples have more regular morphology, relatively concentrated particle size distribution, and smooth surface, indicating that the growth environment of these two directions is more uniform and stable during the synthesis or processing. The particles of the south and north have a more complex morphology, wider particle size distribution, and surface characteristics showing strong roughness and cracks, which may be due to crystal orientation, local temperature gradient, or stress concentration. These phenomena indicate that the microstructure of the samples in different directions is affected by the synthesis or processing conditions, thus showing obvious anisotropic characteristics.

3.2.2. Particle Size Distribution Analysis

The particle sizes of the four rammed earth walls were statistically analyzed, and the results are as follows (Table 1):

Table 1. Particle size statistics.

Wall Orientation	Average Particle Size (nm)	Particle Size Standard Deviation (nm)	P10 (nm)	P50 (nm)	P90 (nm)
East	281.149	128.367	153.563	399.528	645.492
South	315.276	142.235	160.345	417.562	683.459
West	276.593	120.498	148.218	388.452	610.315
North	298.462	134.876	162.774	410.389	657.923

Source: statistics by the author.

There are significant differences in the morphology and particle size of the samples at different wall orientations, indicating that the samples may have been affected by different external environments (such as temperature, reaction rate, or stirring rate) during the past ramming synthesis or processing, forming anisotropic structural characteristics. The average particle size of the east (Table 2, Figure 11) and west (Table 3, Figure 12) walls are similar, 281.149 nm and 276.593 nm, respectively, with a narrow distribution range, indicating that the synthetic conditions of rammed earth production were more uniform and stable in these directions in the past. The average particle size of the south (Table 4, Figure 13) and north (Table 5, Figure 14) walls are larger, 315.276 nm and 298.462 nm, respectively, and the standard deviation is larger, indicating a wider particle size distribution and diverse particle morphology. The P50 (median particle size) of the east and west walls is similar, 399.528 nm and 388.452 nm, respectively, while the P50 of the south and north walls is 417.562 nm and 410.389 nm, with more obvious agglomeration and cracks, which may be caused by crystal orientation or processing stress, and also indicates that particles are more likely to form larger sizes in the south and north walls Figure 15).

Table 2. Particle size statistics of the east rammed earth wall samples.

Total Particles		186
Particle size information	Average particle size	185.341 nm
	Particle size standard deviation	74.159 nm
	P10	118.378 nm
	P50	276.216 nm
	P90	434.054 nm

Source: Statistics by the author.

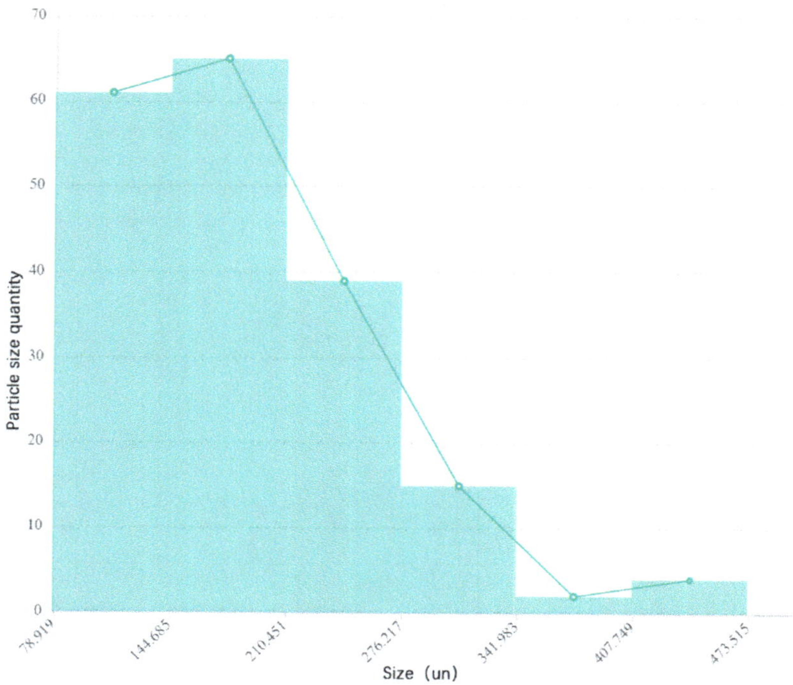

Figure 11. Particle size statistics of the east rammed earth wall samples. (Image source: drawn by the author).

Table 3. Particle size statistics of the west rammed earth wall samples.

Total Particles		**113**
Particle size information	Average particle size	0.431 µm
	Particle size standard deviation	0.253 µm
	P10	0.38 µm
	P50	1.039 µm
	P90	1.698 µm

Source: Statistics by the author.

Table 4. Particle size statistics of the south rammed earth wall samples.

Total Particles		**67**
Particle size information	Average particle size	1.488 µm
	Particle size standard deviation	1.388 µm
	P10	1.215 µm
	P50	4.048 µm
	P90	6.881 µm

Source: Statistics by the author.

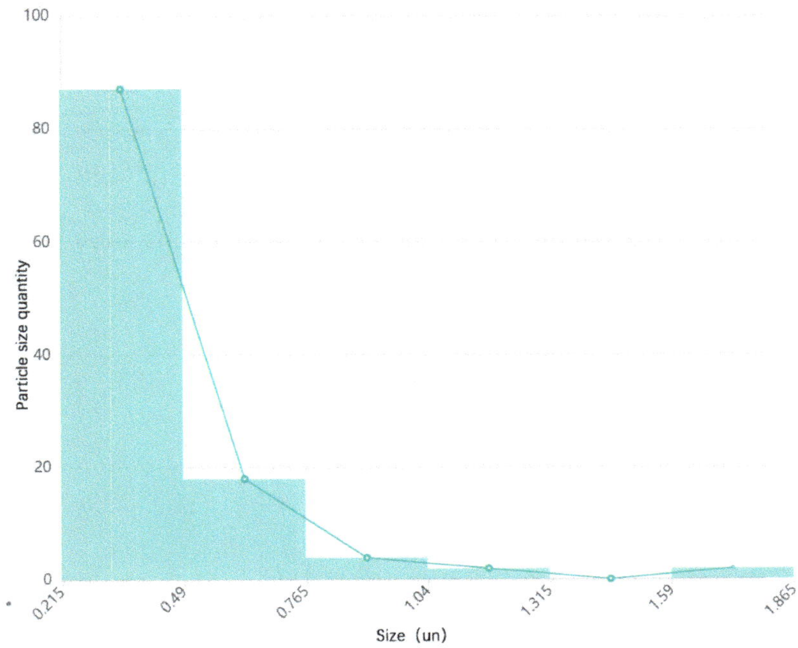

Figure 12. Particle size statistics of the west rammed earth wall samples. (Image source: drawn by the author).

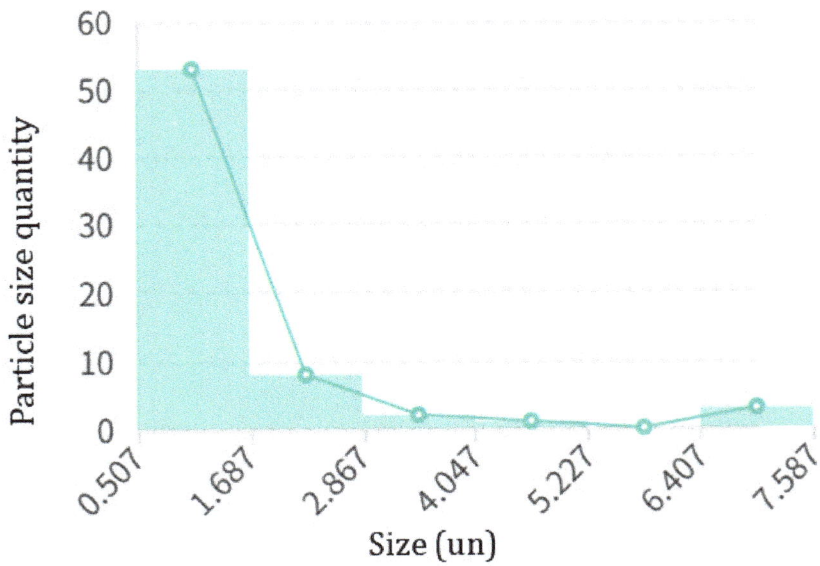

Figure 13. Particle size statistics of the south rammed earth wall samples. (Image source: drawn by the author).

Table 5. Particle size statistics of the north rammed earth wall samples.

Total Particles		**143**
Particle size information	Average particle size	245.012 nm
	Particle size standard deviation	129.568 nm
	P10	144.684 nm
	P50	434.054 nm
	P90	723.424 nm

Source: Statistics by the author.

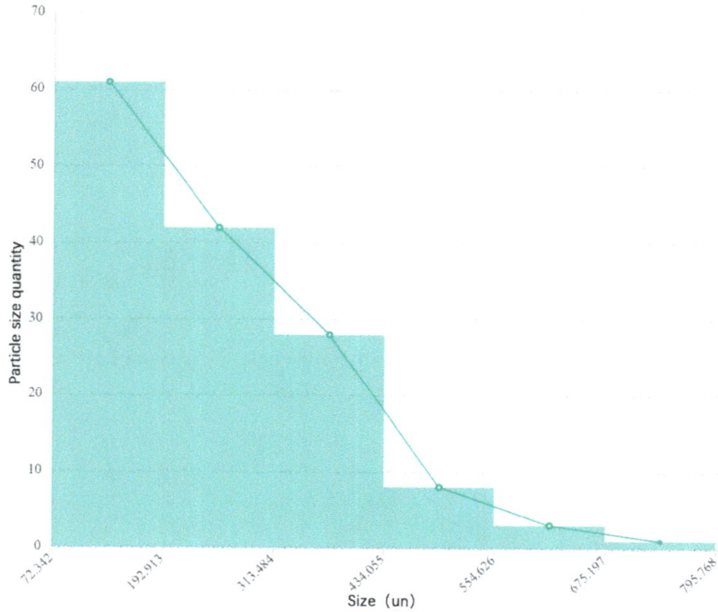

Figure 14. Particle size statistics of the north rammed earth wall samples. (Image source: drawn by the author).

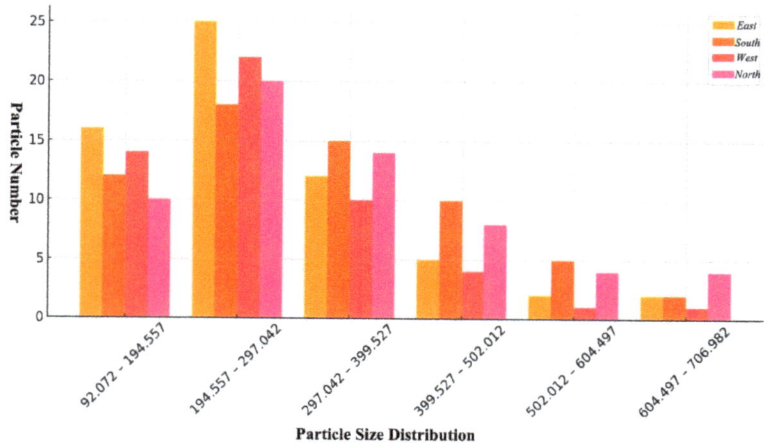

Figure 15. Particle size distribution histogram at different locations. (Image source: drawn by the author).

3.3. Comparative Analysis of Mineralogical Composition (XRD)

By analyzing the XRD spectra of the surface of rammed earth walls in the four directions of east, south, west, and north, the mineral composition and stability of the surface materials were understood. Four main (Figure 16) minerals were marked in the figure: quartz, mullite, lepidocrocite, and nontronite. Quartz is a very stable mineral with high hardness and durability, which can enhance the overall stability of the material. It can be seen from the figure that the diffraction peaks of quartz are relatively obvious in all samples, indicating that it occupies a certain proportion of the material. Mullite is a high-temperature phase mineral with good thermal stability and chemical stability. Lepidocrocite is easily affected by the environment and changes. Nontronite is a layered silicate mineral with certain water absorption and expansion properties. Its presence can cause volume changes in the material when the humidity changes, thereby affecting the stability of the structure. The composition of the north sample is 5.6% lepidocrocite, 45.6% quartz, and 48.8% mullite. In the north sample, the content of mullite is relatively high, indicating that the material in this direction has good thermal stability and chemical stability. Quartz also occupies a considerable proportion, further enhancing the hardness and durability of the material. The composition of the west sample is 8.1% lepidocrocite, 88.7% quartz, and 3.2% mullite. In the west sample, the content of quartz is significantly higher than that of other minerals, with extremely high hardness and durability, and good overall stability. The content of mullite is relatively low. The composition of the east sample is 72.8% quartz and 27.2% nontronite. The composition of the south sample is 70.4% quartz and 29.6% nontronite. Quartz is still the main mineral, ensuring the hardness and durability of the material. Nontronite is a layered silicate mineral, and its water absorption and expansion properties may cause volume changes in the material when the humidity changes.

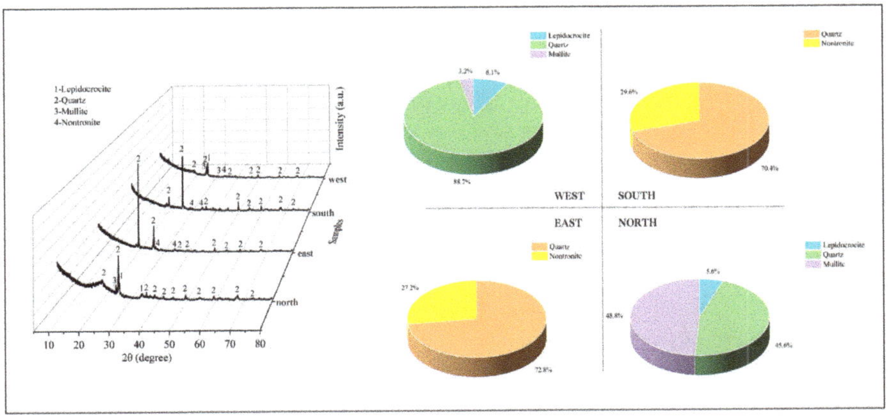

Figure 16. Results of characterization (XRD). (Image source: drawn by the author).

4. Discussion: Long-Term Impact on the Wall and Protection Strategy

By analyzing the anion distribution of the surface soil of the four walls of the ancient, rammed earth village in Zhuhai, the effects of different environmental factors on the long-term stability of rammed earth buildings were revealed. The study found that the long-term exposure of the walls to external environments such as acid rain, humidity changes, salt spray, and biological activities led to significant differences in the chemical composition and stability of walls in different directions. These differences determine which walls are more susceptible to damage and which can maintain good stability under certain conditions. Based on this phenomenon, this study summarizes the main influencing factors and their mechanisms of action on rammed earth walls and puts forward corresponding protection and restoration suggestions.

4.1. South Wall: The Most Obvious Impact of Acid Rain

According to the statistics of the changing trend of acid rain frequency and acid rain days of different intensities in Zhuhai from 2000 to 2015, in the past, the number of acid rain days and strong acid rain days in Zhuhai reached a peak in 2008 and then declined, and the number of acid rain days (pH < 5.6) and strong acid rain days (pH < 4.5) showed similar trends [36]. The number and frequency of acid rain and strong acid rain days showed an increasing trend from 2000 to 2008, a decreasing trend from 2008 to 2015, and the acid rain frequency was less than 50% from 2012 to 2015. After 2008, the acid rain situation in Zhuhai improved significantly, and the annual average precipitation pH value also showed an increasing trend. In such an environment, the current situation of rammed earth walls in Zhuhai's traditional villages has been affected to a certain extent. The south wall is most severely eroded by acid rain and has the highest content of sulfate (SO_4^{2-}). Sulfur dioxide (SO_2) emitted by local industries oxidizes in the air, combines with precipitation and deposits on the surface of the wall, and reacts with mineral components to form gypsum ($CaSO_4 \cdot 2H_2O$). This process may enhance the wall density in the short term, but as the gypsum gradually accumulates, absorbs water, and expands, the wall is prone to cracks and even peeling. Therefore, in terms of the protection of the south wall, measures should be taken to reduce the impact of acid rain, such as using waterproof lime slurry or calcium carbonate coating to reduce the accumulation of sulfates, and regular inspections and maintenance should be carried out to prevent the wall from being damaged by gypsum expansion [37,38].

4.2. West Wall: Salt Spray Has a Greater Impact

Since the west wall faces the nearby river, it has been affected by river wind and salt spray for a long time. This wall seems to be the most affected according to current observations. The salt in the air will adhere to the surface of the wall and continue to deposit with the dry-wet cycle. These salts may accelerate the weathering of the surface of the wall, making it more prone to powdering and peeling, reducing the durability of the wall. For the protection of the west wall, hydrophobic materials can be used to reduce the retention of salt, while optimizing the drainage system, reducing the humidity inside the wall, reducing the accumulation of salt, and thus delaying the weathering process.

4.3. North Wall: The Influence of Humidity and Biological Activity

Because the north wall is in the shade and has high humidity, moss and microorganisms are easy to grow. These microorganisms release organic acids (such as oxalic acid) during metabolism, which accelerates the dissolution of wall minerals and increases the concentration of bicarbonate (HCO_3^-), further affecting the stability of the wall. High humidity may also aggravate the softening and loosening of the wall, reducing its durability. Therefore, during the restoration of the north wall, the growth of microorganisms should be controlled, the moss on the wall surface should be cleaned regularly [39], and a breathable but waterproof protective coating should be used to reduce humidity and reduce the impact of biological weathering [40].

4.4. East Wall: Carbonate Balance Has a Greater Impact

The east wall has a higher content of bicarbonate (HCO_3^-), indicating that it is greatly affected by the dynamic balance of carbonates. This situation may be related to the temperature difference between day and night, the penetration of condensed water, and the carbonate composition of the wall itself. In the short term, the enrichment of HCO_3^- helps the precipitation of calcium carbonate, enhances the density of the wall, and improves its resistance to weathering. However, in the long term, the dissolution-precipitation cycle of calcium carbonate may increase the porosity of the wall, making it more susceptible to weathering. Therefore, in the protection process of the east wall, coating with stable carbonates can be used to reduce the changes in calcium carbonate and maintain the long-term stability of the wall.

5. Conclusions

The four walls of rammed earth buildings show different deterioration trends under the influence of different environmental factors. The south wall is most seriously affected by acid rain, and sulfate accumulation may cause cracks and peeling; the west wall is affected by marine salt spray, and the surface weathering is aggravated; the north wall is affected by high humidity and vigorous biological activity, and the wall minerals are easily dissolved, resulting in a loose structure; the east wall is affected by the dynamic changes of carbonates, which strengthens the structure in the short term, but may increase porosity in the long term, making the wall more susceptible to weathering. These research results have important practical significance for the protection and restoration of rammed earth buildings. In the actual restoration process, the characteristics of different walls should be observed with emphasis. For example, considering the corner areas where walls meet, due to the accumulation of rainwater and poor ventilation, a high-humidity micro-environment is prone to form. It is necessary to focus on designing the drainage slope or adding drainage channels to prevent water accumulation and seepage [41]. Overall, targeted measures should be taken according to the characteristics of different walls, such as reducing acid rain erosion, reducing humidity effects, controlling microbial growth, and stabilizing carbonating structures, to improve the durability of rammed earth buildings and extend their service life.

Through SEM analysis, it was found that the rammed earth walls in different orientations (east, south, west, and north) of the samples had significant differences in morphology and particle size distribution, indicating that the samples had anisotropic characteristics during growth and processing. The particles on the east and west sides have more regular morphology and concentrated particle size distribution, which may be since the synthesis conditions (such as temperature, solvent concentration, etc.) are more uniform and stable in these areas. The particles on the south and north sides have irregular morphology, wider particle size distribution, and significant cracks and pores, which may be due to the local temperature gradient or stress concentration during the growth process. The difference in surface agglomeration and roughness may be related to the electrostatic effect or van der Waals force on the surface of the material, which can be improved in the future by regulating the reaction environment (such as solvent polarity, stirring speed, or catalyst concentration). The results show that by optimizing the reaction conditions and cooling process, it is expected to obtain more uniform particle morphology and particle size distribution in different orientations, thereby improving the overall performance and stability of the material.

Prospects for subsequent research: given the exploratory nature of the initial research phase, SEM (Scanning Electron Microscopy) imaging is currently being conducted on samples collected from the east, south, west, and north faces of the wall. In the subsequent analysis, the number of sampling points on each wall will be increased to five, symmetrically distributed across the upper, middle, and lower sections, as well as both sides. This adjustment will expand the total number of samples to 20, enhancing both the representativeness and accuracy of the analysis.

Author Contributions: Conceptualization, Y.W. and S.Y.; methodology, Y.W.; software, Y.W.; validation, Y.W., Y.H. and S.Y.; formal analysis, J.S.; investigation, J.S.; resources, J.Z.; data curation, J.Z.; writing—original draft preparation, Y.W. and S.Y.; writing—review and editing, Y.W. and S.Y.; visualization, Y.H.; supervision, J.S.; project administration, J.S.; funding acquisition, S.Y. All authors have read and agreed to the published version of the manuscript.

Funding: This research was funded by Zhejiang Provincial Philosophy and Social Sciences Planning Project (grant number: 24NDQN150YBM). The funders had no role in the study conceptualization, data curation, formal analysis, methodology, software, decision to publish, or preparation of the manuscript. There was no additional external funding received for this study.

Institutional Review Board Statement: Not applicable.

Informed Consent Statement: Not applicable.

Data Availability Statement: Readers and all interested researchers may contact Yang Shuai (Email address: samyang@zju.edu.cn) for details.

Acknowledgments: We would like to express our sincere gratitude to the students and the staff who assisted during the field survey. The authors extend their gratitude to Ma Xin from Shiyanjia Lab (www.shiyanjia.com, accessed on 24 February 2025) for providing invaluable assistance with the SEM analysis.

Conflicts of Interest: The authors declare no conflicts of interest.

References

1. Zhou, Z.W. Analysis on the application of contemporary rammed earth materials in the design of the buildings. *Urbanism Arch.* **2023**, *20*, 143–146. [CrossRef]
2. Liu, X.; Bai, W.F. Strategic research on the promotion and development of modern rammed earth building in China. *Urbanism Arch.* **2018**, *2*, 22–25. [CrossRef]
3. Bai, W.F.; Chi, X.N.; Zhou, L. Research on preservation and activation of Kaiping Diaolou with rammed earth wall. *New Archit.* **2021**, *4*, 144–147.
4. Liu, H.Q.; Li, Z.W. Research on the value of Traditional Chinese traditional building materials—Take rammed earth architecture as an example. *Resid. Real Estate* **2020**, *32*, 165–166.
5. Zhang, H.Y.; Li, X.M.; Ren, K.B.; Yi, S.; Wang, Y.L. Stability analysis of rammed soil and masonry structure deterioration. *J. Zhongyuan Univ. Technol.* **2021**, *32*, 48–54. [CrossRef]
6. Cui, K.; Chen, W.W.; Wang, X.D.; Han, W.F.; Zhu, Y.P. Tentative microscopic on degradation of rammed earth relics under coupled function of salinized and wetting-drying in arid region. *J. Cent. South. Univ.* **2013**, *44*, 2927–2933.
7. Peng, X.Q.; Shi, W.J.; Hua, C.C. Study on the mechanism and loss of erosion by wind driven rain for rammed earth buildings. *Adv. Eng. Sci.* **2015**, *47*, 105–111. [CrossRef]
8. Gu, T.Y.; Yang, X.Q.; Shi, D.H.; Yang, Y.G.; Ma, Y.P.; Zheng, X.X.; Wu, J.X.; Zhang, Y.X.; Song, G.D. Scientific analysis and research on rammed earth materials of the Shang Dynasty in Zhengzhou area:taking Zhengzhou Shang City and Zhengzhou Xiaoshuangqiao sites as examples. *Sci. Conserv. Archaeol.* **2024**, *36*, 89–100. [CrossRef]
9. Zhou, Q.; Chen, W. Conservation intervention experience of the rammed earth historic monuments in spain. *Archit. Herit.* **2024**, *1*, 54–59. [CrossRef]
10. Wang, M.L.; Li, S.H. Research on the Applicability of Passive Technology in Traditional Rammed Earth Buildings in Chaoshan. *China Archit. J.* **2022**, *S2*, 42–48.
11. Zhang, Q.; Guo, Q.; Zhang, H.; Wang, Y.; Li, P.; Zhang, B.; Tang, C.a. Surface weathering process of earthen heritage under dry-wet cycles: A case study of Suoyang City, China. *J. Rock. Mech. Geotech. Eng.* **2024**. [CrossRef]
12. Losini, A.E.; Woloszyn, M.; Chitimbo, T.; Pelé-Peltier, A.; Ouertani, S.; Rémond, R.; Doya, M.; Gaillard, D.; Force, M.S.; Outin, J. Extended hygrothermal characterization of unstabilized rammed earth for modern construction. *Constr. Build. Mater.* **2023**, *409*, 133904. [CrossRef]
13. Du, X.Y.; Lin, H. Research on the characteristics of rammed earth buildings in the villages of Guangxi: A Case of Daxin County in Chongzuo. *Urban. Arch.* **2022**, *19*, 75–77. [CrossRef]
14. Manzano-Fernández, S.; Mileto, C.; Vegas López-Manzanares, F.; Cristini, V. Conservation and In Situ Enhancement of Earthen Architecture in Archaeological Sites: Social and Anthropic Risks in the Case Studies of the Iberian Peninsula. *Heritage* **2024**, *7*, 2239–2264. [CrossRef]

15. Petcu, C.; Dobrescu, C.F.; Dragomir, C.S.; Ciobanu, A.A.; Lăzărescu, A.V.; Hegyi, A. Thermophysical Characteristics of Clay for Efficient Rammed Earth Wall Construction. *Materials* **2023**, *16*, 6015. [CrossRef]
16. Losini, A.E.; Grillet, A.-C.; Vo, L.; Dotelli, G.; Woloszyn, M. Biopolymers impact on hygrothermal properties of rammed earth: From material to building scale. *Build. Environ.* **2023**, *233*, 110087. [CrossRef]
17. Luo, Y.; Zhou, P.; Ni, P.; Peng, X.; Ye, J. Degradation of rammed earth under soluble salts attack and drying-wetting cycles: The case of Fujian Tulou, China. *Appl. Clay Sci.* **2021**, *212*, 106202. [CrossRef]
18. Ou, C.Y. Research on the application of new rammed earth construction in the restoration and reconstruction of ancient villages. *Residence* **2023**, *16*, 159–162.
19. Jiang, B.; Lu, R.; Jiang, M.; Wang, L.; Chun, L.; Wan, L. Experimental study on thermal and humidity properties of modified rammed earth buildings in winter. *Build. Environ.* **2024**, *258*, 111583. [CrossRef]
20. Sánchez-Calvillo, A.; Rincón, L.; Hamard, E.; Faria, P. Bibliometric Analysis on Earthen Building: Approaches from the Scientific Literature and Future Trends. *Buildings* **2024**, *14*, 3870. [CrossRef]
21. Lu, G.A.N.; Heng, C.; Yi, Y. The Livability Perception of Residents in Rural Areas and Its Influencing Factors: A Case Study of Doumen Town of Zhuhai. *J. Landsc. Res.* **2019**, *11*. [CrossRef]
22. Zheng, B.; Liu, G.; Wang, H.; Cheng, Y.; Lu, Z.; Liu, H.; Zhu, X.; Wang, M.; Yi, L. Study on the Delimitation of the Urban Development Boundary in a Special Economic Zone: A Case Study of the Central Urban Area of Doumen in Zhuhai, China. *Sustainability* **2018**, *10*, 756. [CrossRef]
23. Song, J.; Chen, Y.; Zheng, L.; Tang, Q. The evolution of traditional ancient towns in South China: Tangjiawan Ancient Town in Zhuhai city. *J. Asian Archit. Build. Eng.* **2024**, 1–19. [CrossRef]
24. Li, S.; Wang, M.; Shen, P.; Cui, X.; Bu, L.; Wei, R.; Zhang, L.; Wu, C. Energy Saving and Thermal Comfort Performance of Passive Retrofitting Measures for Traditional Rammed Earth House in Lingnan, China. *Buildings* **2022**, *12*, 1716. [CrossRef]
25. Wang, W.; Luo, B. Lingnan Architecture Design Based on Ocean Climate Adaptability. *J. Coast. Res.* **2020**, *107*, 226–229. [CrossRef]
26. Yan, L.; Chen, Y. Study of Roof Tiles in Lingnan Traditional Buildings and Roof Drainage Technologies. IOP Conference Series: Earth and Environmental Science. *IOP Publ.* **2021**, *768*, 012151. [CrossRef]
27. Ji, H.; Wu, S.; Ye, B.; Wang, S.; Chen, Y.; Deng, J.-Y. Exploring the Implementation Path of Passive Heat-Protection Design Heritage in Lingnan Buildings. *Buildings* **2023**, *13*, 2954. [CrossRef]
28. Tsai, S.-C.; Zhang, X.-F.; Chang, Y.-T. Toward Nearly Zero-Waste Architecture: Innovation, Application, and Practice of Construction Methods Using Natural Materials. *Buildings* **2024**, *14*, 1584. [CrossRef]
29. Guo, Q.; Pei, Q.; Wang, Y.; Zhan, W.; Yang, S.; Sun, M.; Zhang, J.; Shi, Z. New Advances in the Protection and Research of Earthen Sites in China. *Res. Prot. Grottoes Earthen Sites* **2023**, *3*, 18–31. [CrossRef]
30. Zhang, B.; Wang, G.; Li, W.; Jiao, X. Characteristics and Sources of CBM Well-Produced Water in the Shouyang Block, China. *Appl. Sci.* **2024**, *14*, 4218. [CrossRef]
31. Youssef, M. Passivation Mechanism of Pure Magnesium in Oleum and Sulfuric Acid Solutions/Author Mohamed Youssef. 2024. Available online: https://epub.jku.at/obvulihs/content/titleinfo/10095239 (accessed on 22 March 2025).
32. Alam, B.Y.C.S.; Itoi, R.; Taguchi, S.; Yamashiro, R. Spatial Variation in Groundwater Types in the Mt. Karang (West Java, Indonesia) Volcanic Aquifer System Based on Hydro-Chemical and Stable Isotope (δD and δ^{18}O) Analysis. *Mod. Appl. Sci.* **2014**, *8*, 87. [CrossRef]
33. Rades, S.; Hodoroaba, V.D.; Salge, T.; Wirth, T.; Lobera, M.P.; Labrador, R.H.; Natte, K.; Behnke, T.; Gross, T.; Unger, W.E. High-resolution imaging with SEM/T-SEM, EDX and SAM as a combined methodical approach for morphological and elemental analyses of single engineered nanoparticles. *RSC Adv.* **2014**, *4*, 49577–49587. [CrossRef]
34. Schulz, B.; Sandmann, D.; Gilbricht, S.S. SEM-Based Automated Mineralogy and Its Application in Geo- and Material Sciences. *Minerals* **2020**, *10*, 1004. [CrossRef]
35. Wang, N.; Wang, Q.; Ye, W.M. Research Progress of Microbial Reinforcement Technology. *J. Hydrogeol. Eng. Geol.* **2024**, *5*, 231–244. [CrossRef]
36. Chen, H.Y.; Zhao, X.F.; He, Z.D.; Fan, S.J. Variation characteristics and concept model of acid rain in Zhuhai city during 2000–2016. *Acta Sci. Circumstantiae* **2020**, *40*, 1998–2006. Available online: https://html.rhhz.net/hjkxxb/html/20191105011.htm (accessed on 22 March 2025).
37. Zhang, Y.; Ye, W.M.; Chen, B.; Chen, Y.G.; Ye, B. Desiccation of NaCl-contaminated soil of earthen heritages in the Site of Yar City, northwest China. *Appl. Clay Sci.* **2016**, *124*, 1–10. [CrossRef]
38. Zhu, C.H.; Li, N.; Ma, S. Research on Anti-seepage Measures for Rammed Earth Bases of Ancient Buildings Based on Model Tests. *J. Rock Soil Mech.* **2021**, *42*, 3281–3290. [CrossRef]
39. Teng, F.; Sie, Y.C. Simulating field grouting to improve the shear strength of consolidated clay by microbially induced carbonate precipitation. *Soils Found.* **2025**, *65*, 101. [CrossRef]

40. Ye, C.R.; Tao, Y.; Zhang, Y.R.; Cao, J.; Ke, T.; Wei, S.J.; De Philippis, R.; Chen, L.Z. Monosaccharide composition of primary cell wall polysaccharides as a development level indicator of biologic soil crusts. *Catena* **2020**, *195*, 104782. [CrossRef]
41. Katharina, F.; Marija, M.; Gabriela, K. Preserving earthen architecture: Lessons learned from two case studies. *J. Archaeol. Sci.* **2024**, *55*, 104. [CrossRef]

coatings

MDPI

Article

Experimental Study on the Anti-Erosion of the Exterior Walls of Ancient Rammed-Earth Houses in Yangjiatang Village, Lishui

Yujun Zheng [1], Junxin Song [2,*], Xiaohan Zhang [1], Yake Hu [1], Ruihang Chen [1] and Shuai Yang [1,*]

[1] School of Art and Archaeology, Hangzhou City University, No. 51 Huzhou Street, Gongshu District, Hangzhou 310015, China; 32410041@stu.hzcu.edu.cn (Y.Z.); 32410040@stu.hzcu.edu.cn (X.Z.); 32410045@stu.hzcu.edu.cn (Y.H.); 32410044@stu.hzcu.edu.cn (R.C.)

[2] Institute for Research on Portuguese Speaking Countries, City University of Macau, Avenida Padre Tomás Pereira, Taipa, Macau 999078, China

* Correspondence: a23092100150@cityu.edu.mo (J.S.); samyang@zju.edu.cn (S.Y.)

Abstract: Yangjiatang Village traces its origins to the late Ming and early Qing dynasties. It has evolved over more than 400 years of history. There are 78 rammed-earth buildings left, making it one of the most complete and largest rammed-earth building complexes in East China. This study investigated the traditional rammed-earth houses in Yangjiatang Village, Songyang County, Zhejiang Province. By combining field investigation, microscopic characterization, and experimental simulation, we systematically revealed the erosion resistance of rammed earth in a subtropical humid climate was systematically revealed. Using a combination of advanced techniques including drone aerial photography, X-ray diffraction (XRD), microbial community analysis, scanning electron microscopy (SEM), and soil leaching simulations, we systematically revealed the anti-erosion mechanisms of rammed-earth surfaces in Yangjiatang Village. The study found that (1) rammed-earth walls are primarily composed of Quartz, Mullite, lepidocrocite, and Nontronite, with quartz and lepidocrocite being the dominant minerals across all orientations. (2) Regulating the community structure of specific functional microorganisms enhanced the erosion resistance of rammed-earth buildings. (3) The surface degradation of rammed-earth walls is mainly caused by four factors: structural cracks, surface erosion, biological erosion and roof damage. These factors work together to cause surface cracking and peeling (depth up to 3–5 cm). (4) This study indicates that the microbial communities in rammed-earth building walls show significant differences in various orientations. Microorganisms play a dual role in the preservation and deterioration of rammed-earth buildings: they can slow down weathering by forming protective biofilms or accelerating erosion through acid production.

Keywords: rammed earth; traditional village dwellings; X-ray diffraction (XRD); microbial community diversity composition detection; scanning electron microscopy (SEM); Lishui City

1. Introduction

1.1. Research Background

As one of the oldest forms of architecture in China, rammed-earth architecture can be traced back to the Neolithic Age and is a critical, a vital material carrier of Chinese civilization [1–3]. Rammed-earth architecture ranks among the oldest building forms in China, with origins traceable to the Neolithic Age. It represents a crucial material carrier of Chinese civilization [4–6]. It is a low-carbon and energy-saving building material. Before the emergence of modern building materials such as cement, polymers, and reinforced concrete, rammed earth was a traditional building material widely used in many parts of the world [7,8]. This form of architecture, which uses raw earth as the primary raw material, is made by layering and compacting materials such as clay, sand, and lime and has significant ecological advantages and regional adaptability. The production process of rammed-earth architecture embodies the wisdom of the ancient working people. Its typical process includes soil selection → ingredients (adding organic adhesives such as glutinous rice paste and brown sugar) → formwork installation → layered compaction (each layer is about 15 cm) → maintenance → demolding and other links (Figure 1). This building type offers advantages such as low carbon emissions, environmental protection, low cost, and thermal insulation, while also embodying rich regional culture and historical memory.

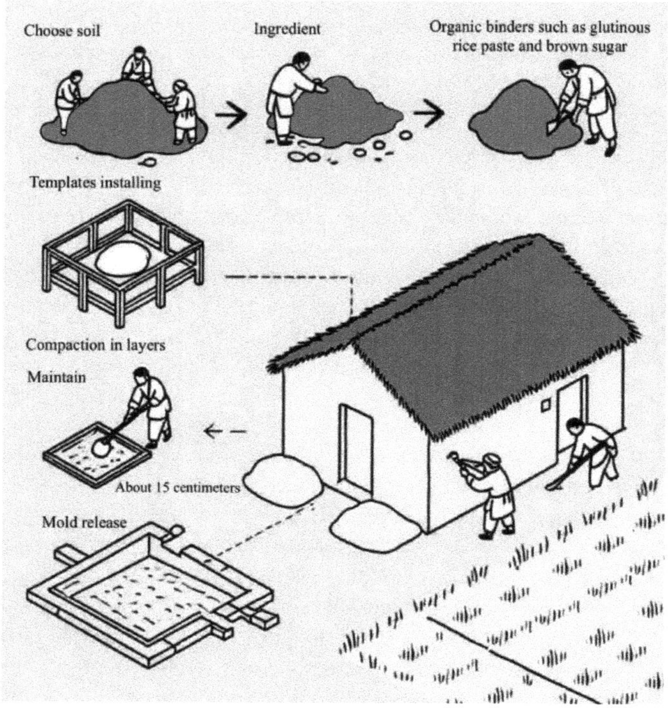

Figure 1. Rammed-earth building construction process (image source: drawn by the authors).

This study selected Yangjiatang Village in Songyang County, Lishui City, Zhejiang Province, as the research object. The village dates back to the late Ming and early Qing dynasties, with a history of more than 400 years. There are 78 rammed-earth buildings, which is one of the most complete and largest rammed-earth building complexes in East China [9,10]. Yangjiatang Village was designated as a Chinese Traditional Village in 2014, as part of the third batch. Its rammed-earth buildings have the following significant features: (1) the wall thickness is 50–80 cm, forming a unique microclimate regulation system; (2) local yellow soil is used, and oyster shell powder and other materials are added to improve durability; and (3) the building layout is based on the terrain, reflecting the traditional construction wisdom of "harmony between man and nature". Driven by the subtropical monsoon climate, this region receives an average annual precipitation of 1837 mm and experiences a high acid rain frequency of 42.3%. These climatic factors collectively contribute to the severe erosion observed in the rammed-earth buildings. According to our survey, 37% of the village's rammed-earth buildings have suffered structural damage in the past ten years, 12 of which are rated as dangerous. This situation poses a serious safety threat to older people and our adults living alone who still live in these buildings (accounting for 68% of the current population) and also puts precious architectural the cultural heritage at risk of irreversible damage.

1.2. Literature Review

The conservation of rammed-earth heritage is now a pressing priority, as accelerated deterioration driven by climate change and environmental factors threatens these structures globally. In Yangjiatang Village, Lishui, ancient rammed-earth houses are particularly vulnerable to rainfall, wind, and biological colonization, reflecting challenges documented across global heritage sites [11].

In recent years, scholars have conducted in-depth discussions on the deterioration mechanism of rammed-earth buildings. In humid subtropical regions, research has established that the primary mechanisms leading to soil disintegration are dry-wet cycles and capillary water migration. These processes are significantly exacerbated by the colonization of biological communities such as mosses and lichens, which accelerate surface softening and structural degradation. On this basis, some studies attempt to reveal the coupling effect of climatic conditions and biological interactions through laboratory simulations and on-site observations, thereby providing a scientific basis for protective measures [12,13].

At the forefront of protective technology research are methods for material modification and surface treatment aimed at enhancing durability. Traditional methods mostly employ lime or silicate stabilizers to enhance the rain erosion resistance of walls. The introduction of novel nanomaterials into rammed-earth conservation has drawn attention owing to their excellent hydrophobicity and minimal alteration to appearance. However, long-term compatibility, reversibility, and potential impact on the original materials remain the focus of academic debate [14].

International research provides a valuable comparative perspective. For instance, in Europe, sacrificial mud coatings are often used as replaceable "sacrificial layers" to withstand wind and rain and maintain the authenticity of the material. In Australia, strategies to mitigate water erosion often prioritize passive design solutions, including extended roof overhangs and enhanced drainage systems. In Latin America, community participation and simple protection measures have proven to be equally effective in arid or semi-arid environments. These experiences indicate that integrating local traditional knowledge with modern scientific methods may be the key approach to achieving sustainable protection [15,16].

Meanwhile, the development of monitoring and diagnostic technologies has also significantly enhanced the level of research. Remote sensing, 3D laser scanning, and digital photogrammetry provide high-precision data support for the deterioration monitoring of rammed-earth buildings, making it possible to track and quantitatively analyze the erosion process dynamically. Such technologies provide a new methodological basis for the systematic protection of large-scale sites [17,18].

In conclusion, significant progress has been made in existing research regarding the analysis of deterioration mechanisms, the development of material modification strategies, and the advancement of technical conservation methods. However, in-depth exploration is still needed in the aspects of long-term effect evaluation, regional adaptability, and cultural inheritance. The sustainable protection of Yangjiatang Village hinges on three critical efforts: advancing eco-material applications, achieving a harmonious integration of traditional and modern techniques, and implementing monitoring-based data management. These efforts are fundamental to guaranteeing both the physical security of structures and the preservation of cultural identity.

1.3. Problem Statement and Objectives

Therefore, the restoration and protection of rammed-earth buildings have multiple values: from a cultural perspective, it is a living inheritance of traditional construction skills; from a technical standpoint, the ecological wisdom it contains is of enlightenment significance to the development of modern green buildings; and from a social perspective, the protection work improves the living conditions of villagers and promotes the growth of rural tourism. A professor from the School of Architecture at Tsinghua University emphasized, "The rammed-earth buildings of Yangjiatang Village are textbooks of living history. Their preservation is tantamount to safeguarding the cultural DNA of the Chinese nation".

The purpose of the study is to deeply analyze the erosion resistance of rammed-earth buildings in East China. Using a multi-technique approach, including advanced drone aerial photography, XRD, microbial community analysis, SEM, and soil leaching simulations, this study systematically elucidates how microbial communities and mineral phase changes influence the durability of rammed earth. The methods selected for this study—XRD, SEM, microbial diversity analysis, and soil leaching simulations—were chosen to comprehensively characterize the material properties and biological factors influencing erosion resistance. These methods are well-established for mineral identification, microstructural visualization, and community profiling. While XRF could provide elemental composition, it was not prioritized as XRD already identified the key mineral phases. Future research could incorporate mechanical strength tests, porosity analysis, and specific surface area measurements to correlate microstructure with macroscopic durability further. At the same time, based on the innovative idea of integrating traditional craftsmanship with modern technology, a set of low-intervention restoration technology systems suitable for the characteristics of mountain villages is planned. This study is not only committed to solving the key scientific problems currently faced in the protection of rammed-earth buildings but also focuses on exploring the potential protection value of natural materials, providing innovative solutions that are both scientific and practical for the sustainable protection of traditional rammed-earth buildings, and establishing a replicable and popularizable technical paradigm for the protection and restoration of similar cultural heritage.

2. Materials and Methods

2.1. Study Area: Yangjiatang Village in Lishui City

Yangjiatang Village is in Sandu Township, Songyang County, Lishui City, Zhejiang Province (Figure 2). It is located between 119°10′ and 119°42′ east longitude and 28°14′ and 28°36′ north latitude. It is surrounded by mountains at an altitude of 400–600 m. Songyang County is the area with the most significant number of ancient villages and the best preservation in East China [19,20]. Songyang County boasts over 100 traditional villages with well-preserved structures and diverse styles, 78 of which are designated as national-level conventional villages. These settlements—including alpine, stepped, waterside, terraced, flat-valley, and Hakka types—are scattered across its mountains. Among them, Yangjiatang Village, established around 1630 AD during the late Ming and early Qing dynasties, has a history of nearly 400 years (Figure 3). It is the most representative traditional rammed-earth building settlement in southwestern Zhejiang. In 1655, the 12th year of the Shunzhi reign of the Qing Dynasty, the Song family settled in this area. They developed it rapidly into a high-mountain village where their clan became dominant. Despite this demographic shift, the original name, Yangjiatang Village, has been retained to the present day. The town has more than 300 residents, with the Song family being the most prominent surname in the area. They are descendants of Song Lian, the first civil official of the Ming Dynasty (1368–1644). The village is terraced along the mountain slope, and its overall layout remains well-preserved. are 78 rammed-earth buildings from the Ming and Qing

Dynasties (1368–1840) to the Republic of China (1912–1949), with a building density of 85%. Zhejiang Province has identified 11 of these buildings as historical and cultural. The most representative buildings, such as "Song's Ancestral Hall" and "Yang's Hall" (Figure 4), have been listed as cultural relics protection units in Zhejiang Province.

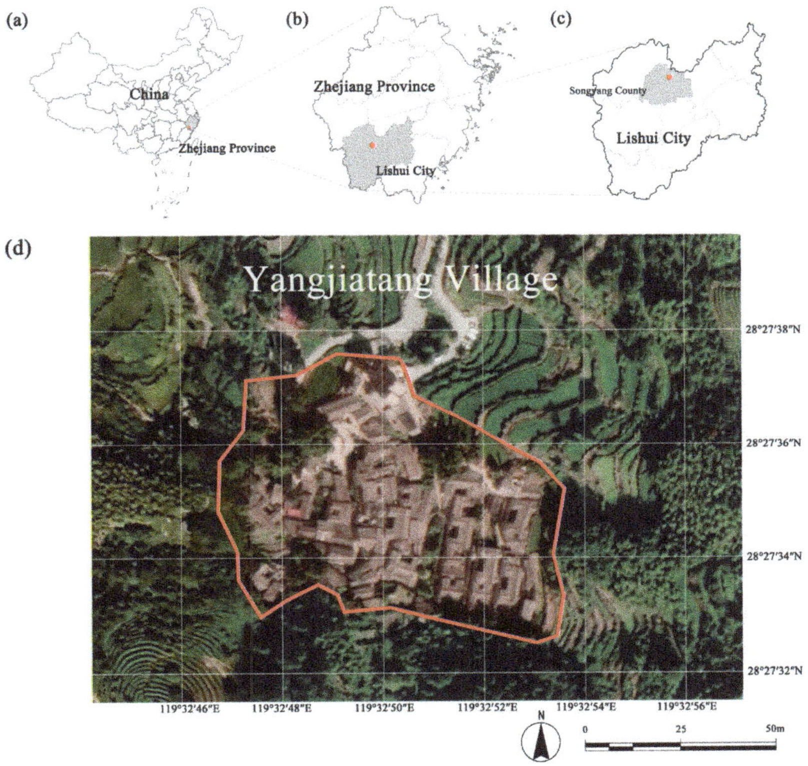

Figure 2. Geographical location of the Yangjiatang Village. (**a**) People's Republic of China; (**b**) Zhejiang Province; (**c**) Lishui City; (**d**) satellite image and scope of Yangjiatang Village (image source: drawn by the authors).

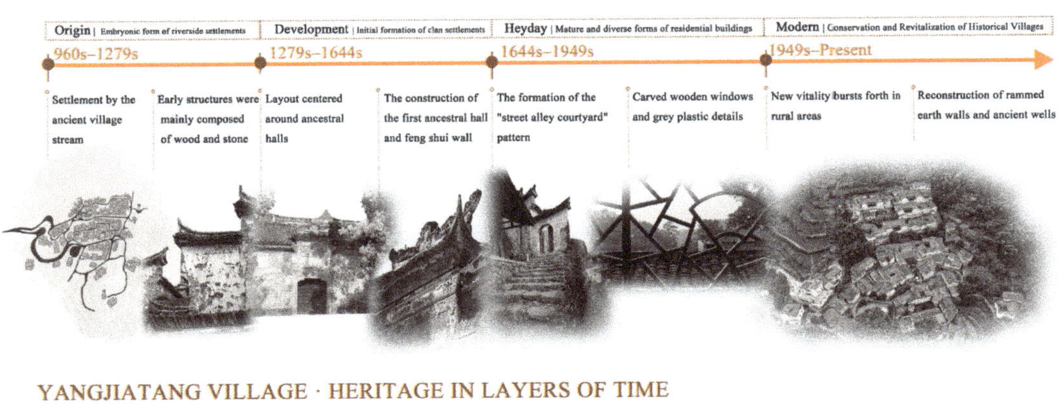

Figure 3. The historical development of Yangjiatang Village (image source: the photographs were prepared by the author or authors).

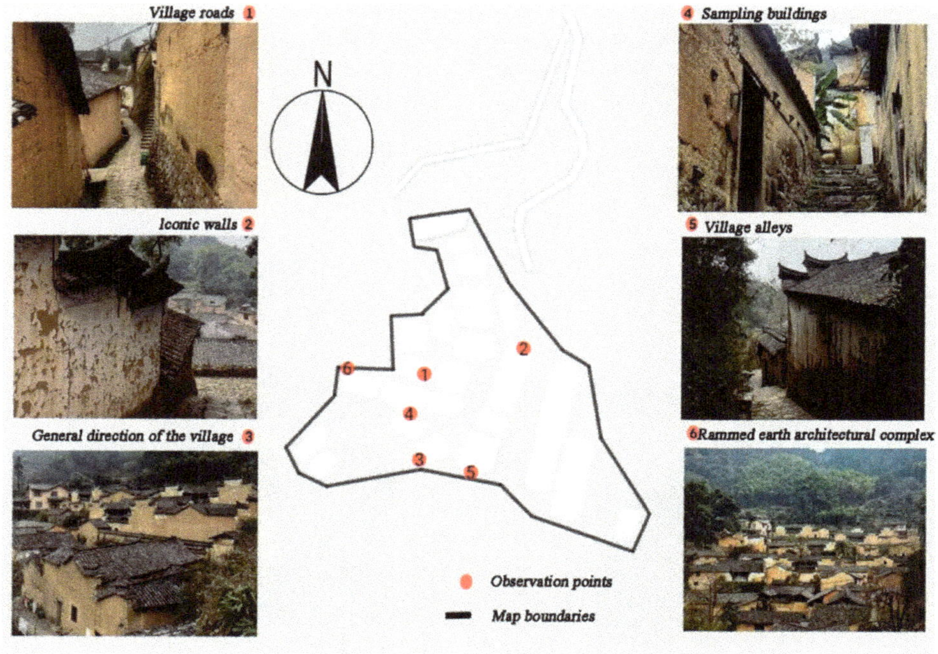

Figure 4. Important rammed-earth buildings in Yangjiatang Village (image source: the photographs were prepared by the author or authors).

The region is characterized by a typical subtropical monsoon climate, featuring an average annual temperature of 17.5 °C, substantial precipitation reaching 1837 mm per year, consistently high relative humidity maintained above 80% throughout the year, and relatively limited sunshine hours totaling only 1750 yearly (Figure 5). Although this warm and humid climate is conducive to agricultural production, it also poses a severe challenge to the protection of rammed-earth buildings. Data indicate that 37% of rammed-earth buildings have been damaged by rainwater erosion in the past decade, 12 of which were subsequently designated as dangerous.

The rammed-earth buildings of Yangjiatang have a special status in the residential system of southern Zhejiang, and their construction techniques integrate the characteristics of She and Han cultures. A specialized "three-in-one soil" formulation is employed in the building's construction. This formulation leverages a base of local yellow soil (≈70%), which is enhanced with fine sand (20%), slaked lime (8%), and oyster shell powder (2%) to improve its properties. Tests confirm that this specific composition achieves a compressive strength of 2.5 MPa and a permeability coefficient lower than 10^{-6} cm/s, underscoring its superior durability. The thickness of the building wall is generally between 50 and 80 cm, which not only has good thermal insulation performance but also forms a unique microclimate regulation system.

However, the continuous high humidity environment and frequent precipitation have caused serious threats to rammed-earth buildings. On-site investigations show that the south-facing wall is most severely eroded by acid rain, with a surface peeling depth of 3–5 cm; the north-facing fence is in a humid environment for a long time, and microbial erosion leads to a loose structure; while the east–west wall is mainly affected by temperature difference stress, and the crack width generally exceeds 1 cm (Figure 6).

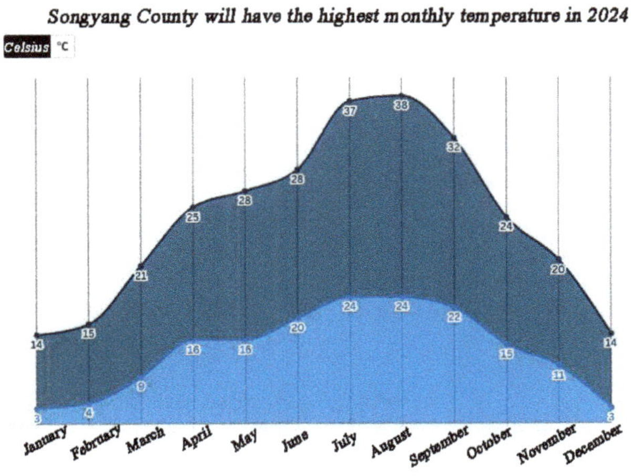

(a) Annual temperature changes in Songyang County.

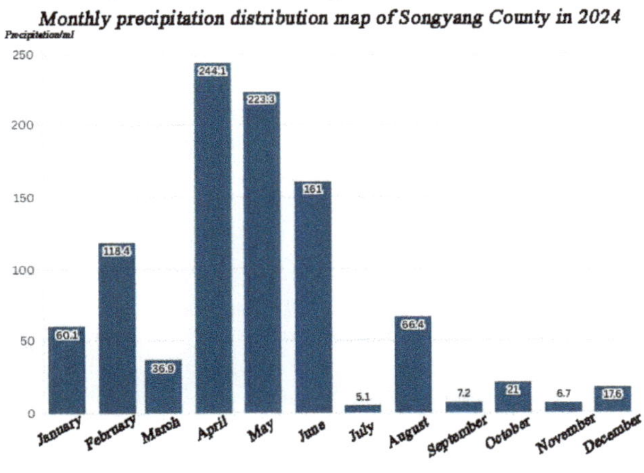

(b) Annual precipitation changes in Songyang County.

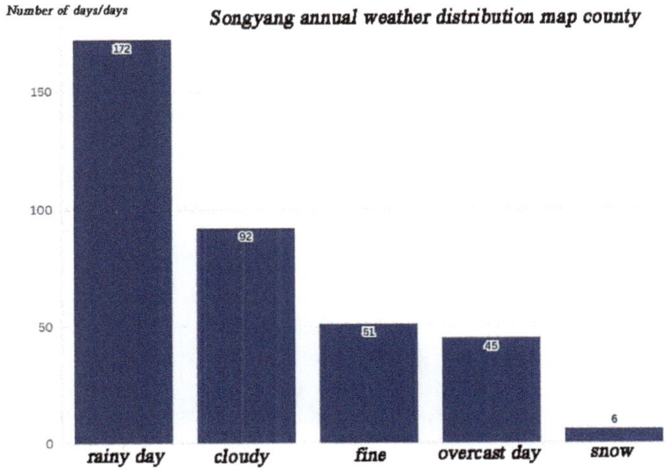

(c) Songyang County Annual Weather Changes.

Figure 5. The climate conditions in Songyang County where Yangjiatang Village is located.

Figure 6. A partial aerial view of Yangjiatang Village (image source: photographed by the authors).

It is touching that there are still 23 households and 38 older adults and older people in the village who stick to these old houses passed down from their ancestors. Grandpa Yang, 78 years old this year, said, "This house was built by my great-grandfather. It is warm in winter and cool in summer, and it is comfortable to live in". This persistence not only reflects the continuity of cultural heritage but also reflects the villagers' profound attachment to their traditional lifestyle. In recent years, the Songyang County Government has attached great importance to the protection of Yangjiatang Village and has invested more than 12 million yuan in special funds to implement projects such as "Save Old Houses" and "Protection and Utilization of Traditional Villages". In 2020, the village was included in the *Important Architectural Heritage Protection List* of Zhejiang Province, and in 2022, it was awarded the "China Traditional Village Digital Museum" construction pilot. These measures not only protect precious architectural heritage but also inject new vitality into rural revitalization.

2.2. Field Survey and Sample Collection

Yangjiatang Village is in a typical mountainous terrain in southwestern Zhejiang. The village is terraced along the mountain slope, with buildings of varying heights blending seamlessly into the natural environment. Sampling was conducted during clear weather conditions with temperatures ranging from 15 °C to 22 °C and relative humidity between 70% and 85%. These parameters were recorded to ensure consistency and to correlate with subsequent laboratory simulations. To fully understand the status of rammed-earth buildings, we conducted systematic aerial photography surveys using a DJI MINI 4 PRO drone (Shenzhen Dajiang Innovation Technology Co., Ltd., Shenzhen, China). To adapt to the cloudy and foggy conditions of Yangjiatang Village, the camera has been set to aperture f/2.8, shutter speed 1/100–1/500 s, and ISO 100-800. The flight altitude was maintained at 11–15 m to capture both building overviews and wall details. The drone's flight altitude is strictly controlled within the range of 11–15 m, which can not only capture the overall shape of the building but also clearly identify the detailed features of the wall. For the high-resolution documentation of severely damaged rammed-earth walls, the image acquisition distance was reduced to 3–5 m. This close-range approach, supplemented by the camera's digital zoom function, enabled the clear capture of millimeter-scale crack details, which is a critical step for subsequent quantitative analysis.

The ground survey adopted a multi-device collaborative working mode: using the DJI POCKET 3 (Shenzhen Dajiang Innovation Technology Co., Ltd., Shenzhen, China) primary camera for 4K high-definition shooting and iPhone Smartphone (Foxconn, Shenzhen, China), Huawei Smartphone (Huawei Technologies Co., Ltd., Shenzhen, China), OPPO Smartphone (Guangdong Oppo Mobile Telecommunications Corp., Ltd., Dongguan, China), and other mobile devices for auxiliary shooting to ensure the integrity and comparability of the image data. Using this ground-to-earth integrated survey method, we created a digital image library featuring high-definition images of rammed-earth buildings (Figure 7).

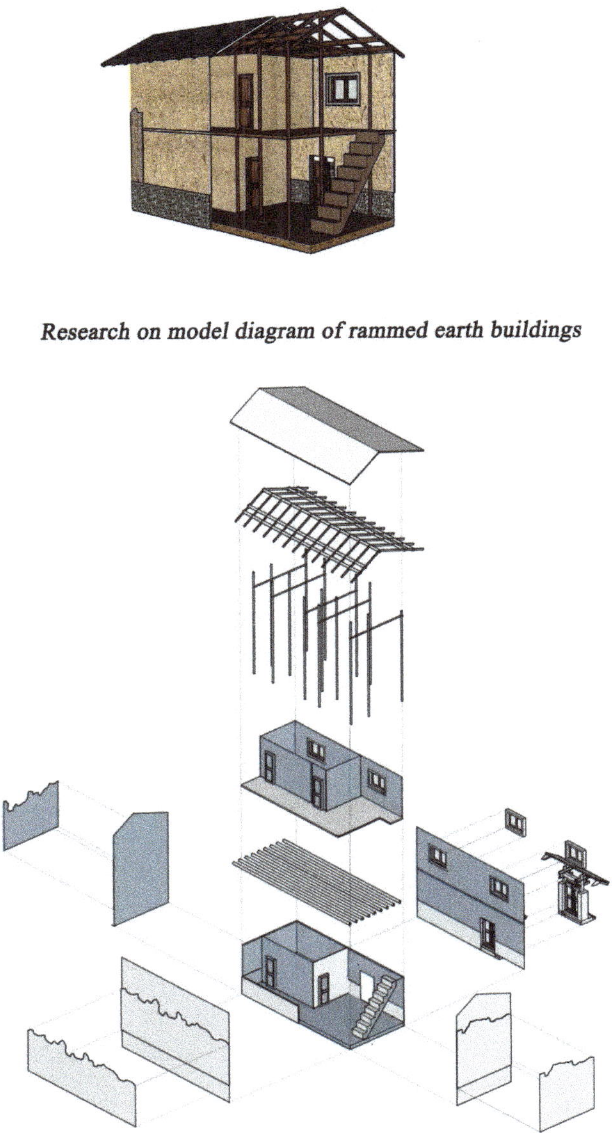

Figure 7. Analysis of typical rammed-earth buildings in Yangjiatang Village (image source: drawn by the authors).

During the field investigation, we systematically recorded various typical damage modes: (1) Structural cracks: mainly distributed at the corners of the wall, with the maximum crack width reaching 12 mm and extending for more than 2 m, showing obvious shear crack characteristics; (2) Surface erosion: The peeling depth of the south-facing wall is generally 3–5 cm, exposing the internal rammed-earth layer; (3) Biological erosion: The moss coverage rate of the north-facing wall is as high as 65%, and the black mold area accounts for 40% of the wall area; (4) Roof damage: About 28% of the buildings have tile slippage, causing rainwater to erode the top of the wall directly. Researchers have shown great interest in the formation mechanisms of these diseases, particularly regarding the differences in damage characteristics of walls facing different directions, which suggests a complex environmental mechanism.

Sampling was conducted with prior consent from homeowners. A scientific sampling plan was designed: (1) Samples were collected from four orientations (E, S, W, N). (2) Sampling was performed 1.5 m above ground, avoiding openings. (3) Using a stainless-steel knife, the surface weathering layer (2 mm) was removed before collecting a 10 × 20 × 10–20 mm cross-section sample (Figure 8).

Figure 8. On-site sampling picture (image source: photographed by the authors).

The sample processing process strictly complies with relevant specifications: (1) Solid samples are wrapped in acid-free paper and placed in polyethylene specimen boxes; (2) powder samples are placed in 15 mL centrifuge tubes. All containers have been marked with unique codes and record metadata, including the location of the sampled house, the direction of the sampled wall, and the sampling time (Figure 9). To ensure the freshness of the samples, the time from collection to laboratory analysis is strictly controlled within 7 days. We use anti-seismic packaging and temperature recorders during transportation to prevent any changes in the sample properties. Through field investigations, we have obtained a wealth of physical samples and image data. This foundational research provides critical support for the informed conservation of traditional rammed-earth buildings and establishes a robust framework for future investigations into the long-term performance of these materials in humid subtropical environments.

Figure 9. Sampling bagging and drone photography records (image source: photographed by the authors).

2.3. *Analysis Methods and Process*

2.3.1. X-Ray Diffraction (XRD) Analysis

Mineral composition was determined by X-ray diffraction (XRD) using a Rigaku SmartLab diffractometer (Rigaku Holdings Corporation, Tokyo, Japan) with Cu Kα radiation (λ = 1.5406 Å), operated at 40 kV and 40 mA. Data were collected over a 2θ range of 5° to 70° with a step size of 0.02° and a counting time of 2 s per step. Quantitative phase analysis was performed by using the Rietveld refinement method. The XRD analysis was performed by using a Rigaku SmartLab high-resolution diffractometer equipped with a Cu Kα X-ray source (wavelength λ = 1.5406 Å) operating at 35 kV and 25 mA. The measurements were conducted in a 2θ range from 5° to 70° with a step size of 0.02° and a scanning speed of 2° per minute. The diffractograms were collected in continuous mode, and phase identification was carried out by comparing the obtained patterns with the International Centre for Diffraction Data (ICDD) PDF-4+ database. Quantitative analysis of mineral phases was performed using the Rietveld refinement method with the software package provided by the instrument manufacturer. By analyzing the XRD diffractograms of the samples, the mineral composition and stability of the surface materials can be understood, which can provide an essential basis for interpreting the durability texture of traditional building materials.

2.3.2. Detection of Diversity Composition of Microbial Communities

This study used high-throughput sequencing technology to systematically study the microbial communities of rammed-earth building walls in Yangjiatang Village. The 16S rRNA gene V3-V4 region of samples from walls in four directions (East, South, West, and North) was sequenced using the Illumina NovaSeq platform, and an average of 78,452 valid sequences (length from 400 to 440 bp) were obtained. Data analysis was performed using QIIME2 (version 2019.4), and representative ASV sequences were obtained after DADA2 denoising. Species annotation was based on the SILVA 138 database, using the Naive Bayes classifier with a confidence threshold of 0.7. The study focused on analyzing microbial alpha diversity (Shannon, Chao1 index), beta diversity (PCoA, UPGMA clustering), and functional prediction (PICRUSt2).

2.3.3. Scanning Electron Microscope (SEM) Analysis

In this study, we performed high-resolution microscopic morphology analysis on rammed-earth samples using a German ZEISS GeminiSEM 360 field emission scanning electron microscope (Carl Zeiss AG, Oberkochen, Germany). The device is equipped with a secondary electron detector (SE) and a backscattered electron detector (BSE), which can achieve an ultra-high resolution of 1.2 nm at an accelerating voltage of 10 kV. The sample was pretreated by coating a 5 nm thick gold film with an Quorum Q150T ES ion sputtering coater (Quorum Technologies Ltd., Laughton, UK) to improve conductivity. During the analysis, the working distance was set to 5 mm, the beam intensity was 100 pA, and the magnification was adjusted from 200× to 20,000×. The particle morphology, pore structure, and interface characteristics of the samples facing the wall in different directions were focused on. Finally, the acquired images were subjected to quantitative analysis in ImageJ 1.54p bundled with 64-bit Java 8 to measure key parameters: particle size distribution (n > 500), porosity (calculated by threshold segmentation), and surface roughness (determined using a grayscale gradient algorithm).

3. Results

3.1. XRD Analysis Results

The XRD diffractograms revealed distinct mineralogical compositions across the four orientations (Figures 10 and 11). Four main minerals are clearly marked in Figures 10 and 11: Quartz (SiO_2), Mullite ($Al_6Si_2O_{13}$), Lepidocrocite (γ-FeOOH), and Nontronite ($Na_{0.3}Fe_2(Si,Al)_4O_{10}(OH)_2 \cdot nH_2O$).

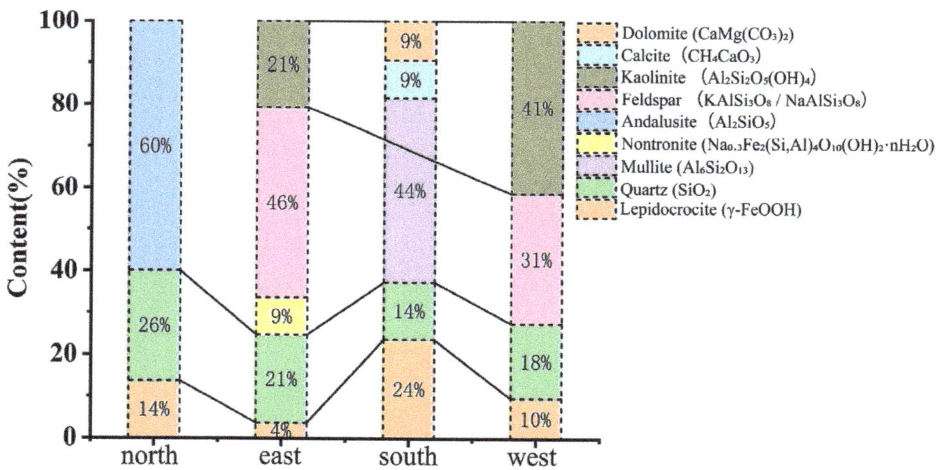

Figure 10. The main minerals and distribution ratios in the wall (image source: the author draws the picture based on the experimental results).

Quartz is a very stable mineral with high hardness, durability, heat resistance, and water resistance, which can improve the overall stability of the material. As can be seen from Figure 10, the Quartz diffraction peaks are apparent in all samples, indicating it is not only a significant component but also the dominant phase in the material. Among them, the proportion in the north wall and the east wall is higher than that in the other two walls (the north wall accounts for 26% and the east wall accounts for 21%). It is noteworthy that while the diffraction peak intensity of quartz appears particularly prominently in the east sample diffractogram (Figure 11), its quantified percentage (21%) is more comparable to other orientations when considering the complete mineral composition (Figure 10). This observation can be attributed to factors influencing peak intensity beyond mere abundance, such as the crystallinity and preferred orientation of quartz crystals within the sample. Therefore, quantitative analysis provides a more reliable measure of the actual mineral content, confirming that quartz is a dominant and critical stabilizing phase across all wall orientations. From the diffraction peaks of the north wall and the east wall, the peak intensity of the east wall is higher, indicating that Quartz is easier to crystallize.

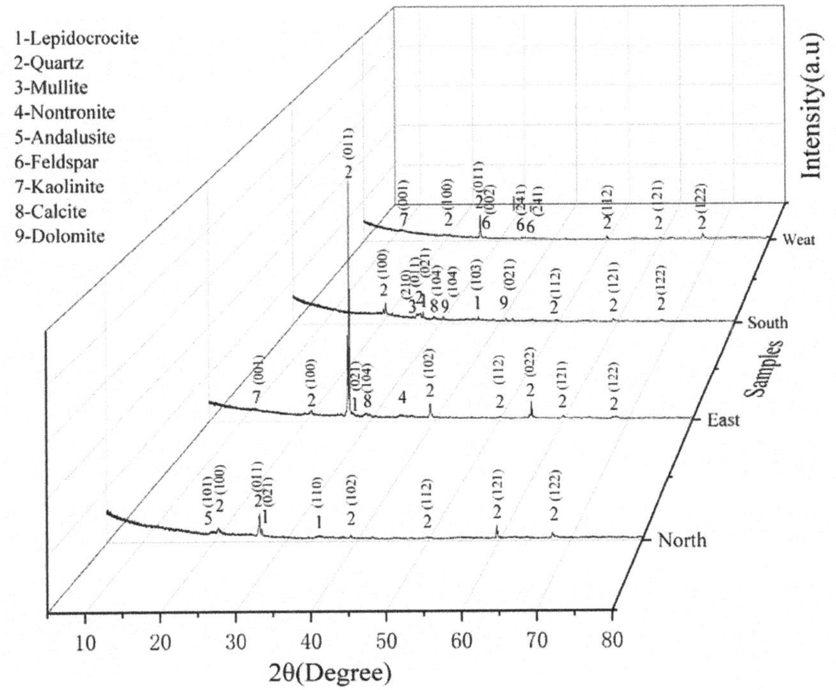

Figure 11. XRD results of 4 samples (image source: the author draws the picture based on the experimental results).

Mullite is a high-temperature phase mineral with excellent thermal and chemical stability. It can be used for thermal insulation of building exterior walls, interior walls, and partition walls, reducing the overall heat loss and energy consumption of buildings. From the figure, Mullite is concentrated on the south wall because it is exposed to more sunlight. Rammed-earth buildings are prone to shrinkage and cracking when drying. The micro-expansion characteristics of Mullite can partially offset the shrinkage force, improving overall stability. Mullite acts as a humidity buffer due to its microporous network, which actively manages moisture by absorbing and slowly releasing it. Synergizing with the natural breathability of rammed-earth construction, this process aids in stabilizing the humidity levels specifically on the south-facing wall, thereby positively influencing the microclimate. Conversely, iron hydroxide is an environmentally sensitive mineral prone to phase changes, potentially causing material weathering and chromatic alterations. The significantly higher content detected on the south wall (24%) versus the north wall (14%) underscores the role of orientation-dependent factors, such as solar exposure and wet-dry cycles, in driving deterioration. The south-facing wall experiences significant diurnal temperature fluctuations, which promote water vapor condensation. This moisture, in turn, accelerates the oxidation of iron compounds within the material. Consequently, protective strategies for the south wall should prioritize enhancing waterproofing capabilities to mitigate the disruptive expansion and contraction cycles associated with iron hydroxide. Conversely, the north-facing wall is characterized by lower temperatures, higher humidity, and reduced sunlight exposure. These conditions suppress the oxidation of Fe^{2+}, allowing iron hydroxide to persist in a colloidal state.

Nontronite is a layered silicate mineral with specific water absorption and expansion properties. Its presence will cause the material to change in volume when the humidity changes, thus affecting the stability of the structure. The east wall is affected by the monsoon climate, and the humidity changes significantly. Rainwater infiltration causes Nontronite to expand, which in turn affects the stability of the wall. Currently, its proportion constitutes only 9%, resulting in a negligible impact on the wall. Furthermore, the high feldspar content (46%) in the east wall effectively restricts these potential expansion spaces. The composition of the north sample is 14% iron hydroxide, 26% quartz, and 60% andalusite. The substantial quartz content in the northern sample contributes to the material's enhanced hardness, durability, and cold resistance. The colloidal state of iron hydroxide can fill the gaps of andalusite and improve impermeability. The composition of the west sample is 10% iron hydroxide, 18% quartz, and 41% kaolinite. In the west sample, the Quartz content is significantly higher than that of others, with extremely high hardness and durability and excellent overall stability. However, in the diffraction peak diagram, it can be seen that the peak intensity of the quartz on the west wall is not high, which may be caused by wind erosion. It is necessary to strengthen the weathering protection of the wall surface. At the same time, the west wall is more humid than the east wall. Kaolinite has strong water absorption and a higher proportion. The composition of the east sample is Quartz accounting for 21%, Nontronite accounting for 9%, Feldspar accounting for 46%, kaolinite accounting for 21%, and iron hydroxide accounting for 4%. Quartz and feldspar can form a skeleton support body with good overall stability. The composition of the south sample is Quartz accounting for 14%, dolomite accounting for 9%, calcite accounting for 9%, Mullite accounting for 44%, and iron hydroxide accounting for 24%. Quartz is still the main mineral, which ensures the hardness and durability of the material. In addition, due to the water absorption and expansion of nontronite, the volume of the material may change when the humidity changes. Calcite and dolomite are carbonate minerals that can neutralize acidic rainwater, thereby improving the erosion resistance of the wall surface.

3.2. Analysis of the Microbial Diversity Detection Data

We analyzed the microbial communities from four wall orientations using 16S rRNA gene sequencing. The data was processed with QIIME2 software to identify bacterial types and their abundances. As shown in Table 1, the original input sizes were 60,951, 76,418, 82,270, and 67,285 sequences. After filtering, these counts decreased to 51,354, 63,719, 68,601, and 56,416, respectively, indicating that preliminary quality control removed approximately 15%–17% of low-quality sequences. Following denoising, the sequence counts slightly varied to 51,220, 60,721, 68,543, and 55,846, respectively. During sequence merge, north retained 50,881 sequences, south significantly reduced to 47,966, west slightly decreased to 68,104, and east dropped to 52,831. After removing chimeras and singletons, the final valid sequence counts obtained for each sample were 50,629, 44,099, 67,048, and 49,823, respectively. These counts will be used for subsequent taxonomic analysis and diversity assessment. Notably, the west sample retained the highest proportion of its sequences (81.5%); in contrast, the south sample suffered the greatest attrition, with a final count representing only 57.7% of the original. This disparity may reflect variations in DNA quality and sequence complexity among different samples, as well as differences in their specific microbial community structures.

Table 1. Analysis of DADA2 sequence denoising results.

Sample ID	Input	Filtered	Denoised	Merged	Non-Chimeric	Non-Singleton
North	60,951	51,354	51,220	50,881	50,630	50,629
South	76,418	63,719	60,721	47,966	44,304	44,099
West	82,270	68,601	68,543	68,104	67,048	67,048
East	67,285	56,416	55,846	52,831	49,873	49,823

The research findings revealed significant differences in microbial community composition richness among the four samples. The south sample demonstrated the highest diversity across all taxonomic levels, with 23 phyla (Phylum), 54 classes (Class), 108 orders (Order), 152 families (Family), 216 genera (Genus), and 55 species (Species) detected. Consequently, the south sample possesses an exceptionally rich and complex microbial community structure, as evidenced by its total of 608 taxonomic units, a count notably higher than that of the other three samples. The east sample exhibited moderate taxonomic diversity, identifying 13 phyla, 25 classes, 46 orders, 61 families, 77 genera, and 40 species, totaling 262 taxonomic units. Notably, the east sample showed a relatively high number of species-level taxonomic units (40), second only to the south sample, suggesting a higher abundance of species-identifiable microbial groups. The north sample displayed relatively low taxonomic diversity, detecting eight phyla, 15 classes, 31 orders, 35 families, 41 genera, and seven species, totaling 137 taxonomic units. The west sample exhibited the lowest taxonomic diversity, identifying seven phyla, 10 classes, 25 orders, 26 families, 31 genera, and six species, totaling 105 taxonomic units. The bar chart visually demonstrates the distribution patterns of microbial taxa across taxonomic levels among the four samples. Sample south exhibits the highest overall abundance, with a significantly greater number of taxa across all taxonomic ranks. This dominance is particularly evident at the order, family, and genus levels, as indicated by the prominent blue, orange, and purple sections in the figure. Samples east and north possess relatively low taxonomic diversity. The west sample, however, is characterized by the least abundant and least diverse microbial community. These findings clearly reveal distinct differences in microbial community composition: The south demonstrates the richest biodiversity, whereas the west shows the lowest diversity (Figure 12).

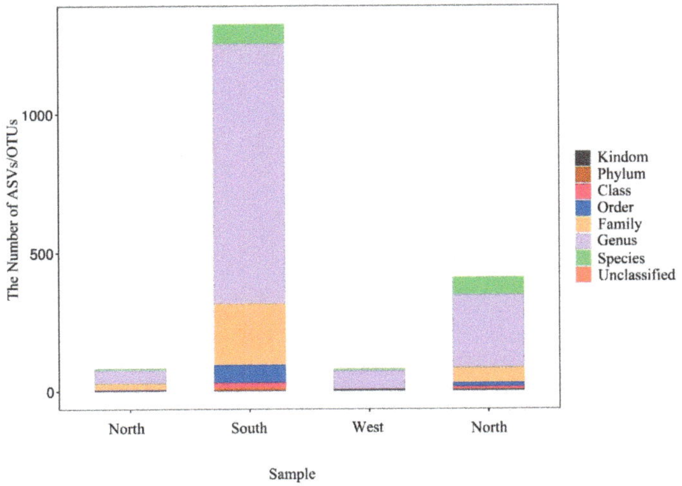

Figure 12. Distribution patterns of microbial taxa in four samples.

This study looked at the tiny living things on the walls of rammed-earth buildings in four directions (east, south, west, and north) in Yangjiatang Village, showing a significant link between the types of microbes present and how well the buildings resist erosion.

As shown in Figure 13, this bar chart displays the relative abundance distribution of microbial communities at the genus level across four distinct samples (North, South, West, and East). The results reveal significant differences in microbial community composition among the sample points. In the North sample, microbial diversity is relatively low, dominated by "Others" taxa with nearly 100% relative abundance, indicating a relatively homogeneous microbial community in this region. The South sample exhibits richer microbial diversity, featuring multiple genera such as *Sphingomonas*, *Pseudonocardia*, and *Rubrobacter*, collectively constituting approximately 40% of the relative abundance, while "Others" taxa account for about 60%. In the West sample, Actinomycetospora was the dominant genus, accounting for approximately 70% of the relative abundance. It was followed by an orange-marked genus (possibly Rubrobacter or another species), which constituted about 25%. The East sample demonstrates the most diverse microbial composition, including *Arthrobacter*, *Pseudonocardia*, *Rubrobacter*, Bacillus, and other genera. These dominant bacterial genera account for the vast majority (80%) of the microbial community, while the "Others" category represents a minor fraction (20%). Overall, microbial diversity gradually increased from the North to the East sample. Concurrently, the community structure transitioned from a simpler to a more complex composition. In contrast to the higher diversity of the East sample, the West sample was marked by a clear predominance of specific bacterial taxa.

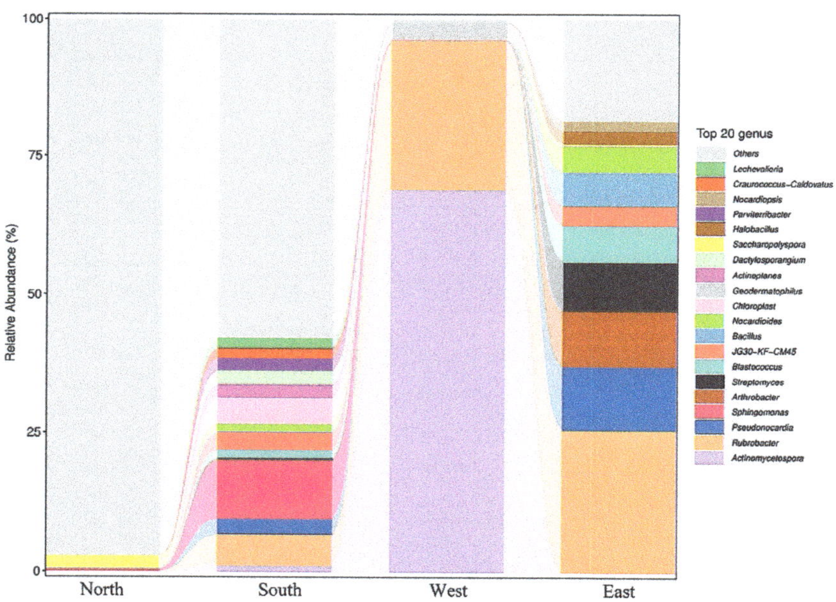

Figure 13. Results of sample microbiota taxonomic annotation.

As shown in Figure 14, the composition of microbial communities (OTU/ASV) across the four sampling sites (North, South, West, and East) exhibited pronounced regional specificity, with limited overlap among them. A network analysis revealed that the South sample demonstrated the highest microbial diversity, containing 1296 unique OTUs accounting for 69.6% of total detected OTUs. The East sample followed closely, with 387 unique OTUs accounting for 20.78% of the total. In contrast, the West and North samples contained significantly fewer unique OTUs, at 74 (3.97%) and 71 (3.81%), respectively. Shared OTUs between any two sites were strikingly limited. The highest overlap was observed between the South and East samples (21 OTUs, 1.13%), while all other pairwise comparisons, such as North–South (5 OTUs, 0.27%) and South-West (2 OTUs, 0.11%), showed even lower numbers. Notably, only one core microbiome OTU (0.05%) was shared across all sites, highlighting significant regional variations in microbial composition.

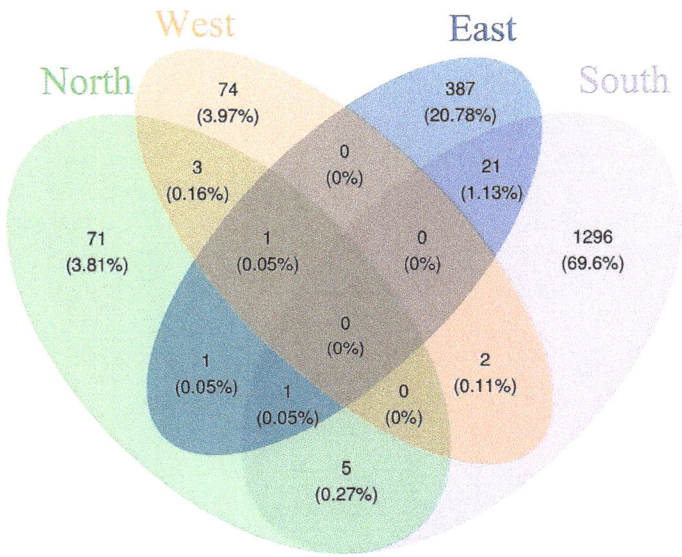

Figure 14. Test results of the Venn diagram for the ASV/OTU analysis.

As shown in Figure 15, this bar chart illustrates the relative abundance distribution of microbial communities at the phylum and genus levels across different sampling points and their combinations. At the phylum level (left panel), *Actinobacteria* (blue-green) and *Proteobacteria* (yellow) are dominant phyla in most samples. *Actinobacteria* was the dominant phylum across all samples, but its abundance varied: it was overwhelmingly dominant in the North sample (~95%), while the predominant phylum remained at a significant level (~75%) in the West and East samples. Notably, Proteobacteria show a significant increase in relative abundance in the South samples and their combinations with other samples. *Firmicutes* (light purple) also exhibit high abundance in East and its combinations. Significantly, *Proteobacteria* dominate the shared microbial communities between the North and the South, indicating that these two sampling sites share predominantly *Microbomonas* bacteria. At the genus level (right panel), differences become more pronounced. *Rubrobacter* (blue-green) constitutes a significant proportion in the North samples, whereas *Actinoplanes* (yellow) and *Sphingomonas* (pink) demonstrate higher abundance in the South samples. In the west sample, *Methylobacterium* (purple) and *Actinomycetospora* (pink) dominate. The East sample's defining feature is its predominance of bifidobacteria (red). In shared microbial communities, *Bacillus* and *Actinoplanes* are the primary shared components between the North and South. *Rubrobacter* predominates in the microbiota shared between the North and West samples, while *Actinomycetospora* and *Blastococcus* constitute the core community shared by the South and East samples. Notably, Bacillus constitutes the main shared microbial component among the North, South, and West, whereas *Delftia* dominates the shared microbiota between the North, South, and East.

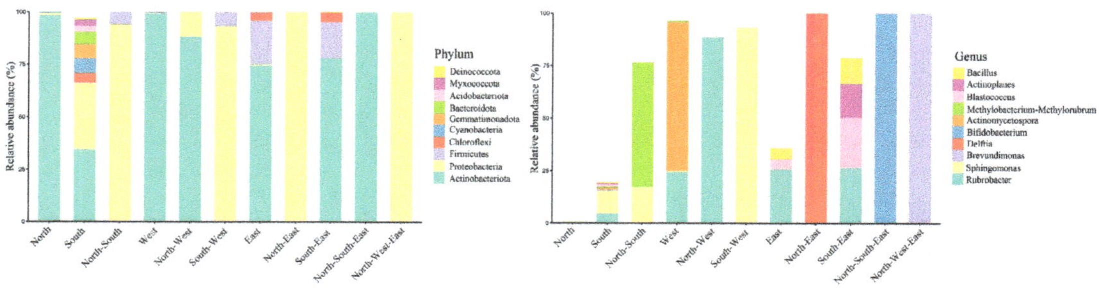

Figure 15. Asv/OTU abundance in different regions of VeenTu.

As shown in Figure 16, the heat map illustrates the relative abundance distribution and clustering relationships of microbial communities at the genus level across four sampling sites (North, South, West, and East). The color gradient from blue (low abundance) to red (high abundance) reflects relative abundance variations in genera across different samples. The heat map analysis reveals significant differences in microbial community structures among the four sites, each exhibiting characteristic dominant bacterial genera.

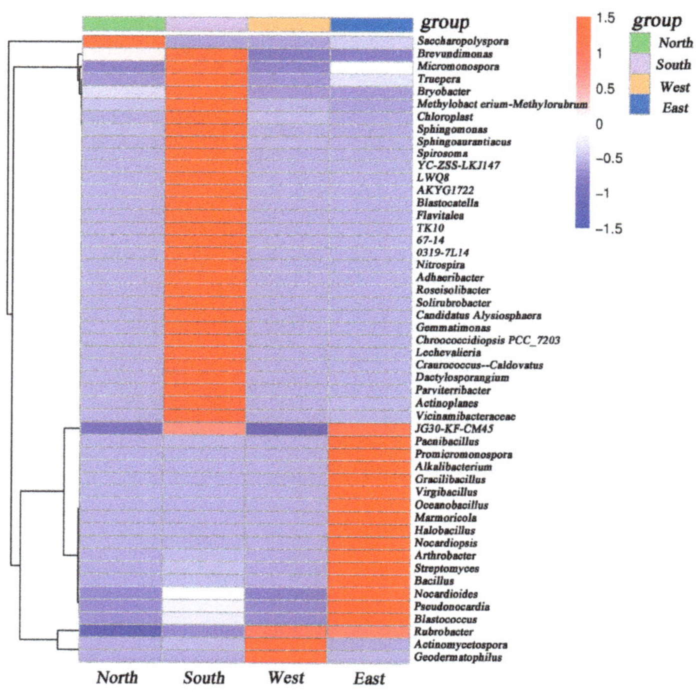

Figure 16. A heatmap showing species composition based on genus-level clustering.

In the South sample, high abundance (red) is demonstrated by over 30 genera, including Brevundimonas, Micromonospora, Truepera, Bryobacter, Methylobacterium-Methylorubrum, Chloroplast, and Sphingomonas, indicating the highest microbial diversity at this site. The West sample, however, shows dominance by *Actinomycetospora*, *Geodermatophilus*, and *Rubrobacter* as key bacterial genera. The characteristic bacterial genera in the East samples primarily include *Paenibacillus*, *Promicromonospora*, *Alkalibacterium*, *Gracilibacillus*, *Virgibacillus*, *Oceanobacillus*, *Marmoricola*, *Halobacillus*, *Nocardiopsis*, *Arthrobacter*,

Streptomyces, *Bacillus*, *Nocardioides*, *Pseudonocardia*, and *Blastococcus*. In the North samples, only *Saccharopolyspora* showed relatively high abundance, while other genera were generally low in abundance. The phylogenetic tree reveals that microbial genera form several major clades: one is predominantly enriched in the South samples, while another is primarily associated with the East samples, with a minor clade (e.g., Actinomycetospora, Geodermatophilus, Rubrobacter) showing high abundance in the West samples. This clustering pattern reflects the selective distribution characteristics of microbial communities under different environmental conditions.

From the perspective of the cluster tree structure, the microbial genera can be divided into several main clusters: one cluster is mainly enriched in the south samples; another is mainly enriched in the east samples; A small number of bacterial genera, such as *Actinomycetospora*, *Geodermatophilus*, and *Rubrobacter*, also showed high abundance in the west samples. This clustering pattern reflects the selective distribution characteristics of microbial communities under different environmental conditions.

In summary, the south-facing wall exhibited the most diverse microbial community, which may contribute to its distinct erosion patterns compared to other orientations.

3.3. SEM Analysis Results

3.3.1. Comparative Analysis of SEM Images

Furthermore, the researchers conducted SEM imaging analysis on samples from the east, south, west, and north, which revealed notable differences in their micromorphology, particle distribution, and surface characteristics (Figure 17).

(1) The particles in the eastern samples are regular polygons, large in size, with significant size differences, relatively concentrated distribution, and a relatively smooth overall surface.

(2) In contrast, the particles in the south are more irregular in shape, mostly polygonal in structure, unevenly distributed, and not obviously aggregated. Its surface is rough and accompanied by cracks, indicating that this orientation may have been affected by greater stress during its formation.

(3) The particle morphology of the western samples is between the eastern and southern ones, mainly round and polygonal, with a large number of total particles; the particle size is relatively consistent, the particle size difference is slight, the distribution is relatively concentrated, and the surface is relatively flat, showing good particle uniformity. The particles in the northern samples are characterized by a fragmented or flaky morphology, a broad size distribution, and an uneven spatial arrangement.

(4) The particle morphology of the northern samples is relatively dispersed, and the particles are mainly distributed on the right side. The particle size is relatively large. There are microorganisms in the north of samples. The organisms may aggregate small particles into larger clusters through secretions or produce fine fragments by decomposition. This microbial activity could lead to the observed heterogeneous particle size distribution, which may result from uneven microbial corrosion or deposition.

In general, the east and west samples exhibit more regular particle morphology, a more concentrated size distribution, and smoother surfaces, suggesting a more uniform and stable growth environment during their synthesis or processing. The south and north samples exhibit more complex morphology, a wider particle size distribution, and rougher surfaces with cracks, which may be attributed to factors such as crystal orientation, local temperature gradients, or stress concentration. Among them, the particle size in the north is more dispersed due to the influence of microorganisms. These phenomena indicate that the microstructure of the sample in different directions is affected by the synthesis or processing conditions, thus showing obvious anisotropic characteristics.

Figure 17. SEM micro graphs of the samples investigated on the four wall surfaces (east, south, west and north).

3.3.2. Particle Size Distribution Analysis

The researchers conducted a statistical analysis of the particle size of the four oriented samples of this rammed-earth wall, and the results are as follows (Table 2):

Table 2. Particle size statistics in Yangjiatang Village.

Wall Orientation		East Wall	South Wall	West Wall	North Wall
Total particles		109	75	187	130
Particle size information	Average particle size	342.636 nm	361.141 nm	284.309 nm	266.304 nm
	Particle size standard deviation	217.108 nm	220.090 nm	159.866 nm	145.531 nm
	P10	243.434 nm	214.087 nm	218.898 nm	203.984 nm
	P50	755.318 nm	608.584 nm	671.127 nm	596.557 nm
	P90	1267.203 nm	1003.082 nm	1123.356 nm	989.13 nm

Source: statistics by the author.

Through statistics, it was found that there were apparent differences in the morphology and particle size of samples at different wall orientations, indicating that the samples may have been affected by different external environments (such as temperature, reaction rate, or stirring rate) during the past rammed-earth synthesis or processing, forming anisotropic structural characteristics (Figure 18). In terms of the total number of particles, the west wall has the most significant number of rammed-earth particles, reaching 187, while the south wall has the fewest, only 75. A larger number of particles may mean that the wall has higher standards in terms of pressure resistance and stability, or more rammed-earth materials were used during construction. The average particle size of the south wall is the largest, at 361.141 nm, and the average particle size of the north wall is the smallest, at 266.304 nm. The difference in average particle size may be associated with localized environmental factors, such as variations in wind force and solar radiation across the different orientations. Larger average particles may make the wall stronger but may also affect the fineness and esthetics of the wall. Using rammed earth with different average particle sizes in different directions may be to balance the physical properties and appearance of the wall. In addition, the standard deviation reflects the degree of dispersion of particle size. The standard deviation of the south wall is the largest, indicating that the particle size of the rammed earth on the south wall varies greatly. The heterogeneity in particle sizes may be attributed to the use of source materials with inherent size variations or to a relatively ad hoc construction methodology. The north wall exhibits the smallest standard deviation, suggesting that its rammed-earth particles have a highly uniform size and may have been subject to more stringent control during construction. P10 means that 10% of the particles are smaller than this particle size. The P10 particle size of the north wall is the smallest, indicating that the north wall has relatively more particles with smaller particle sizes. This may affect the porosity and air permeability of the wall. Finer particles can reduce wall porosity and compromise air permeability, but they may enhance waterproofing performance. P50 is the median, and the P50 particle size of the east wall is the largest. The particle size distribution of the east wall indicates a prevalence of larger mid-range particles, suggesting an overall coarser composition that may contribute to greater wall solidity. The P90 value (the threshold above which only 10% of particles lie) is largest for the east wall, reflecting a higher content of large-sized particles. This particle size distribution likely benefits load-bearing capacity and stability, albeit at the potential cost of greater construction difficulty. These data will help researchers select materials for subsequent repairs and provide a basis for restoration. Figures 19–22 show the particle size distribution of rammed-earth particles from the four wall orientations.

3.4. Spray Water Results

In this experiment, we used a TD model electronic scale produced by Yuyao Jinnuo Tianping Instrument Co., Ltd. (Ningbo, China), and an electric sprayer produced by Delixi Electric (Yueqing, China). Original sample E (unrepaired) on the east side, initial reaction: after encountering water, it showed "rapid penetration" characteristics. Water penetrated the sample within 10 s, and a wet halo quickly appeared on the surface, indicating that it has high porosity and a loose structure. After 6 h, particles began to fall off at the edge, accompanied by small cracks extending to the center, forming an "edge collapse-center suspension" trend. Subsequently, the loss of primary structural support led to an inward collapse of the central section, resulting in a honeycombed fracture pattern. After 12 h, it completely disintegrated into a muddy state, with no block residue, and was determined to be a precarious structure (Figure 23a–c).

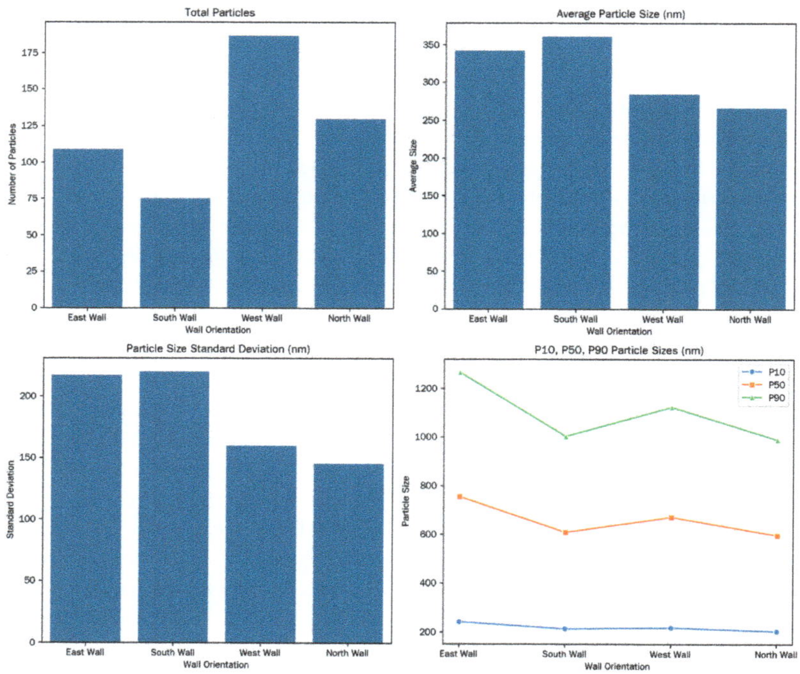

Figure 18. Particle size statistics in Yangjiatang Village (image source: the author draws the picture based on the experimental results).

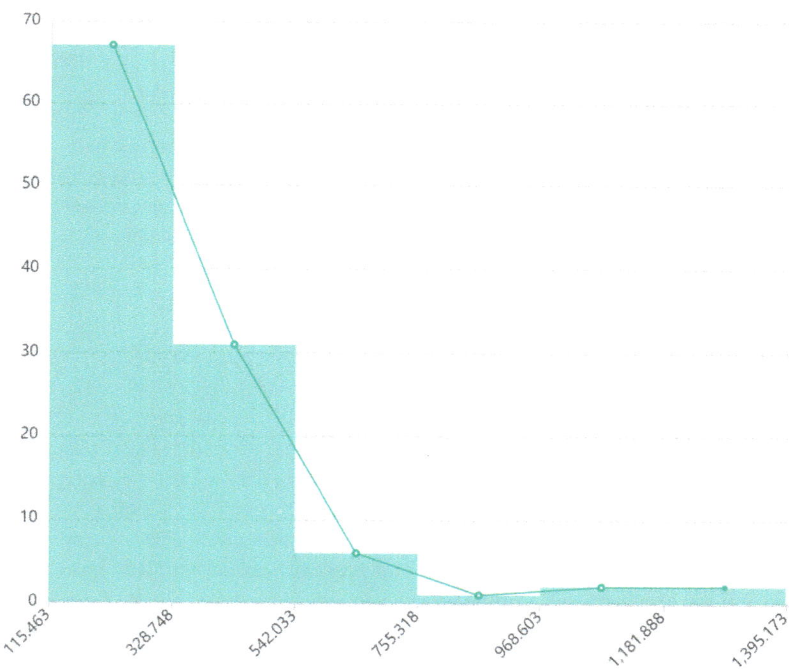

Figure 19. Particle size statistics of the east rammed-earth wall samples in Yangjiatang Village (image source: the author draws the picture based on the experimental results).

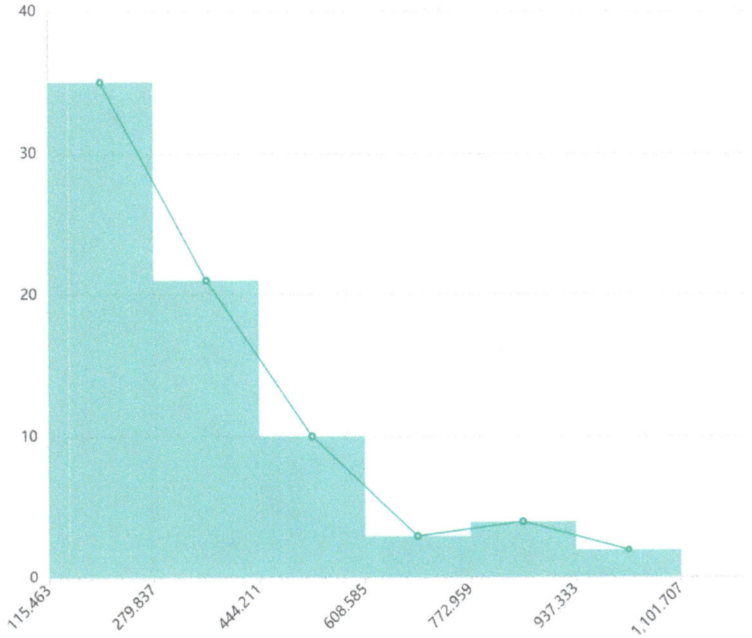

Figure 20. Particle size statistics of the south rammed-earth wall samples in Yangjiatang Village (image source: the author draws the picture based on the experimental results).

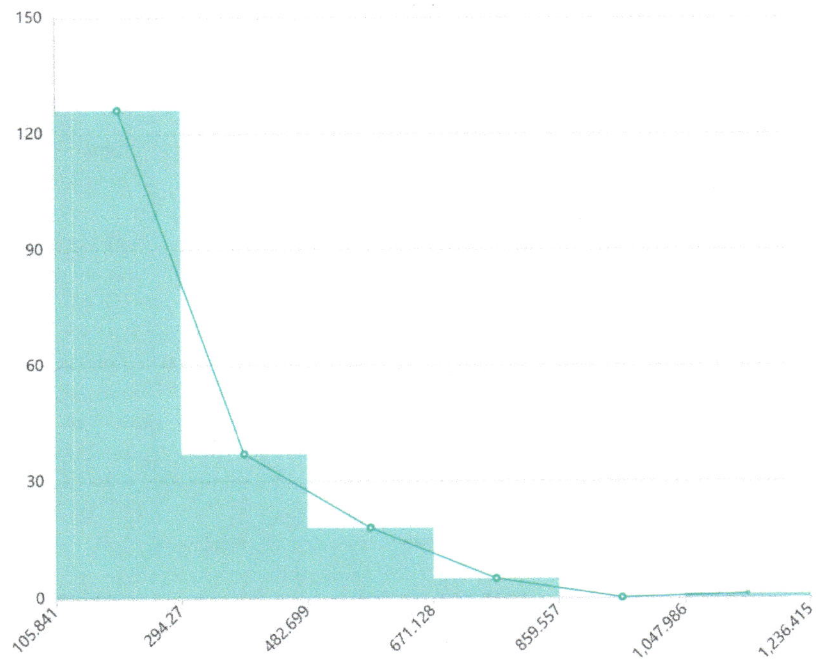

Figure 21. Particle size statistics of the west rammed-earth wall samples in Yangjiatang Village (image source: the author draws the picture based on the experimental results).

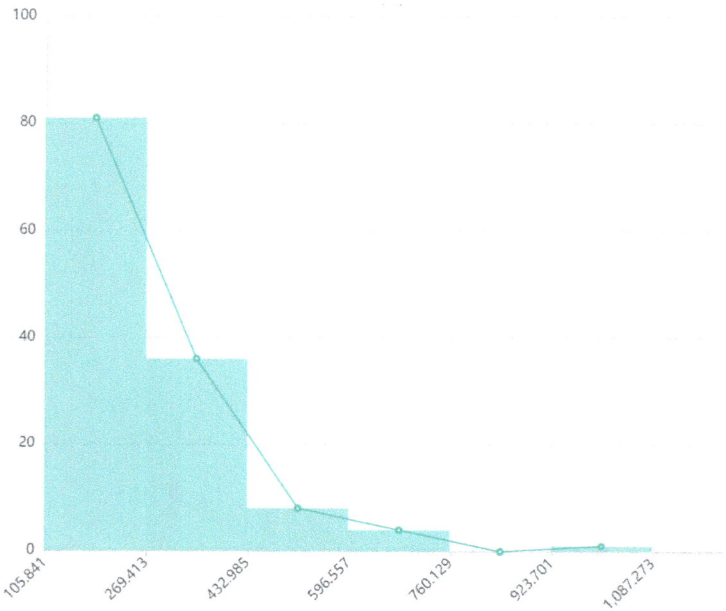

Figure 22. Particle size statistics of the north rammed-earth wall samples in Yangjiatang Village (image source: the author draws the picture based on the experimental results).

During the initial water reaction test, the unrepaired south sample (S) exhibited "surface runoff" behavior: water quickly slid down its slope with minimal infiltration, indicating a dense surface layer with low permeability. After 8 h, scouring erosion caused a "cutting-like" collapse at the water flow edges, forming an approximately 2 mm-deep annular groove, while the central 50% of the area remained intact. Following 12 h of testing, the sample's outer region collapsed, causing stress redistribution and the layered sliding of large soil fragments; the final residual volume was about 35% of the initial volume. Despite this significant material loss, its erosion resistance was greater than that of samples E and N, owing to its dense surface and internal structural support (Figure 23d–f).

West repair sample W (after repair). The repair material (presumably a binder or curing agent) filled the pores and formed a mesh structure, significantly improving the overall bonding strength. The repaired sample demonstrated an approximately 60% reduction in water penetration rate compared to samples E and N, along with the formation of a short-term water film on its surface. After 9 h, small "scale-like" pieces fell off the edge (single piece diameter < 5 mm), and the damage did not extend to the center, showing the characteristics of "local loss-main body stability". After 12 h, the width of the edge fracture zone was about 1–2 cm, and the central area maintained a high degree of integrity. The erosion resistance was significantly improved compared with the unrepaired sample, verifying the effectiveness of the repair process (Figure 23g–i).

The original sample N (unrepaired) on the north side has special geological characteristics. Due to its high sand content and uneven pore distribution, water rapidly infiltrates along cracks during the initial stage. Within 30 s, this leads to the formation of 2–3 small gullies deeper than 3 mm, demonstrating a chain reaction characteristic of "pitting-gully erosion". After 6 h, the structure showed significant degradation through block collapse (1–3 cm^3 per block) and gully expansion (>1 cm width); ultimately, after 12 h, it was completely fragmented, retaining only 10% of its original mass (Figure 23j–l). Erosion resistance is second only to sample E, highlighting the erosion resistance of the loose sand structure.

The overall erosion resistance of the samples can be ranked in descending order as follows: W > S > N > E (Table 3).

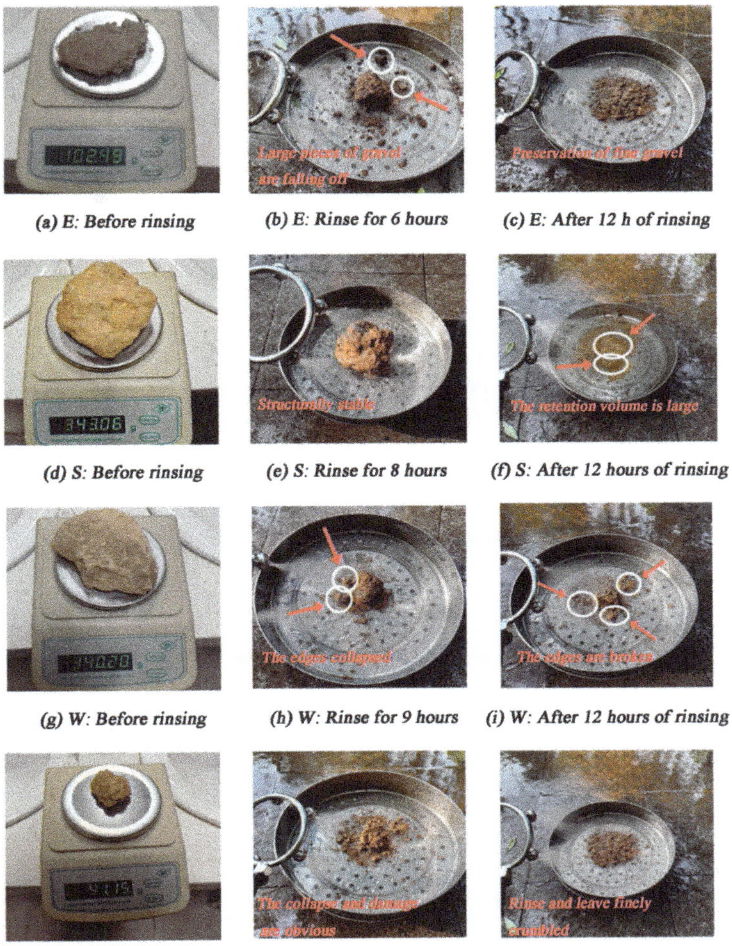

(a) E: Before rinsing ***(b) E: Rinse for 6 hours*** ***(c) E: After 12 h of rinsing***

(d) S: Before rinsing ***(e) S: Rinse for 8 hours*** ***(f) S: After 12 hours of rinsing***

(g) W: Before rinsing ***(h) W: Rinse for 9 hours*** ***(i) W: After 12 hours of rinsing***

(j) N: Before rinsing ***(k) N: Rinse for 6 hours*** ***(l) N: After 12 hours of rinsing***

Figure 23. Experimental process and result record of the spray rammed-earth test block (image source: the author draws the picture based on the experimental results).

Table 3. Experimental process and result record of spray rammed-earth test block changes.

Sample Name	Sample Initial Weight/g	Weight After 6 h of Showering/g	Weight After 12 h of Showering/g
East	102.49	68.7	35.6
South	343.06	96.3	58.4
West	340.2	86.8	43.2
North	41.15	24.5	10.5

Source: The author calculated based on experimental results.

4. Discussion

The experimental results demonstrate that the erosion resistance of rammed-earth walls is influenced by both mineral composition and microbial activity. These findings directly inform the "Save Old Houses" project by highlighting the need for orientation-specific repair strategies and the potential use of microbial regulators in conservation practice. The integration of mineralogical, microbiological, and microstructural data provides a comprehensive explanation for the varying erosion resistance observed in different wall orientations. The high quartz content and regular particle morphology contribute to the stability of the east and west walls (Figures 10, 18, 19 and 21), while the complex microbial community and presence of expansive minerals like Nontronite correlate with the severe erosion patterns on the south and north walls (Figures 13, 17, 20 and 22). The above experimental results show that the "Save Old Houses Action" project identified Songyang County as a pilot county in 2016. The "saving" focuses on cultural relics in traditional Chinese village. Still, it does not include national key cultural relics, Zhejiang provincial cultural relics, or privately owned cultural relics that are part of the general immovable cultural relics listed in the Third National Cultural Relics Survey. The funding for the restoration of old houses in Songyang is shared by three parties: the China Cultural Relics Protection Foundation subsidizes approximately 50%, the Songyang County Government contributes 20% to 30%, and the villagers cover the remaining cost. Let the villagers participate in part of the funds, and the results of repair and protection will be more long-lasting. According to the principle of combining protection and utilization, past repair projects have restored the previously collapsed and dilapidated old houses; these projects not only effectively protected ancient cultural relic buildings but also took into account future use by installing sanitary facilities, re-laying strong and weak electricity pipelines, rectifying the surrounding environment, and demolishing some later-built pig pens and dry toilets, which beautified the living environment and improved sanitary conditions. The guiding principle of maximizing the original appearance mandates the use of traditional craftsmanship and locally sourced materials like soil, wood, and stone in the restoration. The construction workers are mainly composed of the owners of the old house, villagers, and carpenters and stonemasons from neighboring villages. Compared with international conservation strategies, such as the use of sacrificial mud coatings in Europe or community-based maintenance in Latin America, the approach proposed in this study—integrating microbial regulation and local material science—offers a novel, low-intervention alternative. While European methods focus on replaceable protective layers, and Australian strategies emphasize drainage design, our findings highlight the potential of bio-based solutions tailored to subtropical humid climates. In the future, further research into the repair or improvement of rammed-earth walls still has great potential and development prospects. All of this will rely on existing experimental data and microstructural analysis, which will provide a reference for subsequent restoration projects.

5. Conclusions

Through sampling and test analysis of the rammed-earth wall materials of the Yangjiatang Village residence, we can determine its local anti-erosion performance: (1) The minerals on the wall surface are mainly composed of quartz, mullite, lepidocrocite, and nontronite. The minerals contained in all four directions are mainly quartz and lepidocrocite. (2) The particle morphology of the east and west samples is more regular, the particle size distribution is relatively concentrated, and the surface is smooth, indicating that the growth environment of these two directions is more uniform and stable during synthesis or processing. The south and north samples exhibit more complex morphology, a wider particle size distribution, and rougher surfaces with cracks, which may be attributed to factors such as crystal orientation, local temperature gradients, or stress concentration. (3) The overall corrosion resistance shows the order of sample W > sample S > sample N > sample E. (4) The surface degradation of rammed-earth walls is mainly caused by four factors: structural cracks, surface erosion, biological erosion, and roof damage. These factors together caused surface cracking and peeling (depth of 3–5 cm).

While this study provides a comprehensive analysis of the anti-erosion mechanisms of rammed-earth walls through multi-technique integration, it primarily focuses on short-term and micro-scale characteristics. The advantages lie in the detailed characterization of mineral composition, microbial community structure, and microstructural features, which offer immediate insights for restoration strategies. However, the limitations include the lack of long-term environmental monitoring and real-time data on microbial dynamics under varying climatic conditions. Future studies should incorporate continuous environmental sensing and periodic sampling to track changes over decades, thereby enhancing the predictive capacity of microbial regulation technologies. To improve the protection of traditional rammed-earth buildings, future efforts could integrate intelligent monitoring systems (e.g., IoT-based sensors for humidity, temperature, and crack propagation) with eco-friendly materials such as bio-based consolidants or geopolymer coatings. These systems would enable real-time damage assessment and early warning, while environmentally friendly materials would ensure compatibility with the original structure and reduce ecological impact. Such an approach would support sustainable conservation by combining traditional wisdom with modern technology. In the future, work can be integrated with intelligent monitoring systems to evaluate the long-term effectiveness of restoration technology and explore more eco-friendly materials, thereby advancing the sustainable conservation of traditional buildings.

While this study focuses on Yangjiatang Village, the methodological framework—combining multi-scale characterization techniques to decipher the complex interplay between material composition, microbiology, and environment—applies to the conservation of rammed-earth heritage in other humid subtropical regions worldwide. These insights into the dual role of microorganisms and the orientation-specific degradation mechanisms provide a scientific basis for developing targeted, sustainable conservation strategies beyond this specific case study, contributing to the global effort to preserve earthen architectural heritage.

Author Contributions: Conceptualization, Y.Z. and S.Y.; methodology, Y.Z., J.S., X.Z., Y.H., R.C. and S.Y.; software, Y.Z., J.S., X.Z., Y.H., R.C. and S.Y.; validation, J.S., and S.Y.; formal analysis, Y.Z., J.S., X.Z., Y.H., R.C. and S.Y.; investigation, Y.Z., X.Z., Y.H., R.C. and S.Y.; resources, J.S. and S.Y., Y.H., R.C. and S.Y.; data curation, Y.Z., X.Z., Y.H., R.C. and S.Y.; writing—original draft preparation, Y.Z., J.S. and S.Y.; writing—review and editing, Y.Z., J.S. and S.Y.; visualization, J.S. and S.Y.; supervision, J.S. and S.Y.; project administration, S.Y.; funding acquisition, S.Y. All authors have read and agreed to the published version of the manuscript.

Funding: This research was funded by Hangzhou City University's 2025 Provincial College Students' Innovation and Entrepreneurship Training Program Approved Project (Grant Number: S202513021110). The funders had no role in the study conceptualization, data curation, formal analysis, methodology, software, decision to publish, or preparation of the manuscript. There was no additional external funding received for this study.

Institutional Review Board Statement: Not applicable.

Informed Consent Statement: Not applicable.

Data Availability Statement: The datasets used and analyzed during the current study are available from Shuai Yang (samyang@zju.edu.cn) on reasonable request.

Acknowledgments: We would like to express our sincere gratitude to the students and the staff who assisted during the field survey. The authors extend their gratitude to Ma Xin from Shiyanjia Lab (www.shiyanjia.com accessed on 28 June 2025) for providing invaluable assistance with the XRD analysis.

Conflicts of Interest: The authors declare no conflicts of interest.

References

1. Xie, L.; Wang, D.; Zhao, H.; Gao, J.; Gallo, T. Architectural energetics for rammed-earth compaction in the context of Neolithic to early Bronze Age urban sites in the Middle Yellow River Valley, China. *J. Archaeol. Sci.* **2021**, *126*, 105303. [CrossRef]
2. Lu, P.; Tian, Y.; Storozum, M.; Chen, P.; Wang, H.; Wang, X.; Xu, J.; Jing, L.; Yan, L.; Zhang, L.; et al. Shifting Patterns of House Structures during the Neolithic-Bronze Age in the Yellow River Basin: An Environmental Perspective. *Land* **2021**, *10*, 574. [CrossRef]
3. Tang, X.; Shen, S.; Su, X. From rammed earth to stone wall: Chronological insight into the settlement change of the Lower Xiajiadian culture. *PLoS ONE* **2022**, *17*, e0273161. [CrossRef] [PubMed]
4. Akbulut, D.E.; Alibeyoglu, R.N. *Rammed Earth Construction: From Tradition to a Sustainable Future*; IntechOpen: London, UK, 2025. [CrossRef]
5. Ciancio, D.; Beckett, C. Rammed earth: An overview of a sustainable construction material. In Proceedings of the 3rd International Conference on Sustainable Construction Materials and Technologies, SCMT 2013, Kyoto, Japan, 18–21 August 2013; University of Edinburgh Research Explorer: Kyoto, Japan, 2013; pp. 1–9. Available online: http://www.claisse.info/2013%20papers/data/e053.pdf (accessed on 11 May 2024).
6. Avila, F.; Puertas, E.; Gallego, R. Characterization of the mechanical and physical properties of stabilized rammed earth: A review. *Constr. Build. Mater.* **2022**, *325*, 126693. [CrossRef]
7. Dobson, S. Rammed earth in the modern world. In *Rammed Earth Construction: Cutting-Edge Research on Traditional and Modern Rammed Earth*; Ciancio, D., Beckett, C., Eds.; CRC Press: London, UK, 2015; pp. 3–10.
8. Jaquin, P.A.; Augarde, C.E.; Gerrard, C.M. Chronological description of the spatial development of rammed earth techniques. *Int. J. Archit. Herit.* **2008**, *2*, 377–400. [CrossRef]
9. National Geographic Recommendation: The Last Secret Place in Jiangnan. Available online: https://www.sohu.com/a/776405018_100133113 (accessed on 10 May 2025).
10. Old Houses Are Fully Revived, and the Village Is Reborn. This Is How Yangjiatang Village in Songyang Protects and Utilizes the Ancient Town. Available online: http://www.qstheory.cn/2020-07/22/c_1126270926.htm (accessed on 17 May 2025).
11. Juan, P.V.; Bernardo, C.; Eduardo, J.R.; Laura, I.; Johana, A. Homogenization of rammed earth walls under changing environmental conditions. *Constr. Build. Mater.* **2024**, *425*, 136082. [CrossRef]
12. Liu, Y.; Zhou, R.; Zhao, Q. The weathering effect of moisture on ancient rammed-earth city walls. *J. Cap. Norm. Univ. (Nat. Sci. Ed.)* **2025**, *46*, 12–17. [CrossRef]
13. Joo, L.P.; António, S.S.; Cotrim, M. Mineralogical and microstructural characterisation of rammed earth and earthen mortars from 12th century Paderne Castle. *J. Cult. Heritage* **2020**, *42*, 226–239. [CrossRef]
14. Zhou, Z. The application of contemporary rammed earth materials in architectural design is divided. *Urban. Archit.* **2023**, *20*, 143–146. [CrossRef]
15. Martin-Antunes, M.A.; Espuelas, S. Recent developments in stabilized rammed earth: Testing protocols and the recommendations for standardization. *J. Build. Eng.* **2025**, *106*, 112436. [CrossRef]
16. Luo, L.; Chen, X.; Zhou, J. The Application of modern rammed earth construction technology in landscape architecture. *J. Build. Technol.* **2024**, *55*, 1655–1658. [CrossRef]
17. Jiang, M.; Jiang, B.; Lu, R. Thermal and Humidity Performance Test of Rammed-Earth Dwellings in Northwest Sichuan during Summer and Winter. *Materials* **2023**, *16*, 6283. [CrossRef] [PubMed]
18. Wan, L.; Wu, E.; Chi, X. A New Rammed Earth Construction Strategy Based on the Endogenous Development Model in Rural Areas: Taking the "One Specialization, One Village" Project in Rural Southwest China as an Example. *Archit. J.* **2024**, 105–111. [CrossRef]
19. Looking at the Path of Urban-Rural Integrated Development | Songyang: Ancient Villages and Ancient Charm, the Story of Shared Prosperity in Traditional Villages. Available online: http://www.news.cn/20241012/fbc64ff310a241c58a94ca37d2341714/c.html (accessed on 18 May 2025).
20. Yang, Q.; Davenport, B.; Zhang, L. How do older people conserve cultural heritage in rural areas: Evidence from Zhejiang, China. *J. Inst. Conserv.* **2025**, *48*, 31–50. [CrossRef]

coatings

MDPI

Article

Anti-Erosion Mechanism of Biological Crusts and Eco-Protection Technology Using Composite Biofilms for Traditional Rammed Earth Dwellings in Songyang County

Jiahui Yang [1], Ning Wang [1], Zebiao Huang [1], Yue Huang [2], Weilu Lv [1,*] and Shuai Yang [1,*]

1 Center for Cultural Heritage Studies, School of Art and Archaeology, Hangzhou City University, No. 51 Huzhou Street, Gongshu District, Hangzhou 310015, China; 32210119@stu.hzcu.edu.cn (J.Y.); wangning@hzcu.edu.cn (N.W.); 22304002@stu.hzcu.edu.cn (Z.H.)

2 Hangzhou Animation & Game College, Hangzhou Vocational & Technical College, No. 68 Xueyuan Street, 17 Xiasha Higher Education Park, Qiantang District, Hangzhou 314423, China; vivianwhang@zju.edu.cn

* Correspondence: luwl@hzcu.edu.cn (W.L.); samyang@zju.edu.cn (S.Y.)

Abstract: A typical county for traditional village conservation in China is Songyang County. It is renowned for its ancient rammed earth dwellings, which exhibit a unique microclimate and possess significant historical value. However, high precipitation and acid rain under the subtropical monsoon climate have caused severe surface erosion, including cracking and spalling. This study focuses on traditional rammed earth dwellings in Chenjiapeng Village, Songyang County, combining field surveys, experimental analysis, and microscopic characterization to systematically investigate erosion mechanisms and protection strategies. Techniques, such as drone aerial photography, X-ray diffraction (XRD), field-emission scanning electron microscopy (FE-SEM), and microbial diversity detection, were employed to elucidate the anti-erosion mechanisms of gray–green biological crusts on rammed earth surfaces. The results indicate that algal crusts enhance surface compressive strength and shear resistance through macroscopic coverage (reducing raindrop kinetic energy and moisture retention) and microscopic extracellular polysaccharide-cemented soil particles forming a three-dimensional network. However, acidic environments induce metabolic acid release from algae, dissolving cementing materials and creating a "surface protection-internal damage" paradox. To address this, a "transparent film-biofiber-acid inhibition layer" composite biofilm design is proposed, integrating a biodegradable polylactic acid (PLA) mesh, algal attachment substrates, and calcium carbonate microparticles to dynamically neutralize acidic substances, achieving synergistic ecological protection and cultural heritage authenticity. This study provides innovative solutions for the anti-erosion protection of traditional rammed earth structures, emphasizing environmental compatibility and sustainability.

Keywords: rammed earth dwellings; anti-erosion; biological crust; composite biofilm

1. Introduction

Rammed earth architecture, as the crystallization of China's traditional construction techniques that have lasted for thousands of years, carries important historical and cultural memories [1]. The production techniques of rammed earth architecture are relatively complicated and are divided into two stages, namely soil preparation and ramming. The ramming stage is further divided into substages, such as soil dumping, ramming, layer-by-layer ramming, maintenance, mold removal, wall beautification, and wall solidification [2] (Figure 1). After going through these steps repeatedly, a solid and stable rammed earth wall can be completed. Songyang County is located in the hilly area of southwestern Zhejiang and was built in the fourth year of the Jian'an era of the Eastern Han Dynasty (199 AD), with a history of over 1800 years [3]. Today, Songyang County is a demonstration county for the protection of Chinese traditional villages, and its number has long been among the top in the country. Up to now, a total of 78 villages have been listed in the Chinese Traditional Village List (data cover the first to the seventh batches), accounting for nearly one-fourth of the total in Zhejiang Province, and the county is known as the "county sample of classical China". It is noteworthy that a large number of traditional rammed earth dwellings have been preserved in these villages. The wall thickness of these buildings is generally between 40–60 cm, forming a unique microclimate regulation system. At the same time, the biggest advantage of rammed earth is its small thermal conductivity coefficient and good thermal inertia, which creates the characteristic of warmth in winter and a cool interior in summer for rammed earth architecture, making it a valuable treasure in the history of human architectural development [4]. However, influenced by the subtropical monsoon climate, the regional annual average precipitation exceeds 1800 mm. Meanwhile, with the rapid advancement of national industrialization under the background of a coal-dominated energy structure, the sharp increase in the number of coal-fired power plants, heavy industrial bases, and motor vehicle ownership has led to a continuous rise in emissions of acidic gases, such as sulfur dioxide and nitrogen oxides. This has resulted in frequent acid rain occurrences, causing pathological phenomena, such as cracking and flaky spalling on the surface of rammed earth. According to research, in the past decade, some rammed earth buildings in Songyang County have shown significant structural damage due to environmental erosion. Therefore, it is necessary to pay attention to the research on the anti-erosion properties of rammed earth dwellings.

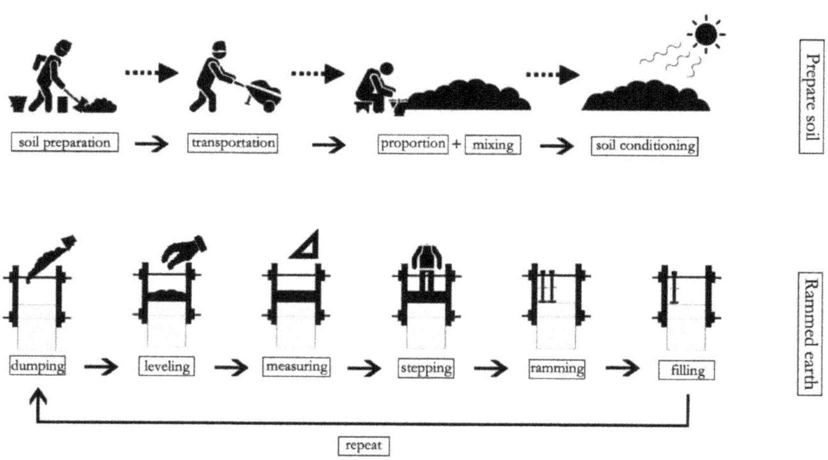

Figure 1. Manufacturing process diagram of rammed earth wall (drawn by the authors).

Traditional villages carry the excellent traditional culture of China and are an important link in the protection of our national culture [5]. Songyang County currently has 96 traditional Chinese villages, making it one of the counties in East China with the most traditional villages and the most abundant styles [6], and it has high research and protection value. Chinese scholars have conducted extensive research on rural areas in the past, mainly focusing on economic issues related to economic crops in these areas [7–10]. With the promotion of the rural revitalization strategy, there have been considerable achievements in the research into these issues. However, the protection of the ancient residential heritage of traditional villages is also worthy of our attention. Among them, traditional rammed earth dwellings, as one of the most typical types of dwellings [11], are widely used in traditional Chinese villages. In the process of historical evolution and inheritance, they have continuously adapted to the environment, forming a distinct and unique rammed earth culture [12,13]. However, with the passage of time, many young laborers have migrated out, leaving only elderly people living alone, which leads to the inability for ancient rammed earth dwellings to receive timely repairs and maintenance. In addition to natural environmental erosion, many ancient rammed earth dwellings have been damaged. In recent years, relevant scholars have carried out research in this regard, exploring the reasons for the erosion of rammed earth materials.

Some scholars have noticed that wind and rain erosion damage of rammed earth materials involves the physical wear and tear of the structure, which is the main form of damage to rammed earth walls. Under the condition of continuous loss of rammed earth materials, the cross-section of the wall becomes smaller and smaller, resulting in the continuous decline of the bearing capacity of the rammed earth structure [14] and leading to in structural instability and other problems of the building. Some scholars have also analyzed the causes of the erosion of rammed earth materials from a biological point of view, such as the root spread of plants in rammed earth materials during the growth process, which will cause cracks in rammed earth [15]. In addition, the erosion of rammed earth by plant litter will change the composition and structure of rammed earth, thus threatening the preservation of ancient dwellings [16].

In the research into the erosion resistance of rammed earth, scholars at home and abroad start from different angles to improve the erosion resistance of rammed earth materials. These angles are mainly divided into physical modification and chemical modification: some scholars choose common admixtures, such as cement and lime, to improve the erosion resistance of rammed earth [17]. In addition, some scholars have added plant fibers (coconut fiber, sisal fiber, sargassum seaweed fiber, etc.) to enhance the crack resistance and durability of rammed earth [18–22].

In addition to this physical and chemical modification, it has also been found that the use of microorganisms to induce calcium carbonate precipitation has increased the water resistance of rammed earth, resulting in cracking repairs (autogenous healing) [23,24], and that algae can form complexes through its photosynthetic metabolites, thereby enhancing the erosion resistance of the surface layer of rammed earth [25]. The authors believe that this method has the advantages of high ecological compatibility, endogenous self-healing function, and low environmental disturbance, but most of the current studies have not designed a scheme for the erosion resistance and crack resistance of the overall walls of rammed earth structures.

Therefore, this study took the traditional rammed-earth residential buildings in Chenjiapu Village, Songyang County, Zhejiang Province (which is a typical example of 78 villages in Songyang County that are included in the list of traditional Chinese villages and which is known as "villages on the cliff"), and systematically carried out research into the erosion and protection mechanisms of rammed earth buildings by combining field investigation, experimental analysis, and microscopic characterization research. First of all, through field investigation and drone aerial photography, researchers conducted a defects survey of representative ancient rammed earth residential buildings in Songyang County and recorded the cracks and eroded surface problems in detail. In terms of microscopic mechanism research, X-ray diffraction (XRD), field-emission scanning electron microscopy (FE-SEM), and microbial diversity detection were used to show that microbial communities and algae have a certain protective effect on the surface layer of rammed earth. Based on this comprehensive research, this analysis proposes a design for the protection of "eroded rammed earth walls and rammed earth with cracks". Specifically, this design is a transparent film–biocellulose–acid inhibition layer composite biofilm scheme. This provides an innovative solution for the protection of traditional rammed earth buildings.

2. Materials and Research Process

2.1. Study Area: Songyang Ancient Village in Lishui

Songyang County is located in the southwest of Zhejiang Province (Figure 2), at the junction of Zhejiang, Fujian, and Jiangxi provinces, with geographical coordinates of 28°14′ to 28°36′ north latitude and 119°10′ to 119°42′ east longitude. The central basin is called "Songgu Plain" because of its openness and flatness, and is also called "Songgu Basin". It preserves a complete local cultural context and has simple mountain characteristics in southwest Zhejiang [26]. Lishui City, where this area is located, has a subtropical monsoon climate, with four pronounced seasons, abundant sunshine, and abundant rainfall. The wind speed in Lishui varies significantly throughout the year, and the wind speed is higher in summer (June–August), especially in July and August, when the wind speed frequently reaches 3–4 m/s. In winter (December to February), the wind speed is relatively low, generally ranging from 1 to 2 m/s. Its temperature changes obviously throughout the year, and the temperature is higher in summer (June–August), usually between 25–35 °C, especially in July and August, when the temperature frequently reaches above 35 °C. In winter (December–February of the following year), the temperature is low, usually between 0–10 °C. High temperatures in summer and cold and humid conditions in winter are the main temperature characteristics. At the same time, the direct radiation in Lishui varies significant throughout the year, and the radiation intensity is higher in summer (June–August), especially in July and August, when the solar radiation intensity frequently reaches 600–800 Wh/m^2. In winter (December–February of the following year), the solar radiation intensity is low, usually between 200 and 400 Wh/m^2—(Figure 3).

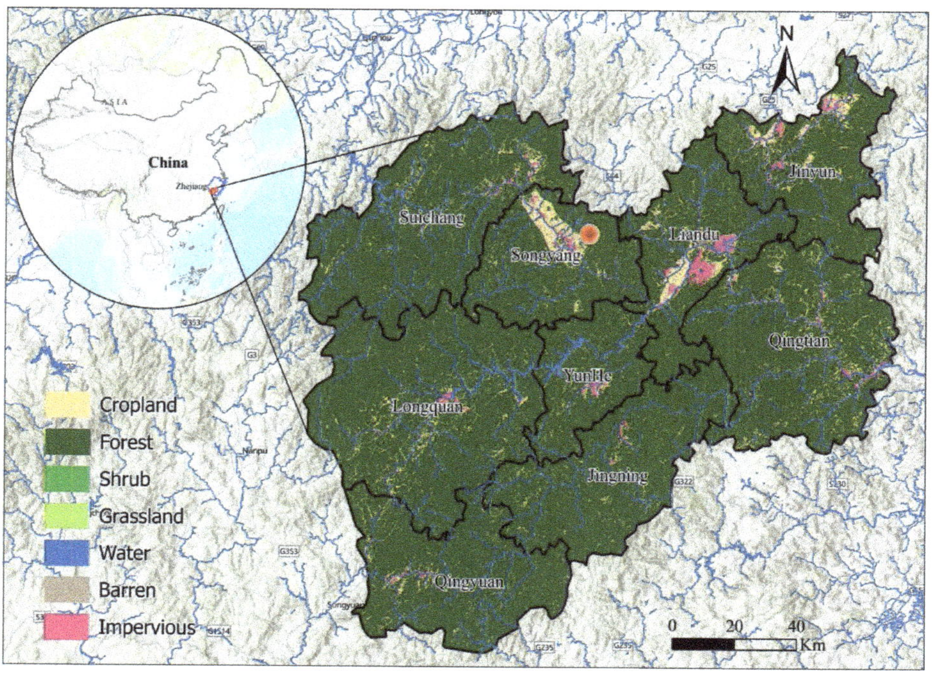

Figure 2. Geographical location of the study area.

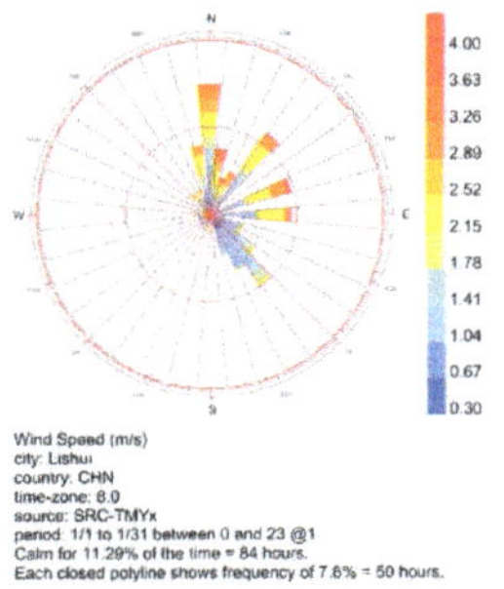

(**a**) January Wind environment simulation

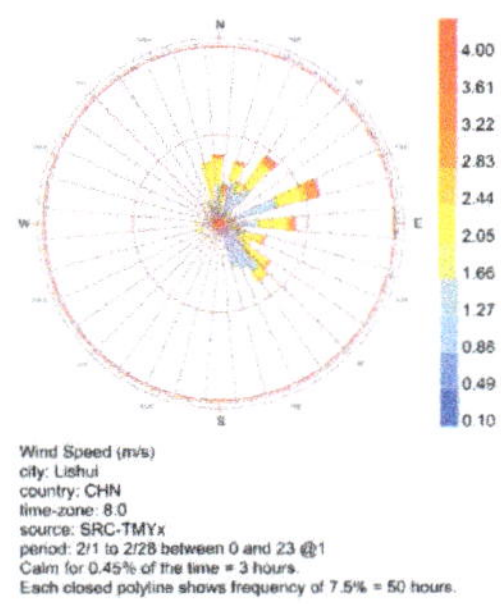

(**b**) February Wind environment simulation

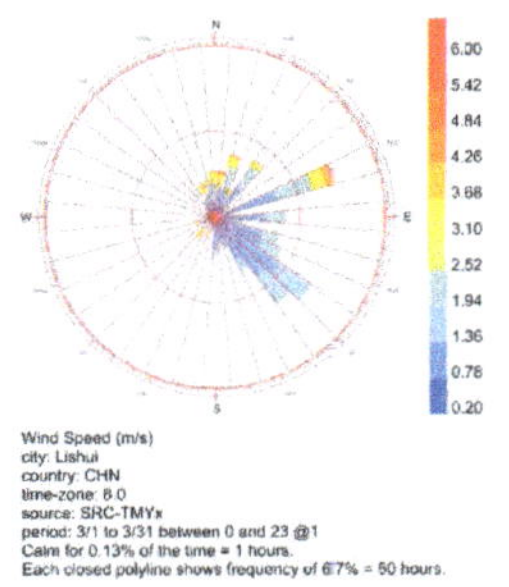

(**c**) March Wind environment simulation

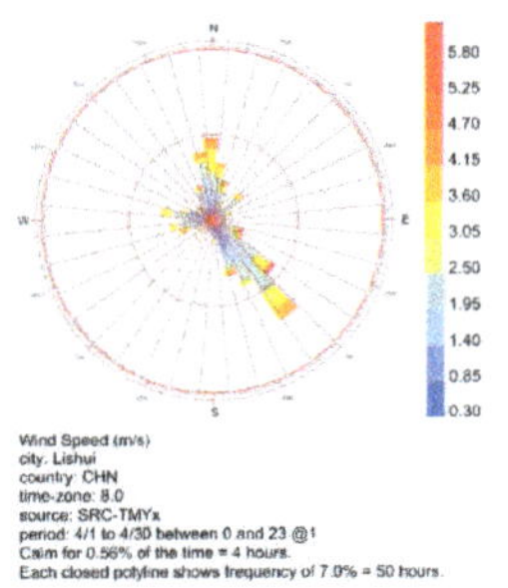

(**d**) April Wind environment simulation

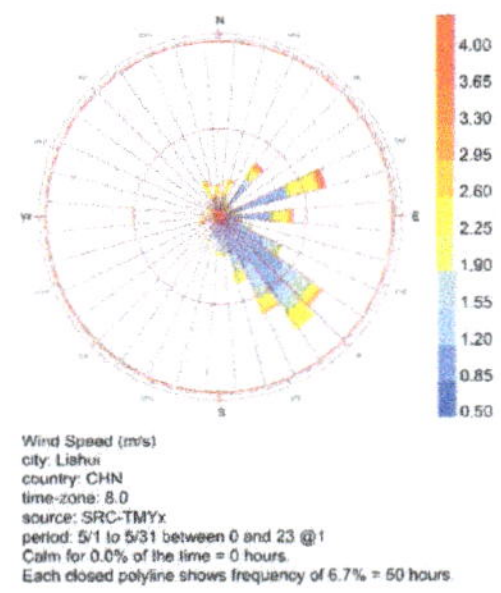

(**e**) May Wind environment simulation

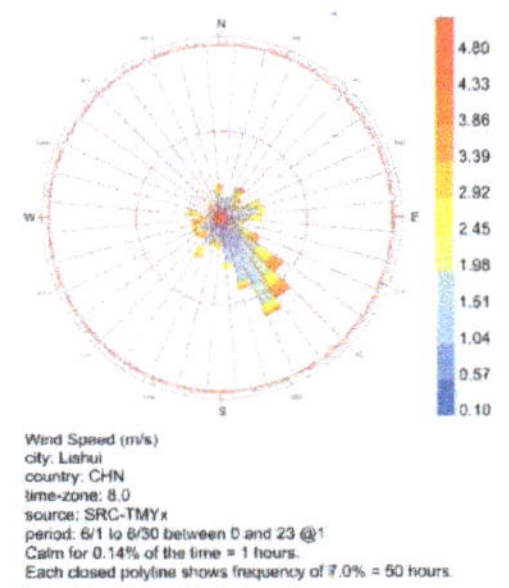

(**f**) June Wind environment simulation

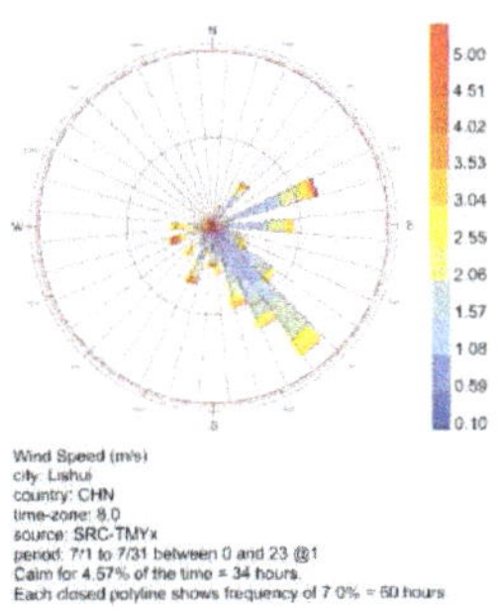

(**g**) July Wind environment simulation

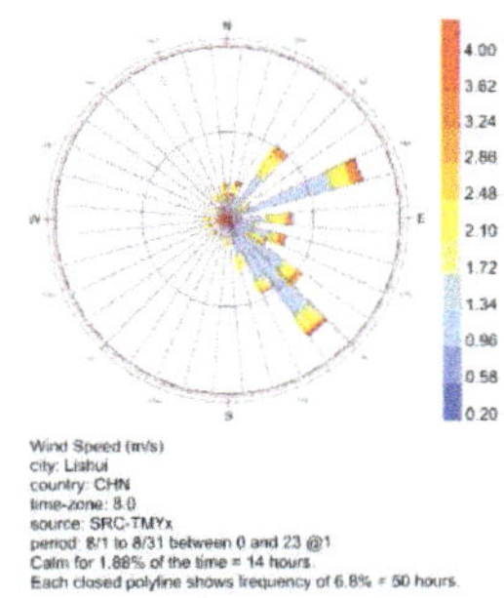

(**h**) August Wind environment simulation

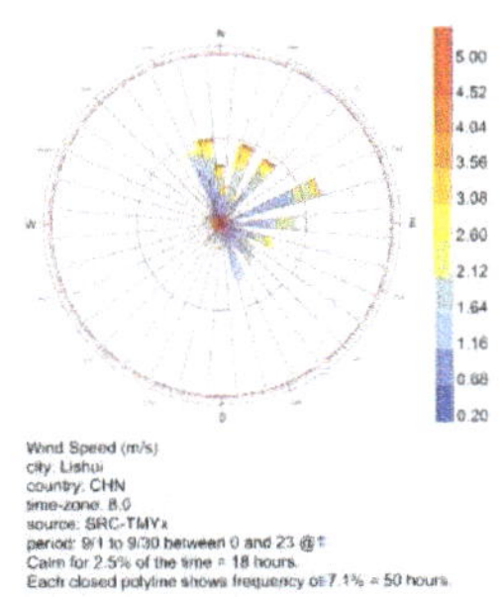

(**i**) September Wind environment simulation

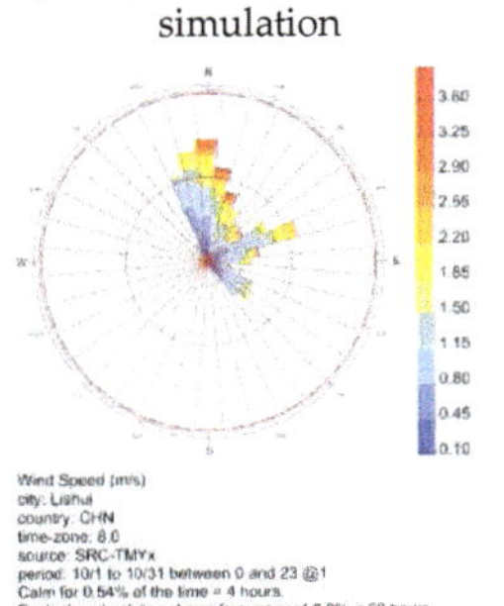

(**j**) October Wind environment simulation

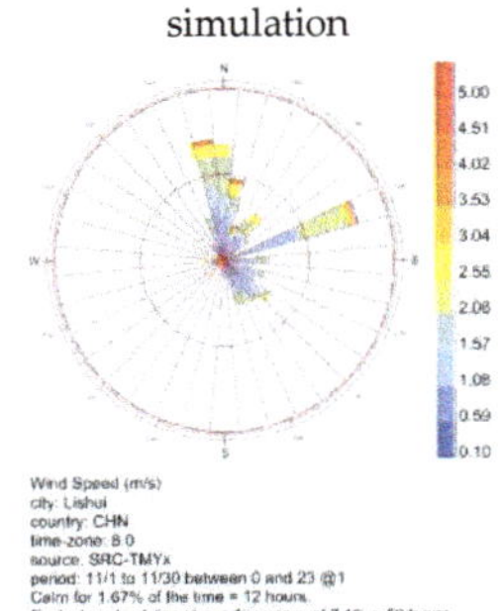

(**k**) November Wind environment simulation

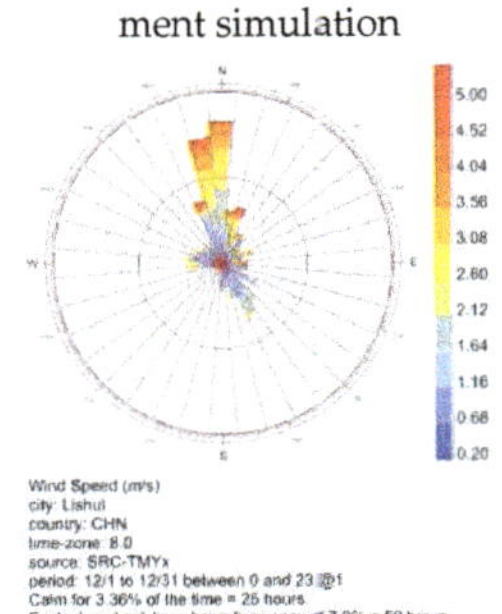

(**l**) December Wind environment simulation

Figure 3. Analysis of annual wind frequency rise in Lishui (source: authors, drawn using Ladybug).

Songyang County in Lishui City, with its unique geographical environment and climatic conditions, preserves a large number of ancient villages and rammed earth residential buildings with characteristics of the mountainous areas in southwestern Zhejiang. These buildings are not only important carriers of local folk culture but also valuable resources for studying the anti-erosion performance of rammed earth buildings. However, due to the significant changes in annual solar radiation and precipitation, the long-term preservation of rammed earth residential buildings is unfavorable, and erosion is prone to occurring. For example, high humidity and frequent precipitation in summer can cause the rammed earth walls to absorb water and expand, and the long-term wet expansion and dry shrinkage cycle will accelerate the cracking and peeling of the walls [27]. At the same time, higher wind speeds in summer will also accelerate the evaporation of moisture on the surface of the walls, leading to dry and cracked wall surfaces. Over time, the stability of rammed earth residential buildings is greatly reduced [28]. In recent years, with the enhanced awareness of traditional village protection, the ancient villages in Songyang County have gradually received attention from the academic community and the government. Therefore, this study selects Chenjiapu Village in Songyang County as the research area (Figure 4).

Figure 4. Aerial photography of Chenjiapu Village, Songyang County (source: photo by the authors).

In recent years, with the growing awareness of protecting traditional villages, the ancient villages in Songyang County have drawn increasing attention from academia and the government. Chenjiapu Village, Songyang County, has many rammed earth dwellings and ancestral hall relics. This study selected Chenjiapu Village in Songyang County as the study area (Figure 4). This choice not only helps us deeply understand the anti-erosion mechanisms of traditional rammed earth buildings but also offers a scientific basis for the conservation and restoration of ancient villages.

2.2. Field Investigation and Sample Collection

Because the terrain of ancient villages in Songyang County is complex, and the buildings are mostly built according to the mountains (Figure 5), the researchers chose to use a DJI Mini 4K, a consumer-grade drone with stability and flexible handling, to perform aerial photography tasks. The UAV is equipped with a high-definition camera and a three-axis mechanical pan/tilt system, which can keep the picture stable in complex environments. It is equipped with 1/2. A 3-inch CMOS sensor supports shooting 12-megapixel photos and 4K/30fps video. Enhanced Wi-Fi tech enables long-distance real-time HD image transmission. Regarding parameter configuration, the researchers synthesized the suggestions of several aerial photography experts and optimized the default settings of UAVs. The researchers set the ISO value of the camera to automatic mode to dynamically adapt to the complex environment of light and shadow interlacing in ancient villages (such as patio shadow and roof intense light). At the same time, according to the cloudy and sunny weather conditions on that day, the researchers fixed the aperture to f/2.8. Researchers set the shutter speed to 1/80 s to 1/2000 s for dynamic adjustment to avoid the overexposure of highlights. Finally, the researchers chose to shoot at 2 pm on 20 December 2024. During implementation, the UAV's flying height was kept at 12–18 m. For the damaged rammed earth residential wall, the shooting distance was reduced to 2–4 m.

Figure 5. Aerial photography of ancient villages in Songyang County (source: photo by the authors).

The traditional earthen wall construction in the Songyang area is usually rammed using the long-slab rafter construction method. Based on this method, the size of the rammed block of the formed earth wall is larger, generally about 1.8 m in height and 1.5 m in length (the specific size can be adjusted according to the actual situation). These large-size rammed blocks are combined with small, staggered masonry, so that there are certain gaps in the earthen wall structure. These gaps not only function as expansion joints but also have a certain structural rationality and can adapt to the natural deformation of the material. However, in the actual use process, due to factors, such as foundation settlement, camber of blocks, improper construction technology, or later maintenance loss, these gaps may gradually expand, which will adversely affect the earth wall structure. Specifically, gap enlargement may cause the following problems: loss of structural stability (the expansion of the gap may lead to uneven settlement of the overall structure of the earth wall, thereby weakening the stability of the structure); increased risk of rainwater infiltration (when a gap expands, rainwater may penetrate into the wall through the gap, causing the soil to soften and further exacerbate the damage to the wall); reduced strength (the widening of the gap may weaken the shear and compressive resistance of the soil wall, reducing its load-bearing capacity); appearance damage: the obvious enlargement of the gap will affect the appearance integrity of the earthen wall, reducing its historical and cultural value.

During the field trip, the researchers found that the residential house at No. 22 Chenjiapu West Road has some typical problems. These included the following: (1) erosion and recessing, mostly at the turning point of the rammed earth wall interface, with an average depth of 5–7 cm (Figure 6(bA)); (2) cracks in the wall (Figure 6(bB)); and (3) obvious traces of water erosion (Figure 6(bC)). At the same time, researchers found that the west wall of the house was less eroded, and the rammed earth wall was grayish–green. The same is true of the other three parts of the wall with a gray–green surface, which makes us curious about their peculiarities.

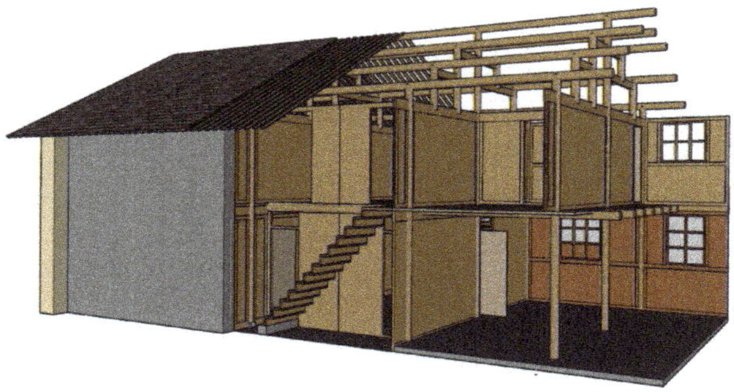

(a)Research on the model diagram of the main part of the dwelling.

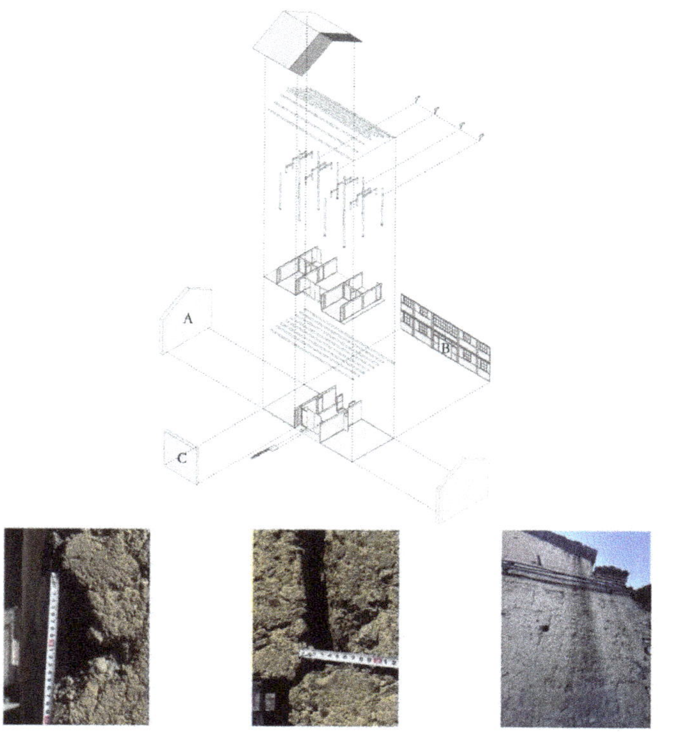

A.Eastern rammed earth wall issue:undercutting recess.

B.South rammed earth wall issue:longitudinal cracking.

C.North rammed earth wall issue:water erosion marks.

(b)Explosion diagram

Figure 6. Explosion diagram of the sampling site problem distribution.

Based on the aforementioned research background, this study conducted systematic sampling of rammed earth walls exhibiting grayish–green surface crust characteristics on four facades (eastern, northern, southern, and western) of the target vernacular dwelling. Using archaeological scalpels, researchers performed stratified scraping along the normal line direction of wall surfaces to collect crust layer powders, with the sampling depth strictly controlled at 10 ± 0.5 mm and individual sample mass was precisely maintained at 10 ± 0.2 g (Figure 7). The samples were coded by orientation as A1 (east), A2 (north), A3 (south), and A4 (west). A dual-sealing preservation protocol was implemented, as follows: primary containment involved sealing compacted earth powders in inert PVC tubes (φ25 × 100 mm), followed by secondary packaging in anti-static aluminum foil bags, effectively isolating environmental contaminants and suppressing moisture exchange. The time interval between sampling completion and laboratory characterization was rigorously limited to 48 h to ensure that test data accurately reflect material behavioral characteristics under in situ environmental conditions. Notably, the research team uniformly applied a specialized glutinous rice glue interface agent for rammed earth walls to the damaged areas incurred during the investigation. The surface was then dyed using a mixture of loess slurry and natural mineral pigments to ensure consistency with the surrounding rammed earth matrix. Finally, a toothed trowel was employed to replicate the original wall texture, with particular emphasis on restoring the horizontal textural characteristics between rammed layers, thereby preserving the visual continuity of the wall structure.

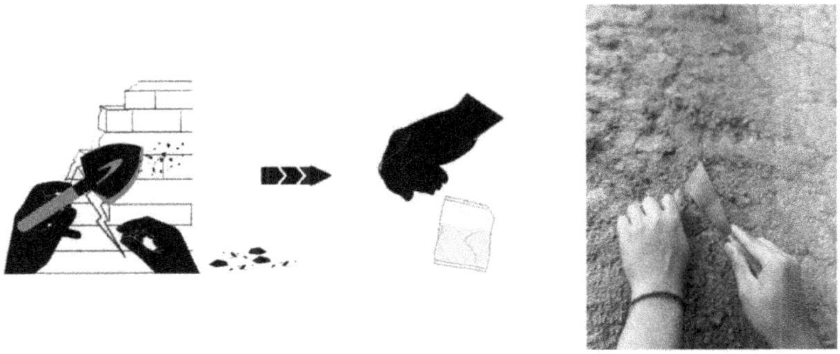

Figure 7. Sampling process for rammed soil.

Before the sampling work was implemented, researchers had preliminary contact with the owner of the house, systematically explaining the research objectives and work procedures. After on-site communication and confirmation, it was found that the building is currently vacant, effectively avoiding potential interference risks from recent human activities to the rammed earth walls. After obtaining verbal authorization from the property owner in accordance with the law, our team conducted a systematic investigation and sampling work on the historical building in accordance with the technical specifications for cultural heritage protection.

2.3. Research Methods and Processes

To elucidate the synergistic interaction mechanisms between rammed earth substrates and algal microbial communities in erosion resistance and crack suppression within the gray–green wall coating system described in Section 2.2, this investigation employed a multidisciplinary experimental approach combining microbial community diversity analysis, X-ray diffraction (XRD) phase characterization, and scanning electron microscopy (SEM) microstructural observation. All o these investigations were carried out using four sets of 10 ± 0.2 g samples.

During the sample preparation process, the rammed earth powder was not subjected to chemical fixation treatment. Since the sample is a complex of inorganic minerals and microorganisms, the integrity of the microstructure can be maintained by directly collecting the sample while avoiding physical damage. After collection, the sample was naturally dried at room temperature for 48 h to ensure complete dehydration and avoid the interference of moisture in a vacuum environment. Subsequently, the sample was sputter-coated with a gold–palladium alloy (Au-Pd, 60:40 ratio), and the coating thickness was 15 nm.

2.3.1. Detection of the Microbial Community Diversity Composition Spectrum

Microbial community diversity detection was used to analyze the microbial community composition on the surface of rammed earth walls. The testing process included total microbiome DNA extraction, PCR amplification of target fragments, magnetic bead purification and recovery of amplification products, fluorescence quantification of amplification products, sequencing library preparation (sequencing libraries were prepared using Illumina's TruSeq Nano DNA LT Library Prep Kit), and high-throughput sequencing. Microbial community diversity detection determines the main microbial species and their relative abundance on rammed earth walls. This offers essential data for understanding microorganisms' role in the erosion process. The complete flow of bioinformatics analysis is shown in Figure 8.

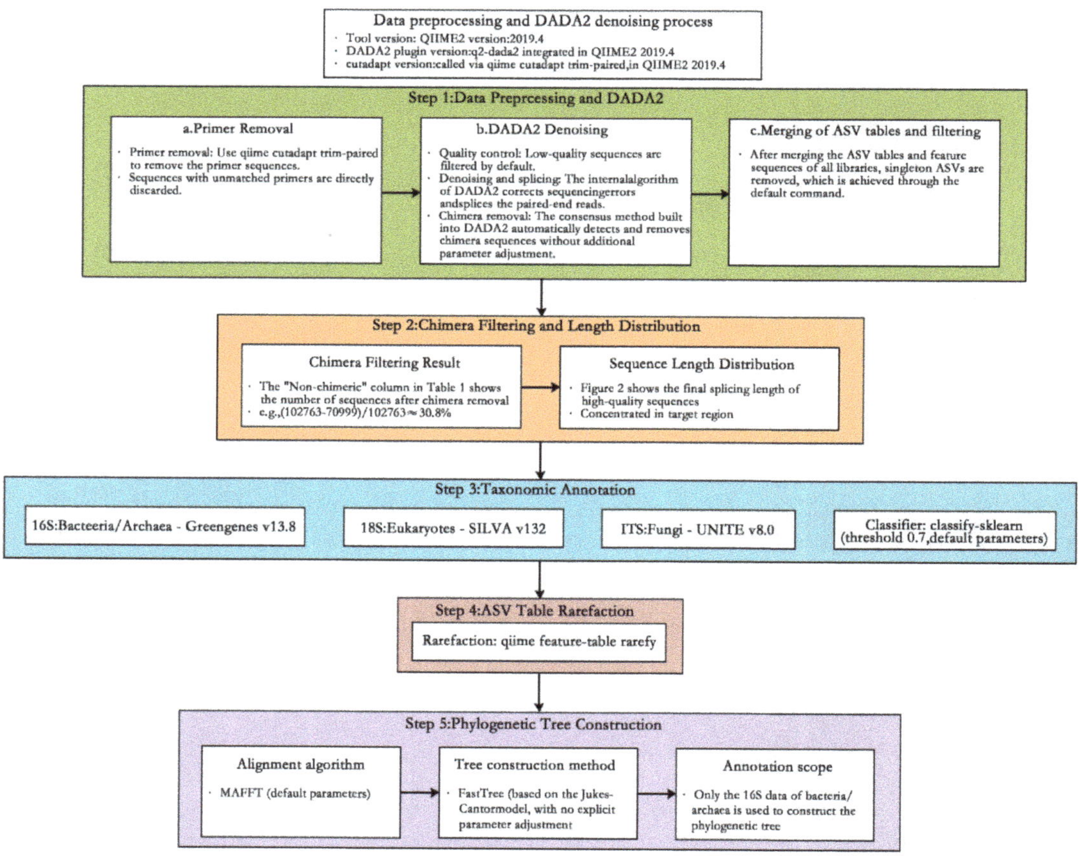

Figure 8. Analysis of the process.

In this study, the Illumina paired-end sequencing platform (MiSeq/NovaSeq) was employed to conduct amplification sequencing targeting the V3-V4 variable regions of the 16S rRNA gene. The primer design was based on the conserved regions of the 16S rRNA gene, and universal primer pairs, namely 338F (5′-ACTCCTACGGGAGGCAGCAG-3′) and 806R (5′-GGACTACHVGGGTWTCTAAT-3′), were used. Moreover, based on the conserved regions of the 16S rRNA gene, the V3-V4 hypervariable regions (~470 bp) could be specifically amplified. Meanwhile, in this study, an amplicon sequence variant (ASV) was used to describe the microbial community diversity.

2.3.2. XRD Test

Samples for XRD testing were collected from rammed earth walls in the east, west, north, and south directions. The thickness of each sample was 10 mm and the weight was about 10 g. The detection instrument used was the SmartLab high-resolution the X-ray diffractometer produced by Rikaku Co., Ltd. of Akishima, Japan, with the reference standard being SY/T 5163-2018 "X-ray Diffraction Analysis Method for Clay Minerals and Common Non-clay Minerals in Sedimentary Rocks". XRD testing identifies the main minerals in rammed earth (such as quartz, feldspar, calcite, montmorillonite, etc.) through X-ray diffraction patterns, clarifying their crystal structure and relative content. The type of mineral directly affects the soil's pH value, porosity, ion exchange capacity, and other factors that determine the colonization environment of microorganisms and algae.

2.3.3. Scanning Electron Microscope (SEM) Testing

Scanning electron microscopy (SEM) testing was used to study the chemical elements and the microstructure of four different samples to determine the growth and morphology of algae on the surface of the rammed earth. The instrument used was a ZEISS Gemini SEM 360(Carl Zeiss AG, Jena, Germany) with a magnification range of 500 to 5000×. In this study, the test parameters were carefully set to balance the requirements of resolution and sample protection. A low accelerating voltage of 1.5 kV was selected to minimize damage to organic components, such as algal extracellular polysaccharides, caused by the electron beam. The working distance was set to 6.5 mm to optimize the collection efficiency of the secondary electron signals and imaging clarity. An in-lens secondary electron detector was chosen for signal detection.

3. Data Analysis

3.1. Analysis of the Microbial Diversity Detection Data

The microbial community diversity results of the four samples (Figure 9a) show that Sample 4 is dominated by cyanobacteria, actinomycetes, and α-proteobacteria. This indicates that the gray–green crust layer of the rammed earth wall is a composite crust layer dominated by algae and supplemented by fungi. It is worth noting that the raw sequencing data (raw reads) of all samples averaged approximately 134,030 reads per sample, with the specific distribution as follows: A1 (128,762 reads), A2 (143,257 reads), A3 (141,631 reads), and A4 (122,470 reads). After quality control processing (removing chimeric and singleton sequences), the average depth of the sequences used for analysis (post-QC reads) was 96,580 reads per sample, specifically A1 (70,685 reads), A2 (117,501 reads), A3 (108,811 reads), and A4 (89,324 reads). The above data indicate that the sequencing depth fully covers the microbial community diversity, providing a reliable data basis for subsequent analysis.

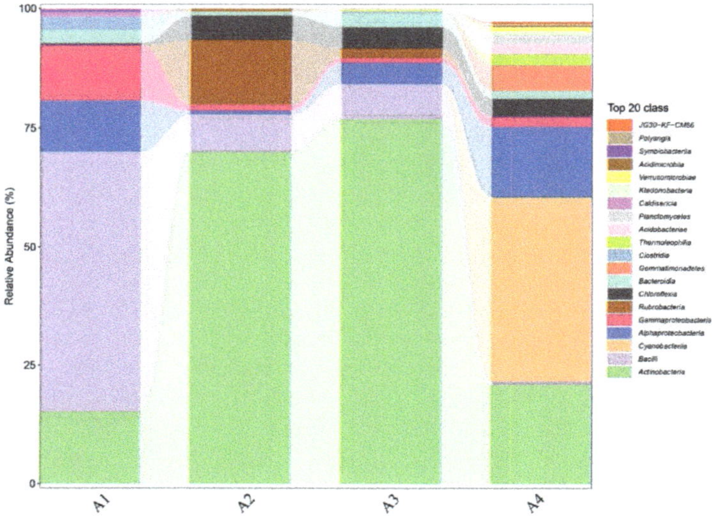

(**a**) Top 20 Bacterial Classes in Soil Microbiome by Relative Abundance

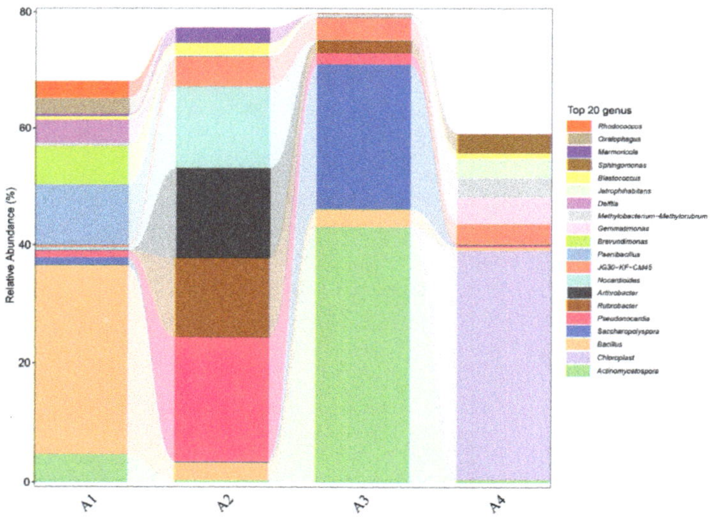

(**b**) Top 20 Bacterial Genera in Soil Microbiome by Relative Abundance

Figure 9. Test results of microbial community diversity.

The photosynthesis of cyanobacteria fixes carbon dioxide and produces organic matter that provides nutrients to other microorganisms. At the same time, actinomycetes and α-proteomycetes promote the circulation of soil organic matter and enhance the stability of the crust layer by decomposing organic matter. This algae-based, fungal-supplemented composite crust layer not only protects the rammed earth walls but also reduces water evaporation and wind erosion.

Figure 9b shows that Sample 4 has more acid-producing microbes. These may generate a large amount of acid, which can erode rammed earth walls.

According to the analysis results presented in Figure 10, there are significant differences in the microbial community compositions of the four samples, and a certain degree of overlap exists among the samples. Sample A4 contains the largest number of ASVs/OTUs (1163, accounting for 33.49% of the total), with a relatively large proportion of unique microbial communities, which might originate from an environment rich in organic matter, facilitating the growth of various microorganisms. Samples A1, A2, and A3 contain 927 (26.69%), 751 (21.62%), and 526 (15.15%) ASVs/OTUs, respectively. The overlapping area between A3 and A2 is relatively small, while A1 shares more common elements with the other samples. Particularly, sample A2, although it has 751 ASVs/OTUs, has very limited overlap with the other samples, with only 8 (0.23%) ASVs/OTUs being shared with A3. These differences suggest that different samples possess distinctiveness in the microbial community structure, and their shared parts might reflect the influence of certain environmental or treatment conditions on the microbial community.

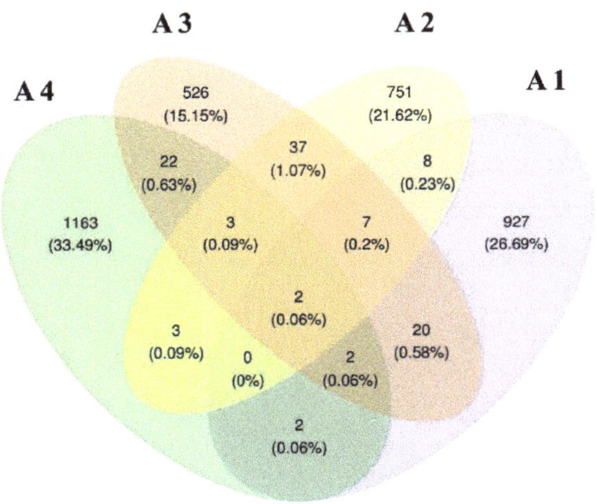

Figure 10. Test results of the Venn diagram for the ASV/OTU analysis.

In Figure 11, the abundant distribution of different bacterial taxonomic groups in each sample is presented. Based on the analysis of bacterial composition at the phylum level, the percentages of OTU occurrence frequency for each sample are displayed, classified by phylum. The main bacterial phyla include Firmicutes, Bacteroidota, and Actinobacteriota, etc. Among them, Firmicutes occupies a relatively high abundance in most samples, particularly in sample A4, demonstrating a prominent characteristic of the dominant bacterial group. Additionally, the abundances of Bacteroidota and Actinobacteriota exhibit certain variations among different samples, which might be associated with the discrepancies in different environments or treatment conditions.

Figure 12 presents a heatmap depicting the relative abundance of various species groups in diverse samples. This heatmap underwent clustering analysis via the UPGMA (unweighted pair group method with arithmetic mean) tree, and the Euclidean distance was utilized to compute the distances of the distribution of species groups within the samples. The samples in the figure were sorted in accordance with the similarity of species groups and further underwent UPGMA clustering based on the Pearson correlation coefficient between samples, ultimately presenting the abundance information of species in each group of samples. It can be discerned from the figure that there are remarkable differences in the structure of species groups among different groups. Red regions signify a higher abundance of the species in the sample, while blue regions imply a lower abundance. Specifically, certain species in A1 (such as Alicyclobacillus and Hymenobacter) possess relatively high abundances, as shown in the red areas of the heatmap, indicating the predominant position of these bacterial groups in this group of samples. In contrast, the distribution of species groups in A4 tends to be more homogeneous, with some species (such as Methylobacterium and Methylorubrum) having dominant abundances in the blue areas, suggesting that these species might be in a lower abundance state in this group of samples. Additionally, some species, such as Pseudorhizobium, Acinetobacter, and Georgenia, have higher abundances in A1 but lower abundances in other groups, which may be related to the environmental characteristics or specific ecological factors of this group. Bacterial groups, such as Bacillus and Thermocymbion, exhibit strong advantages in both A1 and A2 but are relatively scarce in other samples, which is closely associated with specific growth environments or nutrient supplies.

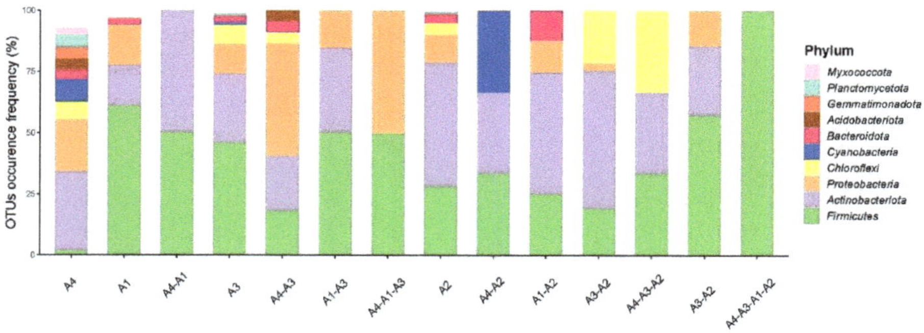

Figure 11. Test results of the ASVs/OTUs in different regions of the Venn diagram.

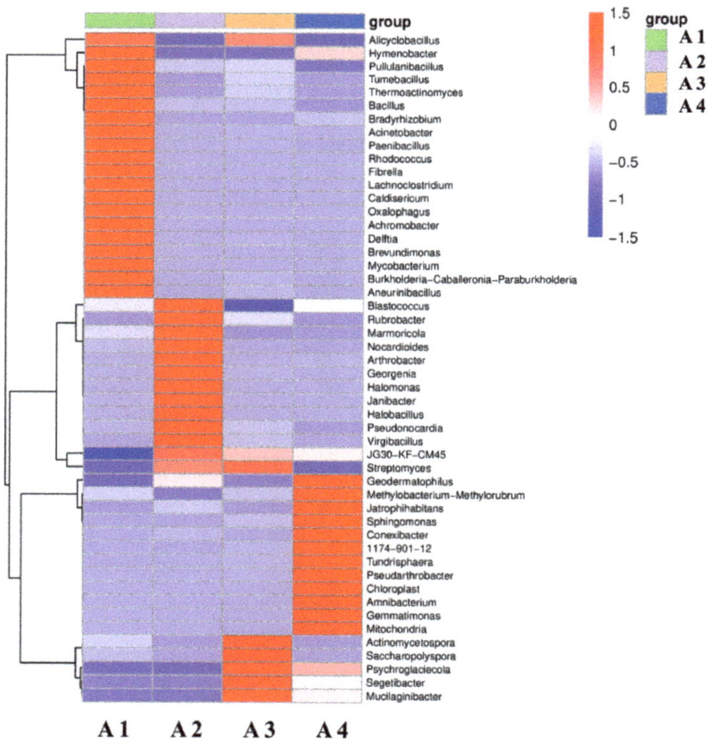

Figure 12. Test results of the heatmap of species composition at the genus level for species clustering.

The statistical data of sample alpha diversity and beta diversity are shown in Table 1 and Figure 13.

Table 1. Index of microbial diversity of bacterial flora.

Sample	Chao1	Simpson	Shannon	Pielou_e	Observed_Species	Faith_pd	Goods_Coverage
A1	968.74	0.983988	6.86743	0.692413	967.5	46.4861	0.999625
A2	844.566	0.966454	6.60856	0.683689	812.4	49.6321	0.998894
A3	637.006	0.931195	5.41482	0.584105	617.5	49.965	0.999273
A4	1211.75	0.947574	6.97516	0.682314	1195	65.8746	0.998868

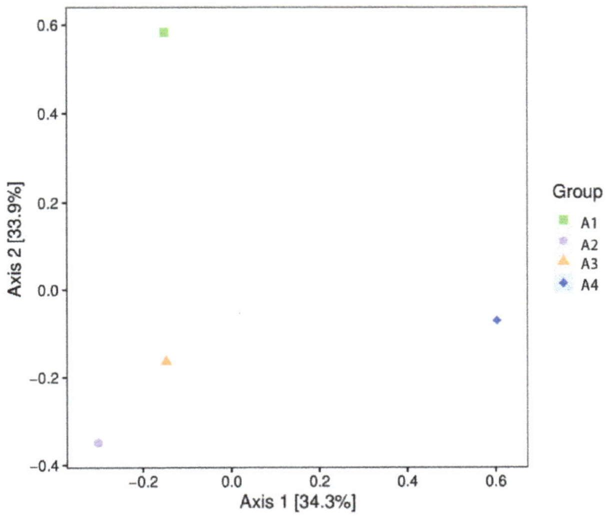

Figure 13. Two-dimensional ordination plot of samples from the PCoA analysis.

3.2. XRD Test Data Analysis

Based on the whole-rock mineral composition analysis of the rammed earth material (Figure 14), there was no significant difference in the soil mineral composition of each sample. The main components included quartz (SiO_2), potassium feldspar ($K_2O{\cdot}$alumina$\cdot 6SiO_2$), and serpentine ($Mg_6[Si_4O_{10}](OH)_8$) (Figure 9). XRD spectra showed that the quartz and feldspar minerals had significant strength in the characteristic peaks of the crystal plane, which showed obvious differences in mineral composition and crystal structure from the rammed earth with cement and lime, indicating that the rammed earth building was formed by natural soil compaction.

It is worth mentioning that the nutrients K, Mg, and Si, released by mineral weathering, can promote the growth of microorganisms and algae, especially diatoms and other silicon-dependent algae. The physical structure of the mineral may also provide a habitat for the microbial community. This also explains the presence of gray–green algae on the surface of the rammed earth walls.

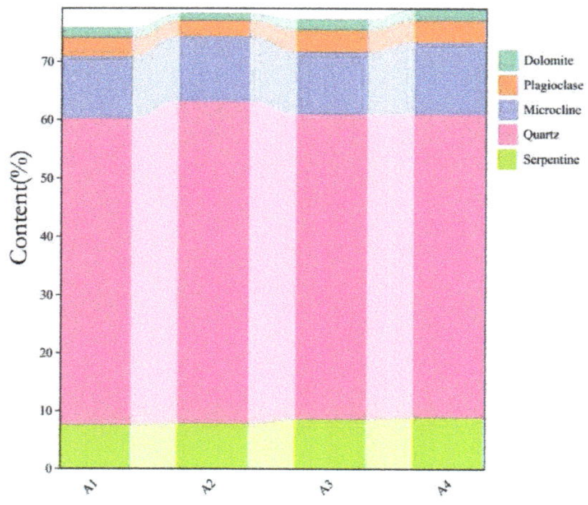

Figure 14. XRD test results.

3.3. SEM Test Data Analysis

The SEM analysis results (Figure 15) demonstrate high consistency in the elemental composition across the four samples. Notably, an abnormally elevated calcium signal was detected in Sample 2 from the northern residential area sampling site. Field investigation revealed this anomaly was attributed to secondary calcium carbonate which had migrated to the sampling area along water flow paths, originating from the lime-based plaster layer of adjacent drainage pipes subjected to long-term water erosion.

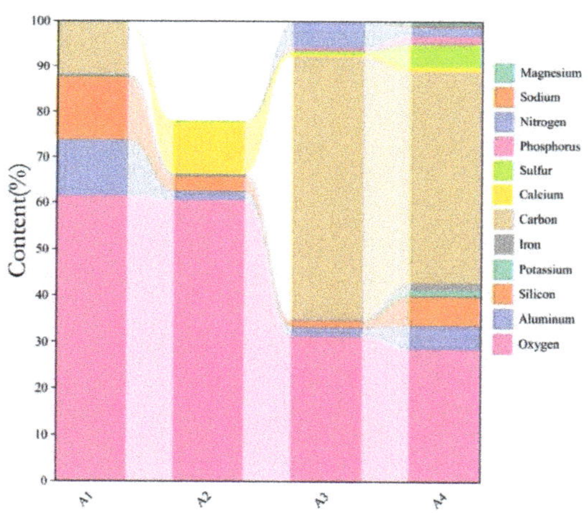

Figure 15. SEM test results.

Figure 16 illustrates the structural characteristics of the A4 connection layer at microscopic scales ranging from 500× to 5000×. The micro-morphology in Figure 16a reveals algae exhibiting a typical filamentous growth pattern, where their three-dimensional network structure physically entraps matrix particle surfaces through entanglement. Figure 16c further demonstrates that algal filaments form cross-scale biological bridges between particles, enhancing interparticle binding forces through bioadhesion. This consolidation mechanism shows significant homology with the protective effects of biocrust layers documented in the Shanhaiguan Great Wall study [29].

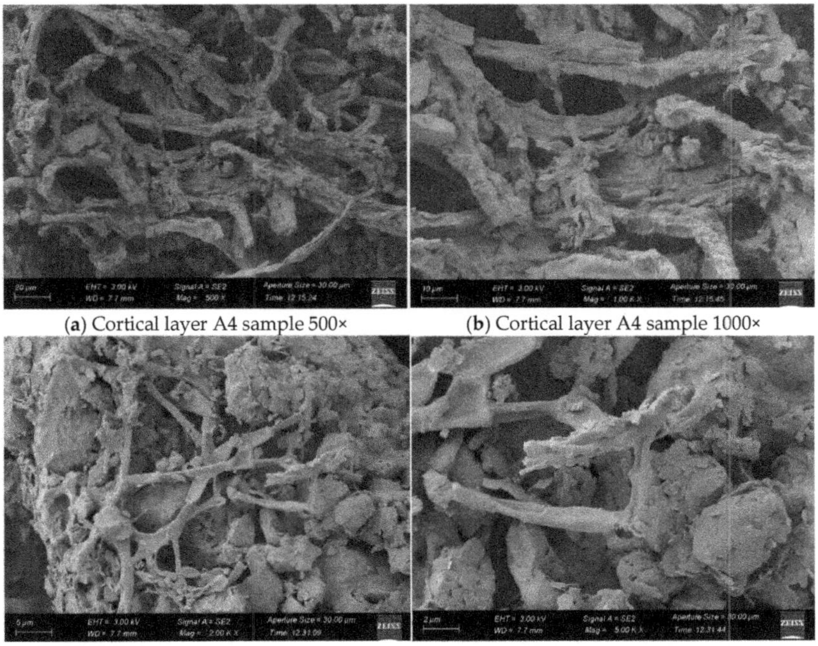

(**a**) Cortical layer A4 sample 500×
(**b**) Cortical layer A4 sample 1000×
(**c**) Cortical layer A4 sample 2000×
(**d**) Cortical layer A4 sample 5000×

Figure 16. SEM photo of the A4 sample.

4. Results and Discussion

The experimental results revealed that algal crusts exhibited a dual anti-erosion mechanism in soil protection. At the macro-protection level, the algae skin can significantly reduce the impact effect of raindrop kinetic energy on the soil surface and reduce the intensity of the splash erosion of rammed earth walls by forming a biological covering. Algal biofilms' water-holding properties enable them to quickly absorb residual water on rammed earth walls after rain, reducing moisture that could otherwise cause structural damage.

Although algae may be damaged during sampling, preventing them from normal secretion, existing studies have shown that algal photosynthetic metabolites (including biopolymers, such as polysaccharides, lipids, and proteins) form stable complexes with soil particles through cementation [30,31]. More noteworthy, the algal filaments and symbiotic hyphae construct a three-dimensional network structure in the soil pores, combined with the viscoelastic matrix formed by exopolysaccharides, to construct a multi-scale stable system with hierarchical characteristics. This structure greatly enhances the topsoil's mechanical strength and shear resistance, thereby improving the rammed earth wall's surface stability.

However, during the study, the researchers found that although the west wall is more intact than the other walls, it still has erosion problems. Therefore, the researchers hypothesized that the protective mechanism of the algal biocrust layer on the rammed earth wall is two-sided. Based on previous scholars' research on the protection mechanism of biocrusts on the rammed earth wall of the Great Wall of Shanhaiguan, the researchers can conclude that when in a dry and rainless environment, just like the hot temperature conditions in Songyang County in summer, the anti-erosion effect of algal biocrusts is more significant. This is because the stable structure formed by the exopolysaccharide it secures is more stable. However, due to the cold and humid winters in Songyang County and the frequent occurrence of acid rain in recent years, the algae on the crust layer accelerate the release of acidic substances in an acidic environment, which dissolve the cemented materials (such as calcium carbonate) inside the wall, resulting in a loose wall structure and small cracks, which may cause internal damage in the long term [32]. Algal biofilms are not found all over the wall but are present in localized areas. For example, they may be present where rain washes repeatedly, such as the wall under the eaves (where the rainwater flows down the eaves) and the parts of the wall that are sunken or poorly drained (wet areas where water has accumulated for a long time). Therefore, in future conservation work, researchers should focus on the areas where algae find it easy to grow, inhibit the erosion of rammed earth walls by the acidic substances secreted by algae under certain conditions through chemical means, and optimize the compound materials of natural fiber and lime, to make algae biofilm play a better anti-erosion role.

Compared with the existing anti-erosion coatings for rammed earth buildings, the biocrust layer has the following significant advantages (Table 2).

Table 2. Comparison of biological crust and artificial erosion control coatings.

Characteristic	Biological Crust Layer	Artificial Anti-Erosion Coating
Environmental friendliness	Completely natural, promotes ecological circulation	Some contain chemicals which may pollute the environment.
Protection dimension	Multi-path dynamic protection (against wind erosion, water erosion, and salinity damage)	Single physical or chemical protection
Maintenance costs	Almost zero	Requires regular maintenance, which is costly
Adaptability	Naturally regulates temperature and humidity, adapts to climate change	Dependent on material properties, susceptible to aging effects
Cultural compatibility	Preserves the original historical appearance, in line with the needs of cultural relics protection	May alter the appearance of the building; careful selection of materials is required

5. Design Strategy

5.1. Design Scheme

It is found that the MICP (microbially induced calcite precipitation) technology has been extended to the field of rammed earth building protection, and the idea of a microbial coating for tulou wall anti-erosion and microbial repair of tulou cracks is proposed, while its application effect is evaluated by experimental means. The results of the study demonstrate that MICP technology can be used as an effective method to alleviate the weathering and crack repair of the surface layer of rammed earth buildings [23,24].

Based on the two main problems found in the rammed earth wall in this paper, namely (1) the existence of cracks and (2) wind and rain erosion, we determined the dual erosion resistance mechanism of algal biocrusts, as revealed by experimental analysis combined with MICP technology. Thus a composite biofilm design scheme of a "transparent film-biofiber-acid suppression layer" was proposed in this study (Figure 17).

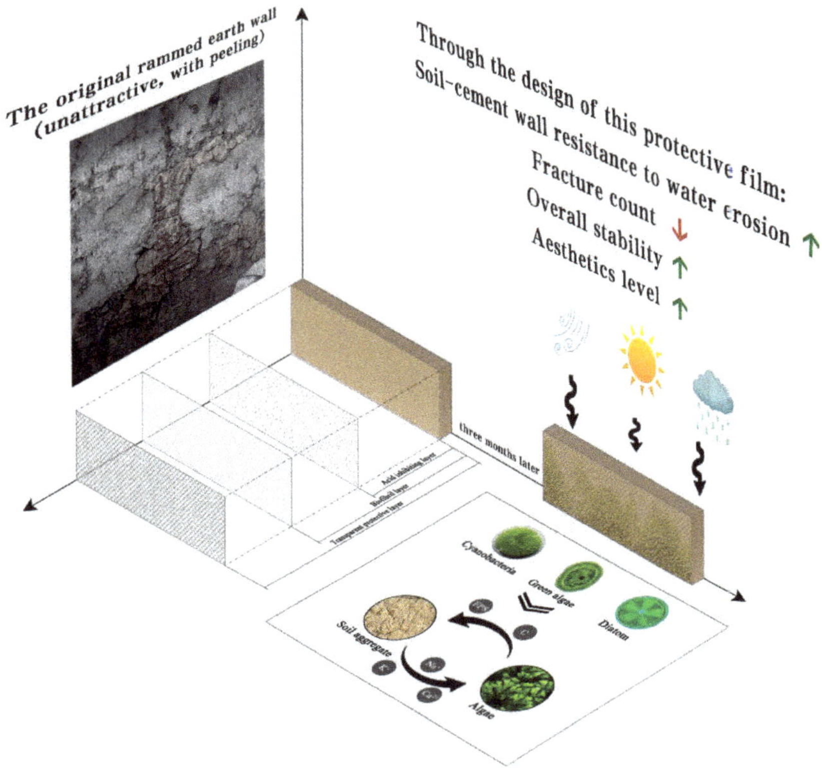

Figure 17. Design concept diagram.

The transparent protective layer is made of a degradable polylactic acid (PLA) mesh (mesh size 2–5 mm), which takes into account the water permeability and raindrop kinetic energy dispersion function to avoid erosion. The biofiber layer is woven into a grid (mesh 5–10 mm) with hemp fiber and coconut shell fiber, which is pre-impregnated with a nitrogen and phosphorus nutrient solution and then paved on the surface of rammed earth, and the algae attachment base (with cyanobacteria as the core, accounting for 60%–70%, supplemented by green algae and diatoms, accounting for 30%–40%) is attached to the pore structure of its rough surface, forming a three-dimensional network complex cemented with extracellular polysaccharide (EPS) soil particles. At the same time, the researchers designed an acid suppression layer to solve the problem of the dissolution of rammed earth wall cementing materials caused by the release of acidic metabolites from algae, that is, calcium carbonate microparticles were embedded in the fiber grid to dynamically neutralize the acidic substances metabolized by algae and alleviate the contradiction of "external protection and internal damage".

Glutinous rice syrup is used between the layers for bonding. First, glutinous rice flour and deionized water were mixed at a 1:4 mass ratio. Subsequently, the colloidal mixture was uniformly coated onto the contact interfaces between the PLA mesh, biological fiber layer, and acid-resistant layer using a spatula, with the coating thickness controlled to 0.5–0.8 mm to ensure no residual bubbles remained. Finally, the layers were laminated layer-by-layer in the order of "PLA mesh → biological fiber layer → acid-resistant layer". The rationale behind the researchers' choice of using glutinous rice syrup as an interlayer adhesive was multifold. Firstly, glutinous rice paste, being a natural organic material, does not pose toxic threats to the microbial community. Secondly, with its weak alkalinity, the glutinous rice paste can work in synergy with the calcium carbonate microparticles in the acid-resistant layer. This synergistic effect enables the neutralization of acidic substances resulting from environmental acid rain and algal metabolism, thereby mitigating chemical corrosion at the bonding interface.

5.2. Design Advantages

This design scheme leverages the experimentally revealed anti-erosion mechanism of algal biofilms to achieve coordinated optimization of ecological compatibility and dynamic protection through a composite structure of a "transparent film-biological fiber-acid inhibition layer". The transparent polylactic acid (PLA) mesh, with its high light transmittance, supports algal photosynthesis while preventing water-induced erosion and washing. Cyanobacteria, as the core component, secrete large amounts of extracellular polysaccharides (EPS), forming a gelatinous protective layer that enhances surface stability. Green algae assist cyanobacteria by contributing to the formation of multi-layer biofilms, thereby increasing the redundancy of protection. Meanwhile, diatom shells increase surface roughness, promoting the attachment of other algal spores, and secrete silicate colloids that aid in filling micro-cracks in rammed earth [25,33]. This synergistic algal attachment base, with cyanobacteria as the primary component and green algae and diatoms as supporting elements, effectively improves the anti-erosion capability, compressive strength, and long-term durability of rammed earth walls.

Moreover, the biodegradable materials (PLA mesh and fiber layer) will naturally biodegrade over time after the algal biofilm stabilizes (after approximately 3 months), creating a self-sustaining protection system that requires no artificial maintenance. This system preserves the historical aesthetic of rammed earth walls through its natural materials and transparent structure, preventing the authenticity of cultural heritage from being compromised by traditional chemical coatings. Compared to conventional physical or chemical protection methods, this innovative design achieves a harmonious balance of efficient erosion resistance, ecological sustainability, and cultural compatibility through a multi-scale cooperative mechanism.

The sandwich-structured composite material developed by Masoud et al. (with an outer layer composed of polyhydroxyalkanoate (PHA) and an interlayer of polylactic acid (PLA) reinforced with sol–gel modified cellulose microfibers (CMF)) significantly enhanced the tensile modulus (approximately 30% increase) and hygrothermal stability of the PLA/CMF composites through chemical modification. Experimental results demonstrated that the modified CMF improved interfacial bonding with the PLA matrix, effectively inhibiting moisture penetration and microbial degradation [34]. Accelerated aging tests under simulated hot–humid conditions (temperature: 25–35 °C, humidity: 70%–85%) revealed that the degradation rate of the PLA layer remained below 15% over 12 months, with mechanical property retention exceeding 80%, indicating its potential adaptability to the subtropical monsoon climate of Songyang County.

The design of this composite biofilm strictly adheres to the principles of minimal intervention and reversibility in the protection of cultural heritage. Its minimal intervention is manifested in non-invasive coverage, which only attaches to the surface layer of the rammed earth to avoid structural damage. Degradable polylactic acid (PLA) mesh, natural plant fibers (hemp, coconut shell), and algal components are used, while irreversible chemical additives, such as cement and resin are completely abandoned. The protective performance is dynamically regulated through the photosynthesis of algae and the acid inhibition layer, reducing the need for artificial intervention. Reversibility is achieved through controllable degradation: PLA and plant fibers gradually degrade into harmless carbon dioxide and water after the algal crust is stabilized (after about 3 months; the degradation rate is less than 15% within 12 months), and a multi-level removal strategy is preset to deal with potential residues—if artificial intervention is required, a soft brush is used to gently brush for physical stripping, or lactic acid bacteria fermentation broth (pH 5–6) is sprayed to accelerate the hydrolysis of PLA, avoiding damage to the rammed earth body. Compared with single physical or chemical protection schemes, this design achieves the unification of a high-efficiency anti-erosion, ecological cycle and cultural compatibility through a multi-scale collaborative mechanism.

6. Conclusions and Outlook

6.1. Research Conclusions

Through the sampling and experimental analysis of the rammed earth wall materials of the dwellings, the conclusions of the local erosion resistance of the rammed earth dwellings in Songyang County can be concluded as follows:

1. Algal biological crusts exhibit a dual anti-erosion mechanism. From a macroscopic perspective, the biological overlay reduces the kinetic energy of raindrops and enhances water-holding characteristics, thereby absorbing surface water and mitigating damage to the wall through dry–wet cycles. From a microscopic perspective, algal photosynthetic metabolites (such as polysaccharides, lipids, and proteins) bind with soil particles to form a stable complex, while myofilament and symbiotic hyphae create a three-dimensional network. This structure significantly enhances the mechanical strength and shear resistance of the surface layer.
2. Environmental conditions dynamically influence the protective efficacy of biological crusts. In dry, rain-free environments, exopolysaccharides form a stable structure, providing robust anti-erosion effects. However, in humid and acidic conditions, algae release acidic metabolites that dissolve cemented materials (e.g., calcium carbonate), leading to a loosened internal structure and the formation of micro-cracks. This creates a contradictory effect of "external protection and internal damage".
3. The composite bio-membrane design of the "transparent film-biological fiber-acid inhibiting layer" proposed in this study achieves the unified protection of rammed earth walls with efficient anti-erosion effects, an ecological cycle, and cultural authenticity through a multi-scale cooperative mechanism.

6.2. Insufficiencies and Prospects

The design of the "transparent film-biofiber-acid-resistant layer" is still in the theoretical stage. For example, based on existing research [35], the behavior of natural fibers (such as hemp fibers and coconut shell fibers) under different environmental conditions is as follows: when exposed to ultraviolet (UV) light, natural fibers are prone to photodegradation, resulting in a decrease in mechanical properties. It is necessary to rely on the outer protective material (such as PLA film) to indirectly mitigate the UV damage, but there is still a risk of aging during long-term exposure. During the humidity cycle, their hydrophilicity leads to moisture absorption expansion and drying shrinkage, which may cause interfacial micro-cracks. However, the water retention characteristics of the fibers can assist in buffering the moisture impact and alleviating the dry–wet stress of the rammed earth. In terms of microbial colonization, the porous structure of the fibers provides an attachment base for beneficial microorganisms (such as cyanobacteria and actinomycetes), promoting the formation of biofilms and the cementing effect of extracellular polysaccharides, and enhancing the anti-erosion ability. However, cellulose-based fibers may be degraded by fungi or actinomycetes, and it is necessary to balance the colonization effect by regulating the microbial community (such as pre-soaking in nutrient solution or using an acid inhibition layer). Therefore, in the future, we will carry out pilot experiments in Songyang County, verify our research through practice and data optimization, and promote its transformation into a scientific and practical ecological protection technology for rammed earth.

Author Contributions: Conceptualization, W.L.; Data curation, Z.H.; Formal analysis, N.W., Z.H. and Y.H.; Investigation, J.Y., Z.H. and W.L.; Methodology, N.W., W.L. and S.Y.; Project administration, S.Y.; Resources, Y.H.; Visualization, N.W. and S.Y.; Writing—original draft, J.Y.; Writing—review and editing, J.Y. and S.Y. All authors have read and agreed to the published version of the manuscript.

Funding: This research was funded by the Humanities and Social Sciences Youth Foundation, Ministry of Education of the People's Republic of China (grant number: 22YJCZH161); the Ministry of Education of the People's Republic of China Industry-University-Research Project (grant number:231105342131517); and the Zhejiang Provincial Philosophy and Social Sciences Planning Project (grant number: 24NDQN150YBM). The funders had no role in the study conceptualization, data curation, formal analysis, methodology, software, decision to publish, or preparation of the manuscript. There was no additional external funding received for this study.

Institutional Review Board Statement: Not applicable.

Informed Consent Statement: Not applicable.

Data Availability Statement: The datasets used and analyzed during the current study are available from Shuai Yang (samyang@zju.edu.cn) on reasonable request.

Acknowledgments: We would like to express our sincere gratitude to the students and the staff who assisted during the field survey. The authors extend their gratitude to Ma Xin from Shiyanjia Lab (www.shiyanjia.com accessed on 22 February 2025) for providing invaluable assistance with the XRD analysis.

Conflicts of Interest: The authors declare no conflicts of interest.

References

1. Chen, L.; Lan, Z.; Wei, C.; Ouyang, D.; Shi, B.; Chen, P.; Wang, M.; Xie, T. Practice and Reflection on Rammed Earth Architecture: The Case Study of Tiles Hill-Xiangshan Campus Reception Centre in China. *Buildings* **2024**, *14*, 4034. [CrossRef]
2. Zhang, Z.; Wang, F.T.; Zhu, Z. The Attempts of Architectural Preservation and Restoration for Traditional Chinese Village "Rescuing Old Houses Action" in Songyang. *Archit. J.* **2021**, *1*, 34–37. [CrossRef]
3. Zhang, B. The value of rammed earth dwellings in Hanzhong country. *Henan Sci. Technol.* **2010**, 87. [CrossRef]
4. Yang, Z.; Chen, H.; Wang, Y.; Lv, Q. Research on the innovation of traditional village protection and utilization mode--Take the Songyang traditional village as an example. *Rural. Econ. Sci.-Technol.* **2023**, *34*, 171–174. [CrossRef]
5. Zheng, Y.; Zhang, G.; Tan, S.; Feng, L. Research on Progress of Forest Fire Monitoring with Satellite Remote Sensing. *Agric. Rural. Stud.* **2023**, *1*, 0008. [CrossRef]
6. Khadka, B. Rammed earth construction: A promising sustainable eco-friendly housing solution. In *Sustainable Materials in Civil Infrastructure*; Woodhead Publishing: Cambridge, UK, 2024; pp. 25–43. [CrossRef]
7. Prudente, L.T. Life-cycle assessment of the environmental impact of earthen builds versus conventional builds on human health. In *Sustainability and Toxicity of Building Materials*; Woodhead Publishing: Cambridge, UK, 2024; pp. 289–306. [CrossRef]
8. Zhang, J.; Lu, J.; Deng, W.; Beccarelli, P.; Lun, I.Y.F. Thermal comfort investigation of rural houses in China: A review. *Build. Environ.* **2023**, *235*, 110208. [CrossRef]
9. Zeng, Z.; Pan, Y.; Chen, Y. Study on construction techniques of traditional rammed earth dwellings in hongling village of bashu area. *Dev. Small Cities Towns* **2023**, *41*, 112–119. [CrossRef]
10. Yuan, R.; Song, X.R.; Li, J.K.; Ding, C. Research on Rammed Earth Composition and Building Craftsmanship in ancient emplacement at Jiaoshan. *Sci. Conserv. Archaeol.* **2010**, *22*, 20–24. [CrossRef]
11. Manzano-Fernández, S.; Mileto, C.; Vegas López-Manzanares, F.; Cristini, V. Conservation and In Situ Enhancement of Earthen Architecture in Archaeological Sites: Social and Anthropic Risks in the Case Studies of the Iberian Peninsula. *Heritage* **2024**, *7*, 2239–2264. [CrossRef]
12. Ou, C. Research on the application of new rammed earth construction in the restoration and reconstruction of ancient villages. *Residence* **2023**, 159–162.

13. Liu, L. Experiential Research and Application of Modified Rammed Earth Wall Materials in Northwest Sichuan Region. Master's Thesis, Southwest University of Science and Technology, Mianyang, China, 2021.
14. Zhao, X.W. The Influence on Cracks and Reaction Mechanism of Plants on the Conservation of Earthen Sites. Master's Thesis, Northwest A & F University, Xianyang, China, 2014.
15. Liu, T. The Fluence of Plants on the Rammed Earth of Earthen Sites in Xi'an. Master's Thesis, Northwest A & F University, Xianyang, China, 2019.
16. Shi, J.; Li, M.; Wang, Y.; Wang, C. The test study of the engineering properties of the earth material of the earth rammed construction. *Sichuan Build. Sci.* **2006**, *32*, 86–87. [CrossRef]
17. Prabakar, J.; Sridhar, R.S. Effect of random inclusion of sisal fiber on strength behavior of soil. *Constr. Build. Mater.* **2002**, *16*, 123–131. [CrossRef]
18. der Kalu, J.; Kraus, I.; Peri´c, A.; Kraus, L. Shear Strength of Reproduced Soil Mixtures Based on Samples from Rammed Earth Walls from Eastern Croatia. *Appl. Sci.* **2021**, *12*, 11708. [CrossRef]
19. Sabbà, M.F.; Tesoro, M.; Falcicchio, C.; Foti, D. Rammed Earth with Straw Fibers and Earth Mortar: Mix Design and Mechanical Characteristics Determination. *Fibers* **2021**, *9*, 30. [CrossRef]
20. Tripura, D.D.; Gupta, S.; Debbarma, B.; Deep, R.S.S. Flexural strength and failure trend of bamboo and coir reinforced cement stabilized rammed earth wallettes. *Constr. Build. Mater.* **2020**, *242*, 117986. [CrossRef]
21. Raj, S.; Mohammad, S.; Das, R.; Saha, S. Coconut fibre-reinforced cement-stabilized rammed earth blocks. *World J. Eng.* **2017**, *14*, 208–216. [CrossRef]
22. Affan, H.; Touati, K.; Benzaama, M.-H.; Chateigner, D.; El Mendili, Y. Earth-Based Building Incorporating Sargassum muticum Seaweed: Mechanical and Hygrothermal Performances. *Buildings* **2023**, *13*, 932. [CrossRef]
23. Liu, S.Y.; Yu, J.; Han, L. Experimental study on the water resistance of microbial induced calcium carbonate precipitation on lime-soil surfaces. *Chin. J. Rock Mech. Eng.* **2019**, *38*, 1718–1728. [CrossRef]
24. Liu, S.Y.; Yu, J.; Zeng, W.L.; Peng, X.Q.; Cai, Y.Y.; Tu, B.X. Effect of microbial induced calcium carbonate precipitation on repairing cracks in lime-soil. *Chin. J. Rock Mech. Eng.* **2020**, *39*, 191–204. [CrossRef]
25. Sharma, M.; Satyam, N.; Reddy, K.R. Strength enhancement and lead immobilization of sand using consortia of bacteria and blue-green algae. *J. Hazard. Toxic Radioact. Waste* **2020**, *24*, 04020049. [CrossRef]
26. Wei, J. Study on traditional villages and dwellings in Songyang county, Zhejiang province. *Beijing Inst. Fash. Technol.* **2021**. [CrossRef]
27. Pei, Q.; Zhang, B.; Shang, D.; Guo, Q.; Huang, J.; Zhu, J. Characters of Temperature Field of Rammed Earth Wall in Arid Environment. *Coatings* **2022**, *12*, 735. [CrossRef]
28. Su, N.; Yang, B.; Chen, W.; Xu, L.; Li, Y. Fluence of Prevailing Wind Direction on SappingQuantity of Rammed Earth Great Wall of the Ming Dynasty. *Coatings* **2022**, *12*, 707. [CrossRef]
29. Liu, L.; Zhang, Y.; Guo, L.; Cao, H.; Dai, Z.; Zhao, Z.; Guo, Y.; Li, D.; Zheng, L.; Li, T. Experimental test and mechanism analysis of soil crust erosion resistance of rammed earth Great Wall in rainy season. *Sci. Rep.* **2024**, *14*, 9069. [CrossRef]
30. Mager, D.M.; Thomas, A.D. Extracellular polysaccharides from cyanobacterial oil crusts: A review of their role in dryland oil processes. *J. Arid. Environ.* **2011**, *75*, 91–97. [CrossRef]
31. Li, X.; Cheng, X.B.; Peng, B. Research and Application of Biological Soil Crusts in the Remediation of Sandy Soil. *Mod. Hortic.* **2022**, *45*, 782. [CrossRef]
32. Komar, M.; Nowicka-Krawczyk, P.; Ruman, T.; Nizioł, J.; Dudek, M.; Gutarowska, B. Biodeterioration potential of algae on building materials-Model study. *Int. Biodeterior. Biodegrad.* **2023**, *180*, 105593. [CrossRef]
33. Zhong, X.Y.; Shi, H.; Li, X.; Zhao, Y.J. Water repellency characteristics of physical crusts and biological crusts and their effects on water infiltration. *Chin. J. Appl. Ecol.* **2022**, *33*, 1843–1852. [CrossRef]
34. Chomachayi, M.D.; Blanchet, P.; Hussain, A.; Pepin, S. Development of a novel sandwich-structured composite from biopolymers and cellulose microfibers for building envelope applications. *Sci. Rep.* **2023**, *13*, 21955. [CrossRef]
35. Huang, M.C.; Zhang, X.L.; Chang, X.; Yang, M.H.; Zhang, J.B. Research Progress of Degradable PLA/PBAT/PBS Composites Reinforced with Different Plant Fibers. *Acta Mater. Compos. Sin.* **2024**, *42*, 1–16. [CrossRef]

coatings

MDPI

Article

Biocrusts Alter the Pore Structure and Water Infiltration in the Top Layer of Rammed Soils at Weiyuan Section of the Great Wall in China

Xiaoju Yang [1,2,3], Fasi Wu [3], Long Li [4], Ruihua Shang [5], Dandan Li [3], Lina Xu [3], Jing Cui [3] and Xueyong Zhao [1,*]

1 Naiman Desertification Research Station, Northwest Institute of Eco-Environment and Resources, Chinese Academy of Sciences, Lanzhou 730000, China; yangxiaoju@dha.ac.cn

2 College of Resources and Environment, University of Chinese Academy of Sciences, Beijing 100049, China

3 National Research Center for Conservation of Ancient Wall Paintings and Earthen Sites, Dunhuang Academy, Dunhuang 736200, China; wfs803@126.com (F.W.); lidd1@dha.ac.cn (D.L.); qingqingcaoyuan5@126.com (L.X.); cuijingpara@163.com (J.C.)

4 Cultural Heritage Conservation and Design Consulting Co., Ltd. of Mogao Grottoes, Dunhuang 736200, China; licanoco183@163.com

5 College of Architecture, Taiyuan University of Technology, Taiyuan 030024, China; shangruihua@tyut.edu.cn

* Correspondence: zhaoxy@lzb.ac.cn

Abstract: The surface of the Great Wall harbors a large number of non-vascular plants dominated by cyanobacteria, lichens and mosses as well as microorganisms, and form biocrusts by cementing with the soils and greatly alters the pore structure of the soil and the ecohydrological processes associated with the soil pore space, and thus influences the soil resistance to erosion. However, the microscopic role of the biocrusts in influencing the pore structure of the surface of the Great Wall is not clear. This study chose the Warring States Qin Great Wall in Weiyuan, Gansu Province, China, as research site to quantify thepore structure characteristics of the three-dimensional of bare soil, cyanobacterial-lichen crusts, and moss crusts at the depth of 0–50 mm, by using optical microscopy, scanning electron microscopy, and X-ray computed tomography and image analysis, and the precipitation infiltration process. The results showed that the moss crust layer was dominated by large pores with long extension and good connectivity, which provided preferential seepage channels for precipitation infiltration, while the connectivity between the cyanobacterial-lichen crust voids was poor; The porosity of the cyanobacterial-lichen crust and the moss crust was 500% and 903.27% higher than that of the bare soil, respectively. The porosity of the subsurface layer of cyanobacterial-lichen crust and moss crust was significantly lower than that of the biocrusts layer by 92.54% and 97.96%, respectively, and the porosity of the moss crust was significantly higher than that of the cyanobacterial-lichen crust in the same layer; Cyanobacterial-lichen crusts increased the degree of anisotropy, mean tortuosity, moss crust reduced the degree of anisotropy, mean tortuosity. Biocrusts increased the fractal dimension and Euler number of pores. Compared with bare soil, moss crust and cyanobacterial-lichen crust increased the isolated porosity by 2555% and 4085%, respectively; Biocrusts increased the complexity of the pore network models; The initial infiltration rate, stable infiltration rate, average infiltration rate, and the total amount of infiltration of moss crusted soil was 2.26 and 3.12 times, 1.07 and 1.63 times, respectively, higher than that of the cyanobacterial-lichen crusts and the bare soil, by 1.53 and 2.33 times, and 1.13 and 2.08 times, respectively; CT porosity and clay content are significantly positively correlated with initial soil infiltration rate ($|r| \geq 0.85$), while soil type and organic matter content are negatively correlated with initial soil infiltration rate. The soil type and bulk density are directly positively and negatively correlated with CT porosity, respectively ($|r| \geq 0.52$). There is a significant negative correlation between soil clay content and porosity ($|r| = 0.15$, $p < 0.001$). Biocrusts alter the erosion resistance of rammed earth walls by affecting the soil microstructure of the earth's great wall, altering precipitation infiltration, and promoting vascular plant colonisation, which in turn alters the erosion resistance of the wall. The research results have important reference for the development of disposal plans for biocrusts on the surface of archaeological sites.

Keywords: Great Wall; biological soil crusts; microstructure; water infiltration; cultural heritage conservation

1. Introduction

Raw earth, as the main building material, occupies an important position in the history of China's architectural development. Archaeological records show that the ancient people gradually formed the earth construction process about 6000 years ago, and improved and perfected it many times in the later production activities, and ultimately formed the earth construction techniques mainly based on palletizing mud, tamping and pressing the ground to raise the arches, etc., which were used in the construction of houses, temples, city walls and burial chambers, etc. [1]. As an important part of the "The Belt and Road Initiative", there are a large number of earthen construction sites along the Silk Road, of which the Great Wall is a typical example. Of the 21,000 km of Great Wall and 43,000 sites in China, more than 70% is earthen structured. The Earth Great Wall, a typical raw earth construction, is mainly made of loess rammed with reeds and wheatgrass stalks, etc. The State Administration of Cultural Heritage (SACH) defined the cultural landscape attributes of the Great Wall for the first time in the Great Wall of China Conservation Report released in 2016 that "the Great Wall is the world's largest and spectacular historical and cultural heritage, and at the same time a cultural landscape with unique aesthetic value, jointly composed of a variety of relics and the natural environment in which they are located". They are not only precious historical materials for the study of ancient defense systems and architecture, but also important empirical evidence in response to global changes and human activities [2]. However, due to the fragile physical and mechanical properties of soil, the extreme open-air environment for a long period of time, temperature fluctuation, wind and sand erosion, rainwater erosion, and biological and human disturbances, seriously threaten the inheritance conservation and sustainable use of the Great Wall [3–5].

Due to the special requirements of heritage protection, leading to increasing difficulty of investigation, sampling and monitoring of the earth sites, researches on the deteriorated mechanism mainly focuses on physical and chemical factors [6–8], and the research on deteriorated prevention and control mainly focuses on chemical material reinforcement, exposed layer, protective shed, and backfill protection, etc. [5,9,10], with less attention to the biological factors and protection strategies [11]. Since the 20th century, especially after [12] proposed the concept of bioerosion, scientists from China and abroad have gradually carried out research on the relationship between organisms and cultural artefacts, and plant soft-covering has been regarded as a new type of site protection technology [13], but the relationship between plants and earthen sites is not clearly defined [14], and the soft capping technology, which is mainly based on vascular plants, is only adapted to the architectural buildings in a wetter climate [15], but not in the arid and semi-arid environment, with low precipitation, both plant establishment and protective effects are limited. Therefore, the screening of plant species with the advantages of strong resistance, long duration of protective effect, and small change in the surface morphology of cultural relics is a bottleneck and biological degradation and protection will be a long-lasting issue for cultural heritage preserved in the "biosphere" [16].

Biological soil crusts (biocrusts) is complex composed of cryptogamic plant and microorganism, such as cyanobacteria, green algae, lichens and mosses, as well as related other organisms glued to soil surface particles through mycelium, pseudo-root and secretion, etc. Which live within, or immediately on top of, the uppermost millimetres of soil. Soil particles are aggregated through the presence and activity of these often extremotolerant biota that desiccate regularly, and the resultant living crust covers the surface of the ground as a coherent layer. They account for more than 40% of the living surface cover in arid and semi-arid zones [17–19], and greatly change the surface structures and even terrestrial ecosystems. Through the on-site investigation of ancient sites, this group of earth surface of the Great Wall in Gansu has nurtured a large number of cyanobacteria, lichens and mosses as well as microorganisms, and the cover even reached more than 30%. They have long adapted to the adverse conditions along the Great Wall and has become a typical feature of the natural human-built composite ecosystems. Most of the research results on wind-sand control and soil resistance to wind erosion have reported positive effects on biocrusts. For example, it accelerates the formation of sand surface soil, improves soil physicochemical and biological properties [20]. With the succession of biocrusts, the water erosion resistance of their overlying soils significantly increases, which in turn affects the eco-hydrological processes, while biocrusts can significantly increase the critical sand uplift friction velocity and enhance the resistance of the soil surface to wind erosion [21]. However, there are some negative reports, for example, in windy environments, biocrusts destabilise soil structure by affecting vascular plant seed germination and growth [22] and providing suitable habitats and food for microfauna [23]. In summary, there may be two sides of biocrusts on the preservation of the Great Wall, biocrusts inhibit the occurrence of diseases such as weathering on the surface of the Great Wall to a certain extent due to strong resistance to wind erosion, but at the same time, biocrusts cover alters the water, heat, soil structure, and bio-composition of the surface of the Great Wall, loosening the originally strong rammed soil, which is more conducive to the invasion of herbaceous plants and animals, and accelerating the weathering of the surface. Although the biological degradation of cultural relics, such as cave temples caused by microorganisms, lichens and mosses, has attracted extensive attention from heritage conservation experts [24,25], the microscopic features of rammed earth sites as a special soil and habitat for biocrusts the micro features of the rammed earth-biotic crust interface interactions as well as the main controlling factors of precipitation infiltration have not been confirmed [26,27].

Soil water infiltration is affected by a variety of factors such as bulk density, pore space, water content, temperature and plant root system, and closely related to surface runoff and soil resistance to rainwater erosion [28]. Since the internal structure of soil pores and fissures provides preferential flow channels for rainfall, clarifying the diameter, morphological characteristics and connectivity of soil pores is crucial to the study of soil water infiltration. The study of soil structure has gradually become more refined with the updating instrumentation such as electron microscopes, optical microscopes, scanning electron microscopes and X-ray computed tomography (X-ray CT). CT is an inspection technique that uses the different attenuation of X-rays when they penetrate substances of different densities to reconstruct images of the internal structure of an object based on external projection data. Due to its non-destructive nature of detection and good rendering of microstructure, X-ray CT technology was used in the early 1990 s for the detection of rice roots and soil moisture content [29], and with the development of CT technology, its use in soil research has virtually encompassed the characterization of pore space geometries and fissures, involving bulk density [30], stratification detection [31], spatial correlation and tortuosity [32], and other different variables. At this stage, CT techniques have been widely used for the study of correlations between parameters such as soil spatial distribution, pore structure, water content and fractal properties, as well as for assessing soil pore network systems and obtaining fine-resolution 3 D soil reconstructions [33]. Traditional methods for porosity detection of biocrusts do not adequately characterise the geometric morphology (pore number, pore volume, pore shape and pore size distribution) and spatial distribution (connectivity, node density and pore path length) of biocrusts. The rapid turnover of CT technology has facilitated the study of non-destructive testing of test samples, which has brought a new opportunity for non-invasive characterisation of artefactual samples to determine the soil pore parameters and their spatial network structure are essential to clarify key processes such as soil mass and energy exchange and water transport.

Due to the long-term natural succession of biocrusts on the surface of the Great Wall, which results in the irregularity, instability and complexity of the structure of the artificially compacted soil, and there is an urgent need to quantify the microstructure of the biocrusts-compacted-soil interface, and revolutionise the knowledge of the mechanism of the biocrusts' role in the pore network and the hydro-physical characteristics of the artificially compacted soil. Despite the fact that biocrusts distribute only within a few centimeters of the surface of the soil profile, they are critical for remodeling the surface soil structure [28]. Therefore, the objectives of this study were:

(1) to characterise and quantify the effects of biocrusts on the pore characteristics of rammed soil at sampled sites, and to reconstruct and visualise the three-dimensional structure of the soil pore network;
(2) to clarify the differences in the effects of biocrusts on the pore parameters of rammed soil at different stages of succession; and
(3) to reveal the key factors affecting the water infiltration into the rammed soil. The results of the study have important value for selecting bio-protective species in the soil sites and guiding the development of site restoration engineerings.

2. Materials and Methods

2.1. Study Site

The sampling site was located in Weiyuan County, Gansu Province, China, in the section of Majiashan Village of the Warring States, Qin Great Wall, and the focus of the research is focusing on the Great Wall made with rammed soils (as a building material), (103°44′–104°20′ E, 34°53′–35°25′ N, elevation 2264.1 m, Figure 1), which is a typical loess landform area. Weiyuan County is located in the central part of Gansu Province, with a temperate continental monsoon climate, with rainfall mainly concentrated in July to October, and an average annual precipitation of about 600 mm, and an annual temperature of 10–12 °C.

The Warring States Qin Great Wall of Majiashan Village was constructed in the 28th year of King Zhaoxiang of Qin (279 B.C.), mainly made of loess, with a thickness of about 8 cm and some of the boundaries of the rammed layer exposed, and a height of 1–5 m. Biological weathering, physical weathering and salt accumulation are obvious, and the layers of rammed construction are clear, with a north-south orientation and farmland on the east-facing side, which cropped with maize, wheat, and potatoes. The village is on the west-facing side. Typical vegetation on the sites includes Caragana Korshinskii Kom, *Potentilla bifurca*, and *Achnatherum inebrians*, etc. biocrusts cover 20–30% of the sites, and biocrusts are dominated by cyanobacterial-lichen crust, lichen crust, and moss crust. The dominant moss crusts were dominated by the Bryum argenteum Hedw. and the Didymodon vinealis (Brid.) Zander, and the biocrusts were mainly distributed in the upper part and the top of the site walls.

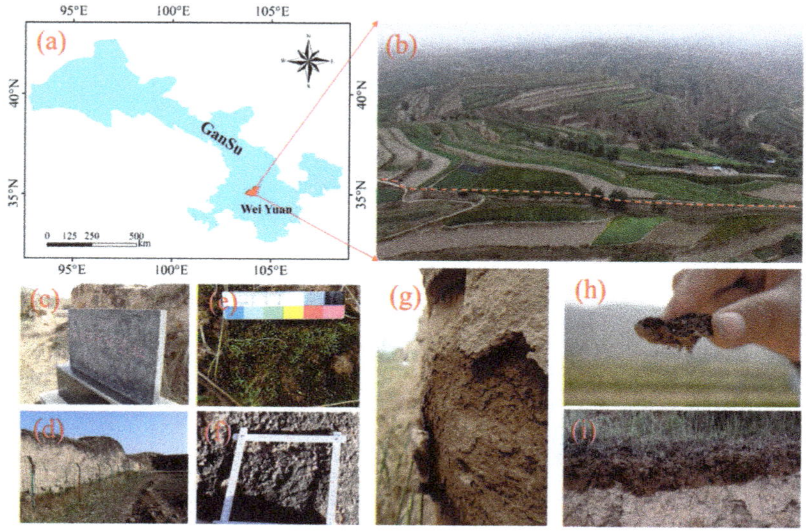

Figure 1. Geographical location of study areas. Location map of Weiyuan County (**a**), the Majiashan section of the Warring States Qin Great Wall (**b**), Great Wall site stele (**c**), distribution of biocrusts on the Great Wall (**d**), moss crusts (**e**), cyanobacterial-lichen mixed crust (**f**), surface weathering defects on the Great Wall (**g**), biocrusts characteristics (**h**) and distribution of moisture on the Great Wall after rainfall (**i**).

2.2. Experimental Design and Sampling

Bare soil (in situ soil), cyanobacterial-lichen crust (lichen crust coverage area > 90%) and moss crust (moss crust coverage area > 90%) were sampled, due to the non renewability of cultural relics resources and the special requirements of laws and regulations such as the "Regulations on the Protection of the Great Wall" (State Council Decree No. 476) and the "Technical Specifications for Soil Site Protection Testing" (WW/T 0039–2012), with only 3 replicates for each sample site, at the top of the wall of a slope < 5°. Before collecting the samples, dual distilled water was sprayed on the surface of the sample points to keep the surface of the biocrusts moist and avoid fragmentation. A polyvinyl chloride cylinder (PVC, inner diameter 50 mm, height 50 mm) was pressed into the soil until the PVC pipe fufilled with soil and then the outsider was cleaned and the top and bottom of the PVC pipe were covered with plastic film to maintain the original moisture content of the soil. The plastic film wrapped around was fixed with adhesive tape and protected with foam board to prevent mechanical damage during transport. A ring cutting (5 cm high and 100 cm^3 in volume) was used to collect in situ soil samples for the determination of bulk density, total porosity and other indicators. Plastic petri dishes (90 mm in diameter and 20 mm in height) were used to collect topsoil from biocrusts and brought back to the laboratory for determination of thickness, chlorophyll a and organic matter content.

2.3. Determination of Soil Physical and Chemical Properties

Biocrusts thickness was measured by vernier caliper, organic matter content (SOM) by potassium dichromate redox titration [34], total biomass by drying method [35], chlorophyll content by UV-V is spectrophotometer (DR 5000, Hash Corp., LoveLand, IA, USA) [28], roughness by the chain method [36], soil bulk density (BD), the total porosity (calculated based on bulk density and particle density, assuming a loess particle density of 2.65 $g{\cdot}cm^{-3}$) [28,37], saturated water content (SSWC), field water holding capacity (FC), saturated hydraulic conductivity (permeability coefficient) by the stainless steel rings (5 cm height, 100 cm^3 volume) [28], and soil mechanical composition by a Malvern laser particle sizer (MS2000, Malvern Panalytical, Malvern Town, UK).

2.4. Characterisation of Soil Microstructure

A super depth-of-field digital microscope (VHX−1000, KEYENCE, Osaka, Japan) and a scanning electron microscope/energy spectrometer (SEM-EDS, JE-OLJSM−6610 LV, JEOL, Tokyo, Japan) were used to characterise the morphological features of moss crusted rhizines and the two-dimensional spatial microstructure of soil.

2.5. Determination of Ct Pore Characterisation Parameters

The pore structure study was conducted in the natural dry state of the Great Wall soil (no rainfall occurred in the month before the simulation test). The sample pores were scanned by a nanoVoxel 4000 CT scanner (Wuhan Sanying, resolution 0.5–100 μm, Wuhan, China) with the following scanning parameters: helical scanning, spatial resolution of 10 μm × 10 μm × 10 μm, voltage of 150 kV, current of 180 μA, layer thickness of 1 mm, layer spacing of 1 mm, 8 layers/revolution, and pitch of 5 mm/revolution. A total of 3000 images with a resolution of 10.4 μm were acquired for different angles of each sample, and all the slices were exported and saved as 8-bit grey scale images for the next analysis.

Image J(2) software was used to binarise the image, remove artifacts and other image corrections, and then segment the binarised image into soil and porosity to obtain a black and white binary image, with the white area representing the soil matrix and the black area representing the soil porosity. The three-dimensional quantitative structural spatial model of rammed soil was established through the process of image filtering and enhancement, threshold segmentation calculation, and three-dimensional visual reconstruction using Avizo software, and the geometric characteristics of soil-rhizoid pores (pore surface area, length, number, and volume), morphological characteristics of the pores (degree of anisotropy, fractal dimension, Euler number, and mean tortuosity), pore diameter and shape distribution (connected porosity, isolated porosity), and pore space network structure analyzed for each parameter [28].

2.6. Simulation Test of Soil Moisture Infiltration

Due to the special provisions of the heritage conservation guidelines, only Mini Disk (USA), which has less impact on the soil, was used for in situ simulation of precipitation to measure the process of water infiltration in bare soil, mixed algal crusts, and moss crusts, and a stopwatch was used to record the change of the water level scale line of the water storage of the infiltrometer over time, and the change was recorded once every 15 s. The test site was selected in the same area where the samples were collected, and three replicates were sampled from each test site, the infiltration study was conducted in the natural dry state of the Great Wall soil (no rainfall occurred in the month before the simulation test) and the water temperature was maintained at 18 ± 2 °C during the experiment. The initial infiltration rate (3 min from the beginning), the stable infiltration rate, the average infiltration rate, and the cumulative infiltration amount (30 min) were selected to analyse the soil infiltration characteristics. Initial infiltration rate = the amount of infiltration in the initial time period/infiltration time, stable infiltration rate = the infiltration rate when the infiltration amount per unit of time is stable, and average infiltration rate = the total amount of infiltration when the stable infiltration is reached/infiltration time.

2.7. Data Analysis and Mapping

SPSS 25.0 software was used to statistically analyse the data. The t-test and ANOVA were performed on the porosity, face porosity, fractal dimension, Euler number, physicochemical properties, and soil water infiltration parameters of each sample, and the significance was tested by the one-way ANOVA ($p < 0.05$), and pearson correlation analyses were carried out on each index. IBM SPSS AMOS 26.0 (Chicago, IL, USA) software was used to construct the structural equation model, and Origin Pro 2024 software was used for graphing.

3. Results

3.1. Characteristics of Biocrusts and Their Impact on Soil Properties

As shown in Table 1, there were significant differences ($p < 0.05$) in the physicochemical properties of the soils of bare soil, moss crust and cyanobacterial-lichen crust cover. The thickness of moss crust was 14.47 ± 0.65 mm, the biomass 0.11 ± 0.02 g·cm^{-2}, and the chlorophyll content 2.29 ± 0.21 μg·g^{-1}. The thickness, biomass, and chlorophyll content of cyanobacterial-lichen crust are only 55.21%, 69.64%, and 6.63% of the moss crust. Compared with bare soil, moss crust conductivity, clay particle content, silt particle content, total porosity, organic matter content, and surface roughness increased by 35.72%, 15.54%, 10.56%, 46.33%, 519.25%, and 201.99%, respectively; And cyanobacterial-lichen crust silt particle content, total porosity, organic matter content, and surface roughness increased by 1.13%, 23.61%, 302.97%, and 373.58%, respectively; Moss crust sand content, soil bulk density and pH decreased by 5.11%, 23.74% and 8.75%, respectively; And cyanobacterial-lichen crust soil bulk density, pH, conductivity, and silt particles content decreased by 12.10%, 4.57%, 6.71%, and 2.79%, respectively, but sand content increased by 1.07%. Biocrusts also significantly changed the hydraulic parameters of the compacted soil, the saturated hydraulic conductivity of cyanobacterial-lichen crust and moss crust were 247.06% and 852.94% of that of the bare soil, and the saturated water content was 196.30% and 385.19% of that of the bare soil, respectively; The field capacity of the above two was 872.73% and 1281.82% of that of the bare soil, respectively.

Table 1. Characteristics of bare soil and biocrusts.

Measurement Indicators	Bare Soil	Cyanobacterial-Lichen	Moss Crusts	F
Crust thickness(mm)	-	7.99 ± 0.30 b	14.47 ± 0.65 a	412.52
Total biomass (g·cm^{-2})	-	0.08 ± 0.02 b	0.11 ± 0.02 a	8.377
Chlorophyll a content (μg·g^{-1})	-	0.15 ± 0.04 b	2.29 ± 0.21 a	412.51
pH	8.48 ± 0.08 a	8.10 ± 0.03 b	7.74 ± 0.01 c	255.7
Conductivity	97.2 ± 0.27 b	90.68 ± 0.68 c	131.92 ± 5.19 a	269.89
Bulk density (g·cm^{-3})	1.75 ± 0.01 a	1.54 ± 0.01 b	1.34 ± 0.01 c	1081.73
Total porosity (%)	33.89% ± 0.56 c	41.89% ± 0.60 b	49.58% ± 0.43 a	1081.73
Percentage of clay (<2 μm, %)	3.71 ± 0.04 c	3.76 ± 0.09 b	4.29 ± 0.03 a	141.57
Percentage of silt (2–20 μm, %)	27.73 ± 0.17 b	26.95 ± 0.27 c	30.66 ± 0.06 a	558.03
Percentage of sand (20–2000 μm, %)	68.56 ± 0.21 b	69.29 ± 0.33 a	65.05 ± 0.01 c	506.89
Organic matter content (g·kg^{-1})	6.45 ± 0.26 c	26.01 ± 3.43 b	39.97 ± 6.52 a	78.30
Surface roughness	2.01 ± 0.03 c	9.54 ± 0.14 a	6.08 ± 0.10 b	7005.53
Field capacity (cm^3·cm^{-3})	0.02 ± 0.02 c	0.19 ± 0.07 b	0.28 ± 0.05 a	8.377
Saturated water content (cm^3 cm^{-3})	0.11 ± 0.07 c	0.21 ± 0.08 b	0.42 ± 0.11 a	15.59
Saturated hydraulic conductivity (cm·min^{-1})	0.03 ± 0.02 c	0.08 ± 0.03 b	0.29 ± 0.10 a	15.59

Note: The data is expressed as mean ± standard deviation. Different letters presented indicate significant differences among soil treatments at the 0.001 probability level. F is the ratio of between-group variance to within-group variance, reflecting the magnitude of between-group differences relative to within-group differences.

3.2. Microscopic Morphological Differences Between Biocrusts and Rammed Earth

Digital microscope results showed (Figure 2a,b) that mosses had 5–8 mm stems and leaves buried in the soil, cemented with soil particles, and the moss crust was dominated by large pores, with distribution of worm pores, herbaceous root system and moss pseudo-root pores, and soil aggregate pores, which had long extensions, good connectivity, and provided preferential seepage pathways for precipitation infiltration; Algae-lichen crust was distributed in the shallow on the thin layer of the surface of soil of 0.5–1 mm, forming a clear boundary with the underlying soil, with no obvious changes in the pores. The pore size of large pores in the cyanobacterial-lichen crust was significantly smaller than that of the moss crust, and the connectivity between the voids was poor, with a large area of fracture penetration and connected to the pores (Figure 2b shows a number of profiles of fracture penetration in the soil samples), forming a pore-fracture network, and these larger fractures were likely to increase the depth of effective infiltration of precipitation and conduct to the pore channels connected to them to enlarge the area of water infiltration.

Figure 2. Digital microscope and scanning electron microscope images of different types of biocrusts. Digital microscope: moss crust ×50 (**a**), cyanobacterial-lichen crust ×50 (**b**); Scanning electron microscopy: moss crust ×50 stem and leaf (**c**), moss crust ×400 (**d**), cyanobacterial-lichen crust ×80 fiber (**e**) and bare soil ×1000 (**f**).

Under the scanning electron microscope (SEM) at different magnifications (Figure 2c–f), the size of soil particles around the moss pseudopods was smaller than that of the pseudopod-free area (Figure 2c showed pore spaces of globular soil aggregates), and the soil particles were closely associated with each other, with high porosity, and a large amount of flocculent material was wrapped on the surface of the particles or filled into the pore spaces, which had a clear "mesh-like" surface structure. Biocrusts algal filaments and moss pseudo-roots can extend through the pore space of soil particles to the neighbouring particles, and have the function of capturing and fixing soil particles. The surface of the bare soil area is not covered by biocrusts, but the site has been subjected to rainfall for a long period of time, and the surface layer of the bare rammed soil is very solid, and the soil particles are tightly arranged. There is a small difference in the compactness of the upper and lower particles, and the particles are tightly nested with each other and show a granular morphology and the boundaries are more obvious. There are a small number of pore fissures distributed on the surface, and the pores are mostly inter-particle mosaic pores and less intra-particle pores, with pore diameters smaller than those of the crust-covered area, and poor pore connectivity.

3.3. Effect of Biocrusts on Soil Pore Characteristic Parameters and Spatial Network Structure

3.3.1. Effect of Biocrusts on Porosity and Pore Size

Due to the limitation of CT scanning resolution, the following pores all refer to the pores with equivalent diameter greater than 50 μm, therefore, there is a difference between the actual soil porosity and the actual soil porosity [28]. As shown in Figure 3a, the porosity of all three soil types showed moss crust > cyanobacterial-lichen crust > bare soil in the 0–50 mm depth range ($p < 0.05$), with the porosity of cyanobacterial-lichen crust and moss crust of 12.84% and 21.47%, respectively, 5 times and 9.03 times higher than bare soil (2.14%). The porosity of the biocrusts samples was significantly higher in the biocrusts layer than in the subsurface layers, and the porosity of the subsurface layers of cyanobacterial-lichen crust and moss crust was significantly lower than that of the biocrusts layer by 92.54% and 97.96%, respectively ($p < 0.05$), and the porosity of the moss crust was significantly higher than that of the cyanobacterial-lichen crust in all the same layers ($p < 0.05$) (Figure 3b).

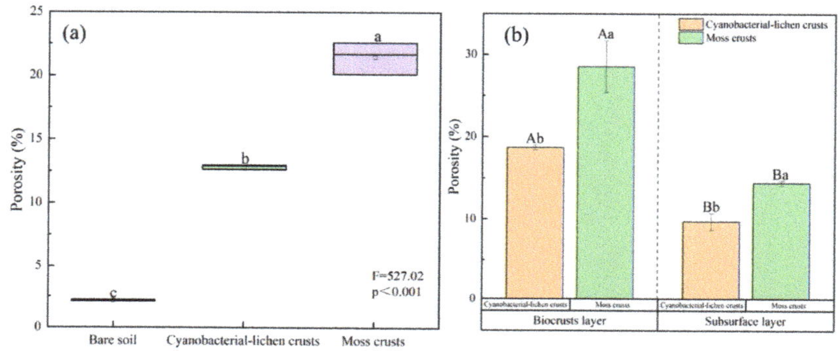

Figure 3. Porosity of bare soil, cyanobacterial-lichen crusts and moss crusts. Total porosity of the entire soil column (**a**), porosity of the biocrusts layer and subsurface layers of the biocrusts (**b**). Different lowercase letters (a and b) indicate significant differences in the porosity of different soil types in the same soil column ($p < 0.05$). Different uppercase letters (A and B) indicate significant differences in porosity between the biocrusts layer and subsurface layers of the same type of soil ($p < 0.05$).

3.3.2. Effect of Biocrusts on Three-Dimensional Soil Pore Characteristics

The surface area density reflects the size of the physically and chemically active surface of the soil, the length density reflects the branching of pores, the network density reflects the interaction ability of pores, the mean pore volume indicates the "size" distribution of pores, and the node density reflects the complexity of the pore network [28]. As shown in Table 2, the surface area density, average pore length density, network density, mean pore volume and node density of biocrusts were significantly higher than those of bare soil in the 0–50 mm depth range. The surface area density, mean pore volume and node density of cyanobacterial-lichen crust and moss crust was 6.54, 2.42, 303.69, 122.43, 105.13 and 5.04, 3.63, 79.13, 82, 44.70 times higher than those of bare soil, respectively. Algae-lichen crusts had a greater effect on soil pore characteristics than moss crusts, and the trend of the effect of the biocrusts layer and the lower layer on soil pore characteristics was basically the same as that of the overall soil column.

Table 2. Pore characteristics of three types of crust.

Soil Depth	Crust Types	Surface Area Density ($mm^2\ mm^{-3}$)	Length Density ($mm\ mm^{-3}$)	Network Density (Number mm^{-3})	Mean Pore Volume ($\times10^{-2}$) (mm^3)	Node Density (Number mm^{-3})
Entire Soil Column	Bare Soil	9.06 ± 2.03 c	0.31 ± 0.11 c	13.92 ± 2.14 c	0.03 ± 0.01 c	13.14 ± 2.65 c
	Cyanobacterial-LichenCyanobacterial-lichen Crust	59.27 ± 19.03 a	37.95 ± 15.84 a	1463.06 ± 211.23 a	0.07 ± 0.04 b	3990.57 ± 538.35 a
	Moss Crust	45.64 ± 0.35 b	25.42 ± 0.12 b	622.10 ± 49.12 b	0.11 ± 0.03 a	1039.79 ± 25.14 b
Biocrust Layer	Cyanobacterial-Lichen Crust	51.64 ± 8.59 a	39.04 ± 10.17 a	741.88 ± 90.04 a	0.06 ± 0.04 b	2938.97 ± 139.21 a
	Moss Crust	43.23 ± 2.9 b	18.27 ± 5.0 b	329.24 ± 40.09 b	0.13 ± 0.08 a	1363.56 ± 60.12 b
Subsurface Layer	Cyanobacterial-Lichen Crust	72.54 ± 20.25 a	143.41 ± 21.52 a	4086.82 ± 991.99 a	0.02 ± 0.01 a	4563.61 ± 553.51 a
	Moss Crust	42.79 ± 1.91 b	61.66 ± 0.51 b	916.17 ± 70.4 b	0.02 ± 0.01 a	2018.51 ± 74.75 b

The different letters presented at the same soil depth indicate significant differences between different types of crust at the 0.05 level.

3.3.3. Effect of Biocrusts on Characterisation Parameters of Porosity

Anisotropy reflects the degree of difference in soil pore structure in different directions, fractal dimension quantifies the complexity and self similarity of pore structure parameters, Euler number describes the connectivity of pore networks, and mean tortuosity characterizes the degree of curvature of fluid migration paths [28]. The results in Table 3 show that in the 0–50 mm depth range, cyanobacterial-lichen crusts increased the degree of anisotropy, mean tortuosity of soil porosity and moss crusts decreased the degree of anisotropy, mean tortuosity of soil porosity compared to bare soil, but the differences were not significant. Fractal dimension was higher in biocrusts than in bare soil, but the differences were not significant. Cyanobacterial-lichen and moss crust Euler number were significantly ($p < 0.05$) higher than bare soil. Overall, cyanobacterial-lichen crusts had a greater effect on soil pore morphological parameters than moss crusts. Between different biocrusts layers, the effects of cyanobacterial-lichen crust and moss crust on soil pore morphological parameters were basically consistent with the trend of the whole soil profile.

Table 3. Pore morphology parameters of the three types of crust.

Soil Depth	Crust Types	Degree of Anisotropy	Fractal Dimension	Euler Number ($\times10^5$)	Mean Tortuosity
Entire Soil Column	Bare Soil	0.22 ± 0.04 a	2.30 ± 0.03 a	−0.12 ± 0.01 c	1.14 ± 0.26 a
	Cyanobacterial-Lichen Crust	0.26 ± 0.07 a	2.73 ± 0.22 a	36.64 ± 7.59 a	1.15 ± 0.07 a
	Moss Crust	0.16 ± 0.03 a	2.67 ± 0.22 a	0.24 ± 0.06 b	1.11 ± 0.11 a
Biocrust Layer	Cyanobacterial-Lichen Crust	0.34 ± 0.11 a	2.67 ± 0.22 a	36.64 ± 0.32 a	1.14 ± 0.09 a
	Moss Crust	0.2 ± 0.03 a	2.67 ± 0.27 a	−5.99 ± 0.30 b	1.12 ± 0.17 a
Subsurface Layer	Cyanobacterial-Lichen Crust	0.35 ± 0.07 a	2.74 ± 0.18 a	9.84 ± 0.64 a	1.16 ± 0.08 a
	Moss Crust	0.23 ± 0.06 a	2.67 ± 0.22 a	3.76 ± 0.36 b	1.16 ± 0.08 a

The different letters presented at the same soil depth indicate significant differences between different types of crust at the 0.05 level.

The mean shape factor characterizes the complexity of pores, sphericity quantifies the degree to which pores approach an ideal sphere, flatness reflects the fluctuation of pore contours relative to a reference plane, and elongation describes the deviation of pore defects relative to a rectangular sample [28]. The mean shape factor is a descriptive statistic of the complexity of the shape of the soil pores after integrating the shape deviations of the sphericity and other parameters. The sphericity is used to describe whether the pores are close to a rounded sphere or not, the flatness can respond to the deviation of the pores above and below the same reference surface, and the elongation, compactness, and gap are descriptive of the deviation of the pore defects with respect to the rectangular samples. As shown in Table 4, the mean shape factor, flatness, and elongation of the pores of the biocrusts were significantly lower than those of the bare soil in the 0–50 mm depth range ($p < 0.05$). Compared with bare soil, the mean shape factor of pores of cyanobacterial-lichen crust and moss crust decreased by 67.54% and 74.19%, flatness by 126.67% and 120.00%, and elongation by 103.23% and 96.77%, respectively ($p < 0.05$). Meanwhile, the biocrusts also reduced soil pore sphericity and compactness, but the difference was not significant, compared to bare soil. The pore gap of cyanobacterial-lichen crust and moss crust increased by 0.44 times and 0.48 times, respectively ($p < 0.05$), respectively, compared with bare soil. The effect of different types of biocrusts surface layers on each shape parameter of the pore space was not significant, but there was a significant difference in the mean shape factor and elongation of the biocrusts lower layers ($p < 0.05$).

Table 4. Pore shape parameters of three types of crust.

Soil Depth	Crust Types	Mean Shape Factor	Sphericity	Flatness	Elongation
Entire Soil Column	Bare Soil	6.47 ± 0.49 a	0.55 ± 0.15 a	0.45 ± 0.19 a	0.31 ± 0.1 a
	Cyanobacterial-Lichen Crust	2.1 ± 0.95 b	0.54 ± 0.25 a	−0.12 ± 0.06 b	−0.01 ± 0.01 c
	Moss Crust	1.67 ± 0.17 c	0.54 ± 0.22 a	−0.09 ± 0.04 c	0.01 ± 0.01 b
Biocrust Layer	Cyanobacterial-Lichen Crust	1.45 ± 0.31 a	0.54 ± 0.25 a	−0.18 ± 0.07 a	−0.07 ± 0.03 a
	Moss Crust	1.45 ± 0.04 a	0.55 ± 0.05 a	−0.18 ± 0.07 a	−0.07 ± 0.02 a
Subsurface Layer	Cyanobacterial-Lichen Crust	1.27 ± 0.38 a	0.54 ± 0.25 a	−0.13 ± 0.07 a	−0.02 ± 0.01 a
	Moss Crust	1.09 ± 0.01 b	0.54 ± 0.11 a	−0.11 ± 0.02 a	−0.003 ± 0.001 b

The different letters presented at the same soil depth indicate significant differences between different types of crust at the 0.05 level.

3.3.4. Effect of Biocrusts on Modelling of Soil Pore Networks

As in Table 5, each parameter differed significantly between bare soil and biocrusts. The connected porosity of cyanobacterial-lichen crust and moss crust was 1.33 and 2.75 times higher than that of bare soil, respectively. Compared with bare soil, Moss crust and cyanobacterial-lichen crust increased isolated porosity (by 40.85 and 25.55 times, respectively), node porosity (by 36.08 and 17.13 times, respectively), and channel number (by 16.64 and 7.71 times, respectively), respectively, but decreased the average throat area both by 99%, the length of channels by 68.71% and 76.24%. The coordination number of cyanobacterial-lichen crust increased by 0.64 times compared to bare soil and moss crust decreased by 61.73%, compared to bare soil. Porosity also varied significantly between the different layers of the biocrusts.

The 2 D and 3 D visualised structural models of the pores of the three soils are shown in Figures 4 and 5, with significant differences in the pore structure of the samples, and the two types of biocrusts showing elongated pore systems at the connection with the ramified soil, which is more complex than the pore network model of the bare soil. The moss crust had the most complex connected pore network system, followed by the cyanobacterial-lichen crust and bare soil. The black areas are less dense and generally fissures, pores or plant roots, etc., while the light-coloured areas are more dense and mostly mineral particles. Through observation, it can basically show that there are more black areas in the image of moss crust, with more microstructures such as pores and cracks, showing complex pores connected or disconnected with cracks, presenting an obvious complex porous structure, and most of these pores and cracks come from moss pseudopetioles or other organic matter decay, indicating poorly soil densification at this time. The moss crust was dominated by the connected large pores formed by rhizines, with fewer medium and small pores, and obviously small aggregate structure could be observed inside the large aggregates. The soil aggregate structure was relatively loose, with a larger number of pores and better pore connectivity, and the difference in the three-dimensional pore structure in the middle portion of the agglomerates was even more significant, and the growth of the moss crust contributed to the development of the large soil pores. The moss rhizines filled in the shallow soil layer of 5–8 mm and the porosity increased with root development, the pore distribution was more uniform in areas without moss root distribution in the 20–40 mm soil depth range, in addition, a small number of larger pores were distributed around the edges of the soil columns, which could be cracks caused by sampling disturbance. The images of cyanobacterial-lichen crusts showed large differences in performance, with black areas being more numerous and unevenly distributed in the surface layer, indicating that their internal porosity and density showed an uneven distribution, dominated by fine pore structures and poor pore connectivity. The small pores are mainly distributed in the 0.5–1 mm of the soil surface layer, and the deep soil pores gradually decrease. The scanning image of bare soil was brighter overall, with a small number of black dots, showing clear homogeneity, with dense microstructure and a small number of internal isolated pores, the profile was more compact, the contours and boundaries of medium and small aggregates could not be observed inside the large aggregates, and the inter-pore connectivity was poor, with a small number of microcracks distributed and a lack of connecting channels, which was unfavourable for water transport, this indicates that the artificially compacted soil is highly dense. The total pores and connected macropores in biological crusts are significantly more than those in bare soil. Skeleton models, node skeleton models, and ball and stick models are more complex and diverse than bare soil. Fresh crusts exhibit higher diversity in three-dimensional pore structures than lichen crusts (Figure 5).

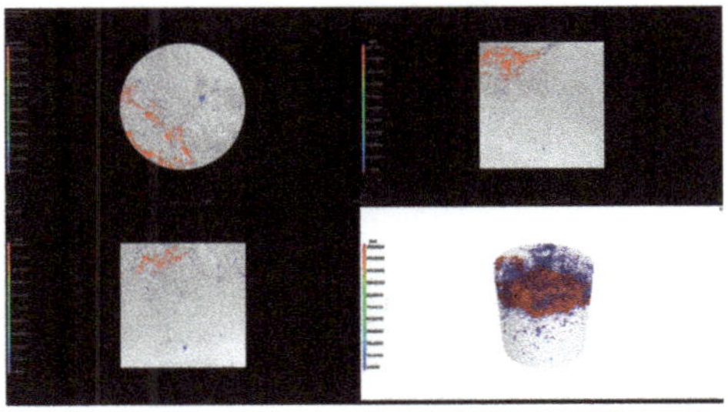

(a)

Figure 4. *Cont.*

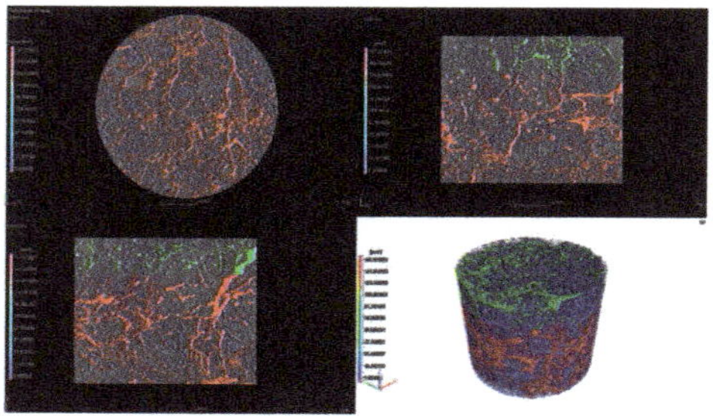

(b)

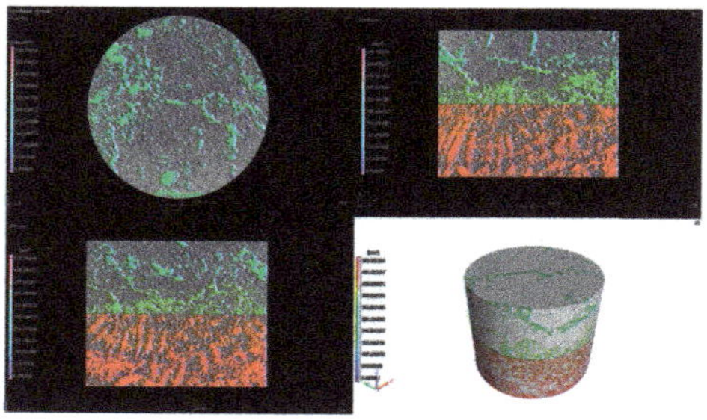

(c)

Figure 4. 2 D and 3 D spatial network pore structures of biocrusts and bare soil. Bare soil (**a**), cyanobacterial-lichen crust (**b**) and moss crust (**c**). Different colors represent the size of pore volume, and the closer it is to red, the larger the pore volume.

Table 5. Characteristics of three types of crust pore networks.

Soil Depth	Crust Types	Connected Porosity (%)	Isolated Porosity (%)	Number of Node Pores	Average Coordination Number	Number of Channels	Average Throat Area (mm^2)	Average Channel Length (mm)
Entire Soil Column	Bare Soil	1.72 ± 0.22 c	0.40 ± 0.14 c	197 ± 23.07 c	4.05 ± 0.32 b	410 ± 20.22 c	1 ± 0.46 a	4.25 ± 1.96 a
	Cyanobacterial-Lichen Crust	2.29 ± 0.26 b	10.62 ± 2.41 b	3571 ± 606.09 b	6.64 ± 1.29 a	3570 ± 671.51 b	0.01 ± 0.01 b	1.33 ± 0.57 b
	Moss Crust	4.73 ± 0.31 a	16.74 ± 0.28 a	7305 ± 96 a	1.55 ± 0.26 c	7232.33 ± 32.50 a	0.01 ± 0.01 b	1.01 ± 0.02 c
Biocrust Layer	Cyanobacterial-Lichen Crust	0.67 ± 0.31 b	18.10 ± 2.98 b	1580.33 ± 426.39 a	7.16 ± 1.56 a	7232 ± 309.71 a	0.01 ± 0.01 a	1.14 ± 0.21 a
	Moss Crust	27.39 ± 0.2 a	1.14 ± 0.12 a	1580 ± 19.52 a	7.15 ± 0.1 a	7235 ± 32.79 a	0.01 ± 0.01 a	1.04 ± 0.12 a
Subsurface Layer	Cyanobacterial-Lichen Crust	0.62 ± 0.18 b	9.06 ± 1.03 b	1232.33 ± 109.50 a	4.45 ± 0.24 a	3346 ± 267.25 b	0.11 ± 0.06 a	1.34 ± 0.34 a
	Moss Crust	2.5 ± 0.4 a	12.21 ± 11.91 a	2987 ± 116.89 b	3.4 ± 0.26 b	5077 ± 51.68 a	0.01 ± 0.01 a	1.31 ± 0.11 a

The different letters presented at the same soil depth indicate significant differences between different types of crust at the 0.05 level.

3.4. Impact of Biocrusts on Precipitation Infiltration

As a key indicator of soil precipitation infiltration, the surface infiltration capacity of soil reflects its ability to contain water. Comparison of water infiltration parameters was carried out by selecting the 30 min representative of the infiltration process, and it can be seen from Figure 6a,b that the cumulative infiltration curves of the biocrusts were basically the same, and all of them showed a steadily increasing trend, but the trend of bare soil changes is more gradual. The cumulative infiltration of bare soil, cyanobacterial-lichen crust, and moss crust shows a linear function relationship with time. Overall, the infiltration rates of the three types of soils decreased sharply in the first 1 min, with greater change between 2–4 min, after which the decreased rate rapidly became smaller and gradually levelled off. The cumulative infiltration rate of bare soil, cyanobacterial-lichen crust, and moss crust shows a power exponential function relationship with time. The initial infiltration rate of moss crust was 3.19 ± 0.48 mm·min^{-1} >

cyanobacterial-lichen crust (1.41 ± 0.08 mm·min^{-1}) > bare soil (1.02 ± 0.04 mm·min^{-1}), and the stabilisation rate of moss crust was 0.07 ± 0.01 mm·min^{-1} > cyanobacterial-lichen crust (0.06 ± 0.03 mm·min^{-1}) > bare soil (0.04 ± 0.03 mm·min^{-1}). Bare soil (0.04 ± 0.01 mm·min^{-1}), as infiltration proceeded, rammed soil reached the stabilised infiltration rate at the earliest and mosses the latest. The average infiltration rate of moss crust was (1.93 ± 0.06 mm·min^{-1}) > cyanobacterial-lichen crust (1.27 ± 0.08 mm·min^{-1}) > bare soil (0.83 ± 0.03 mm·min^{-1}), and the total amount of infiltration of moss crust was 44.97 ± 2.70 mm > cyanobacterial-lichen crust (39.80 ± 1.21 mm) > bare soil (21.63 ± 0.12 mm). The differences between different types of crust soils were significant ($p < 0.05$) for all infiltration parameters. The initial infiltration rate, stable infiltration rate, average infiltration rate and total infiltration rate of the moss crust soil was 2.26 and 3.12, 1.07 and 1.63, 1.53 and 2.33, 1.13 and 2.08 times higher than those of the cyanobacterial-lichen crust and bare soil, respectively (Figure 6c–f), which showed that the biocrusts could significantly change the process of water infiltration traits in the surface compacted soil of the soil sites.

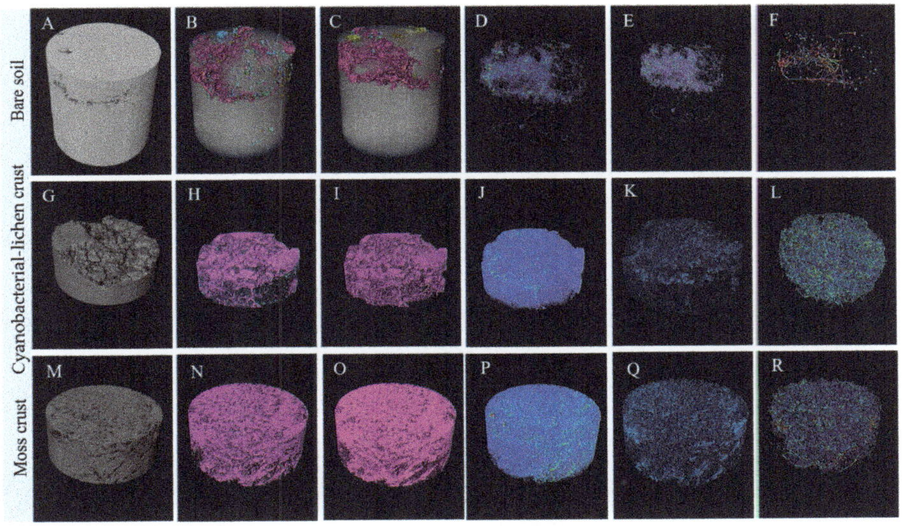

Figure 5. 3 D visualization of bare soil (**A–F**), cyanobacterial-lichen crust (**G–I**), and moss crust (**M–R**), including three-dimensional model (**A,G,M**), total pores (**B,H,N**), connected pores-macropores (**C,I,O**), skeleton models (**D,J,P**), node skeleton models (**E,K,Q**), and ball stick models (**F,L,R**).

3.5. Influence of Soil Properties on Soil Pore Characteristics and Infiltration

Soil properties have a significant impact on soil porosity and its morphological parameters. As shown in Figure 7, Pearson correlation showed that soil bulk density was negatively correlated with CT porosity ($|r| = 0.88$, $p < 0.01$); Soil clay content, organic matter content, field capacity, saturated hydraulic conductivity, and saturated water content were strongly and positively correlated with CT porosity ($|r| \geq 0.69$, $p < 0.05$); And there was a non-significant correlation between CT porosity and soil pore geometry characteristics; Mean pore volume, number of node pores, average coordination number, number of channels, connected porosity, and isolated porosity were significantly correlated with CT porosity ($|r| \geq 0.69$, $p < 0.05$); Except for fractal dimension, which was significantly correlated with PH, soil organic matter content, and soil bulk density ($|r| \geq 0.68$, $p < 0.05$), soil pore characteristics (degree of anisotropy, Euler number, fractal dimension, mean tortuosity) were not significantly correlated with other soil properties; Initial infiltration rate was significantly correlated with soil properties (soil bulk density, clay particle content, soil organic matter content, field capacity, saturated hydraulic conductivity, saturated water content, pH) (|r| $\geq$ 0.74; p < 0.05), and significantly correlated with number of node pores, average coordination number, number of channels, connected porosity, and isolated porosity (|r| $\geq$ 0.70; p < 0.05), and stable infiltration rate showed significant correlation with mean pore volume, degree of anisotropy, and fractal dimension (|r| $\geq$ 0.76; p < 0.05).

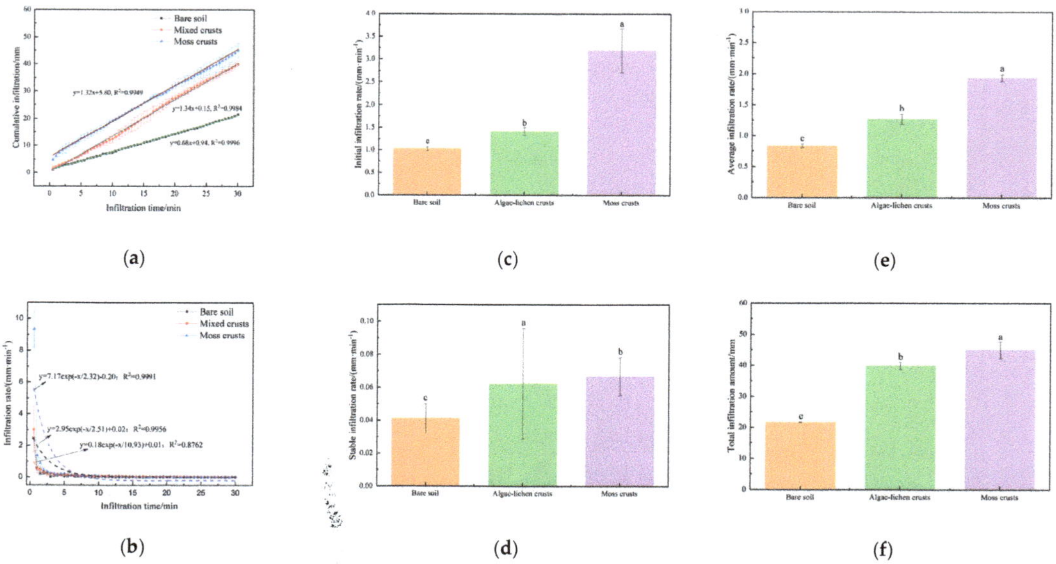

Figure 6. Water infiltration characteristics. Cumulative infiltration (**a**), infiltration rate (**b**), initial infiltration rate (**c**), stable infiltration rate (**d**), average infiltration rate (**e**) and total infiltration (**f**) of biocrusts and bare soil.

The results of the SEM indicate that (Figure 8a), soil type, bulk density, clay particle content, organic matter content, and CT porosity could directly or indirectly affect the initial soil infiltration rate, and all the variables together explained 98% of the initial soil infiltration rate (R^2 = 0.98). CT porosity and clay particle content had a significant positive effect on the initial soil infiltration rate (|r| $\geq$ 0.85, p < 0.001), soil type had a significant direct negative effect on the initial soil infiltration rate (|r| = 2.62, p < 0.001), and the direct negative effect of organic matter content on the initial soil infiltration rate was not significant. Soil type and bulk density had direct positive and negative effects (|r| $\geq$ 0.52) on CT porosity, respectively, but not significant. Although soil density did not have a direct effect on the initial soil infiltration rate, soil bulk density may have a direct effect on porosity and thus indirectly on soil infiltration. Soil clay particle content had a significant direct negative effect on porosity (|r| = 0.15, p < 0.001), with soil type, bulk density, and clay particle content explaining 99% (R^2 = 0.99) of the variation in soil porosity. Crust types significantly affected bulk density, organic matter content, clay particle content, and initial soil infiltration rate (|r| $\geq$ 0.95, p < 0.001). The results of the total effects analysis showed that the initial infiltration rate was mainly affected by the indirect positive effect of crust type, the direct positive effect of CT porosity, and the indirect negative effect of soil bulk density (Figure 8b).

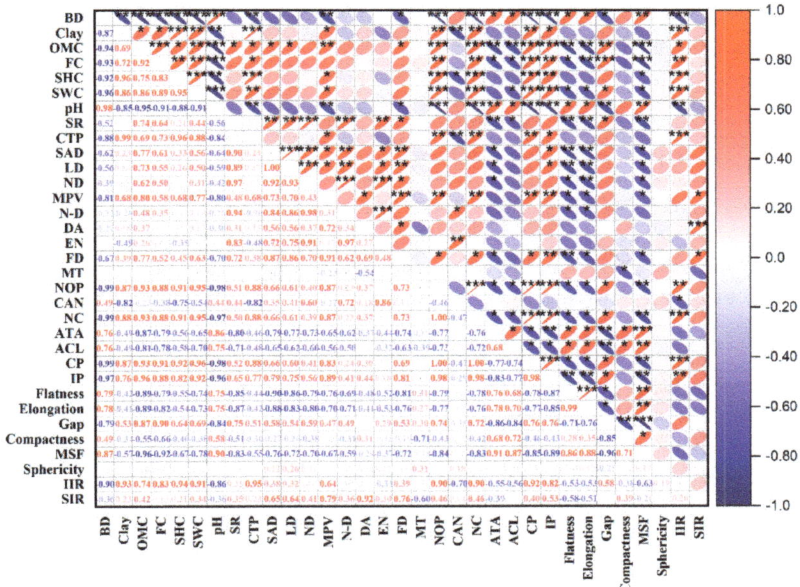

Figure 7. Pearson correlation coefficients matrix of bulk density (BD), clay (Clay), organic matter content (OMC), field capacity (FC), saturated hydraulic conductivity (SHC), saturated water content (SWC), pH, surface roughness (SR), CT porosity (CTP), surface area density (SAD), length density (LD), network density (ND), mean pore volume (MPV), node density (N-D), degree of anisotropy (DA), Euler number (EN), fractal dimension (FD), mean tortuosity (MT), number of node pores (NOP), average coordination number (CAN), number of channels (NC), average throat area (ATA), average channel length (ACL), connected porosity (CP), isolated porosity (IP), flatness (Flatness), elongation (Elongation), gap (Gap), compactness (Compactness), mean shape factor (MSF), sphericity (Sphericity), initial infiltration rate (IIR), stabilised infiltration rate (SIR). Correlation coefficients are listed in the lower left corner of the matrix and significance is listed in the upper right corner, with red indicating positive correlation and blue indicating negative correlation. *, **, *** represent $p \leq 0.05$, $p \leq 0.01$, $p \leq 0.001$ respectively.

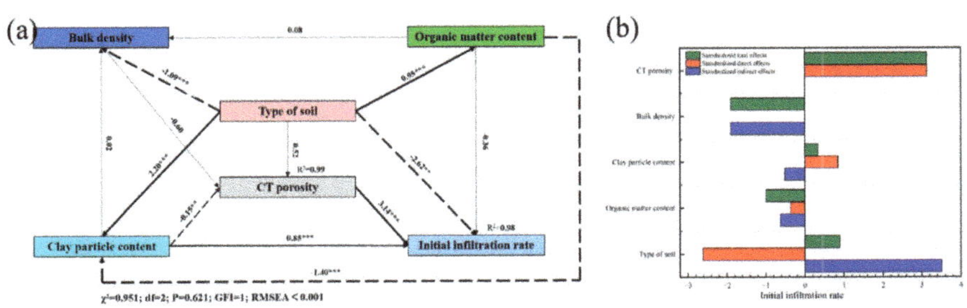

Figure 8. Path analyses (**a**) and effect plots (**b**) of changes in initial soil infiltration rates and their main influencing factors.

The graph shows the relationship between crust type (bare soil, cyanobacterial-lichen crust, moss crust) and organic matter content, bulk density, clay particle content, porosity and initial infiltration rate. Values next to arrows indicate standardised path coefficients, unidirectional arrows indicate direct effects of unidirectional causality, arrow widths indicate strength of correlation, and solid lines indicate positive path coefficients and dashed lines indicate negative path coefficients. *, **, and *** indicate statistically significant paths at $p < 0.05$, $p < 0.01$, and $p < 0.001$, respectively.

4. Discussion

4.1. Biocrusts on the Artificially Compacted Soils Alter Physical Properties

Biocrusts significantly alter the physical, chemical, and biological properties of topsoil through their complex biotic-soil interactions [38,39], which are important for both the functioning and the stability of the ecosystems of earth sites. Soil physicochemical properties differed significantly between biocrusts and bare soil, with sand content and bulk density showing moss crust < cyanobacterial-lichen crust < bare soil, and organic matter, silt, clay, total porosity, initial water content, and saturated water content showing opposite trends, suggesting that biocrusts increased the fine particle content of loess and reduced soil compactness during growth, which in turn affected water-holding capacity, this is consistent with previous research findings [28]. Biocrusts can also increase soil acidity, causing acid erosion of the site, and then form biodegradation, which may be the secretion of acidic metabolites by mosses, lichens, and microorganisms, etc. [16]. Biocrusts can also affects the wind profile owing to the increase in rough-ness length that affects the horizonal wind speed near the ground, and weaken soil wind erosion. The loess in Weiyuan area is mainly composed of particles smaller than 0.25 mm, with the silt content higher than 50%, but the content of sand particles only accounts for 1%. This type of loess in its natural state is very loose and permeable, and prone to disintegration by wind erosion and rain. The bulk density of the Great Wall soil compacted by several manual blows increased, the pore ratio decreased to 0.4–0.6, the maximum dry density increased by 40%−50% (1.8–1.94 $g \cdot cm^{-3}$), and the shear strength increased significantly [37]. As the particle size of the Great Wall rammed soil layer is mainly dominated by silt, the compressive and shear strengths are high in the dry state, but extremely easy to disintegrate when it meets water.

Soil physical and chemical properties are affected by a combination of factors. This study area is distributed in the upper and middle parts of the Great Wall of earth at a height of 2–3 m, with minimised human-disturbance, the physical properties of the soil are mainly affected by the physical, biological and chemical weathering of the surface of the sites. Soil bulk density and porosity are dominant factors affecting soil infiltration [40]). Multiple artificial tamping made the particles of the soil of the Great Wall tightly arranged, the soil pore structure deteriorated, and the bulk density increased.

4.2. Modification of Pore Structure of Artificially Compacted Soil by Biocrusts

Bare soil in a long run of weathering process, the original structure of the larger agglomerate particles within the shallow surface site was damaged and disintegrated into smaller particles, and this change prolonged rain and wind erosion, water and salt transport, and temperature changes caused by drying, wetting, freezing, and thawing cycles, as a result of reduction of the adhesion between soil particles, increased roughness of the bare soil surface, at last decreased the mechanical properties of the soil, and friction between the soil particles, and accelerated the weathering of rammed soil, and caused site degradation [14]. The topography of biocrusts improves the soil microenvironment, leading to colonisation of the wall by shallow-rooted herbaceous plants, shrubs, and microscopic soil fauna, which causes a differential distribution of the internal pores of the wall, where the connecting pores are well ventilated and easily forms a pathway for infiltration of precipitation (Figure 9).

Study of the pore size, distribution morphological characteristics, connectivity and internal network structure of biocrusts is not only an effective way to reveal the ecological function of biocrusts through micro cognition of artificially compacted soil-biological interfaces [28], but also an important scientific issue for the study of biological degradation and protection of soil sites in arid and semi-arid areas. The weathering deterioration process of the Great Wall sites preserved in the open environment is the result of long-term coupling of multiple environmental and anthropogenic factors [41]. The plant-soil interface is a complex porous medium that exhibits a high degree of heterogeneity due to the presence of roots and other organic matter [42]. The heterogeneity of soil pore distribution in areas covered by biocrust is significant, with fine cracks of an important part of the pore space. The formation of the biocrusts changes the mechanical properties of the artificially rammed soil and the formation of the soil pore system, suggesting that the growth of the biocrusts and the process of soil contraction/expansion caused by water flow are the main mechanisms for the formation of the pore space in the artificially rammed soil of the Great Wall of Earth.

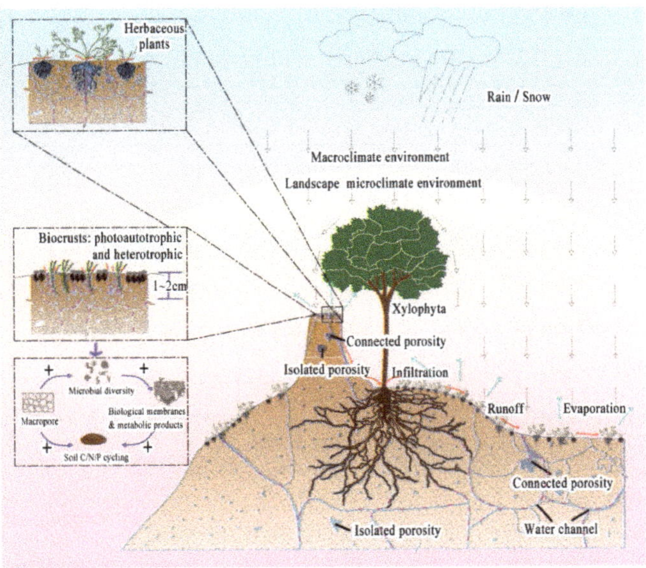

Figure 9. Schematic diagram of soil pore distribution and precipitation pathways formed by moss rhizines and vascular plant roots.

Through the vertical distribution of porosity, it was found that the moss crust had higher porosity (28.53%) above 10 mm in the surface layer of the soil column, and then rapidly decreased to about 14.41%, which was mainly because the moss rhizines was only within 10 mm in the surface layer, which significantly increased the macroporosity of the surface layer of the soil, and altered the soil permeability and air permeability. This "short and thick" root will gradually loosen the soil around the root so that the soil macroporosity near the root increased, the number of macroporosity decreased with the increase of depth, the relative increase of small pores, and the pore distribution being uniformed. This microscopic pore and fissure structure has a decisive effect on the macroscopic seepage and mechanical properties. The subsurface layer soil still retained the compactness of the rammed soil, which was opposite to the results of the study in the wind and water erosion area of the Loess Plateau [28], mainly due to the fact that the Great Wall of the earth was artificially rammed and had a very strong compactness; The lichen crust had a high porosity in the soil column of 10 mm (18.68%), and the porosity of the lower soil rapidly decreased to 9.7%, probably due to the fact that the lichen crusts were only distributed in the surface layer of the compacted soil about 10 mm depth, and fungal mycelium, extracellular polysaccharides, etc. and the formation of soil gels led to the increase in the porosity of the compacted soil; Due to the completion of the rammed soil and not subjected to external extrusion, the bare soil porosity is small (2.14%), and the distribution of pores is relatively uniform, the soil particles arranged as a dense stacked lamellar, the agglomerates dispersed into a smaller particle size and tightly arranged soil particles, and the degree of change in the microstructure is greater than that of the internal soil, and the porosity of the surface layer is slightly greater than that of the lower layer, which may be due to the increase in inter-particle distances caused by surface weathering, and the potential risk of shallow surface stripping at the macroscopic level [43]. Soil cross-sectional porosity can indirectly represent the changes in pore structure caused by plant root growth, root filling of pore space results in a decrease in porosity, and root growth produces external forces that loosen the soil and thus increase porosity [44]. The moss rhizines dominated the shallow compacted soil with loosening action, resulting in an increase in the cross-sectional porosity of the shallow soil. The overall porosity of the bare soil layer and the lower layer of the biocrusts is small because the Weiyuan Great Wall mainly used local loess when rammed, mixed with straw, glutinous rice, etc. to increase the bonding force. The loess has a large proportion of clay particles, and after manual sorting, hammering, and many times of tamping [37], the soil particles are nested with each other, tightly connected, and it is not easy to form a shelf structure between the soil bodies. This kind of dense soil structure is not conducive to precipitation infiltration of, but prone to the formation of surface runoff and causes development of wall cracks, gullies and slumps and other deterioration. Soil pore structure and internal pore distribution are complex and variable, and the use of pore network models to analyse soil pore structure and distribution is essential for describing water transport, solute transport and erosion resistance [45,46]. Fractal dimension, Euler number, tortuosity, sphericity, etc. can reflect the complexity of granular pores within the soil [47,48]. But in this study, the anisotropy and Euler number (interpreted as higher connectivity) of the surface layer of moss crusts are significantly lower than those of cyanobacterial-lichen, which is consistent with the findings of Sun et al. [28], but contrary to the study that "biocrusts reduce water permeability". We speculate that this may be due to differences in the research subjects. Sun et al. [28] mainly focused on loess or sandy soil, which are generally loose. This study focuses on artificially compacted soil, where only the surface layer with biocrusts has a higher porosity, while the soil particles in the compacted soil layer (parent material layer) are very dense, hindering further infiltration of water. The connectivity porosity of moss crusts is significantly higher than that of cyanobacterial-lichen crusts, and the thickness of moss crusts is much greater than that of cyanobacterial-lichen crusts. Therefore, there are more pores inside the moss crusts to quickly accommodate water, while the thickness of cyanobacterial-lichen is smaller and closer to the parent material layer, reducing the rate of water infiltration. In this study, the distribution of pore spatial structure of bare soil was simple, and with the continuous succession of biocrusts, its profile soil pore reticulation tended to be more complex, with increasing mean tortuosity, pore connectivity, and degree of anisotropy in the profile, and the fractal dimension of soil pore in the biocrusts was higher than that in the bare soil, which indicated that the pore space structure was more complicated in the biocrusts. Moss crust soils had lower tortuosity than bare soil, but cyanobacterial-lichen crusts increased pore tortuosity, suggesting that mosses have better connectivity than cyanobacterial-lichen pore structures. If preferential flow is present, moss crusts can prolong the residence time of water in the soil and increase the risk of surface spalling. Moss crust rhizines also create precipitation pathways that allow water flow to act on the soil internal pores along root fissures, expanding the number of macroporous pore throats and further enhancing the connectivity between the soil internal macropores. The decaying root system increases the organic matter content within the soil, and the salinisation of the ramified soil surface (especially the Na^+ content) promotes the development of macropores, the number of inter-porous connecting channels increases significantly, the connectivity between pore throats increases, and the pore network tends to be more complex [28].

4.3. Biocrusts Alters the Erosion Resistance of the Compacted Soil

The Great Wall sites are all preserved in the open air, and during the weathering process caused by long-term natural factors such as precipitation, wind, sand, light, the quality of the earthen building materials is degraded, the mechanical properties are reduced, and diseases such as physical crusting, shedding, gullies, and slumping are formed, all posing a serious threat to the stability and long term preservation of the site property [1,4,24]. The formation and development of these diseases are related to the soil microstructure [49]. Under the condition of heavy rainfall, the top layer of rammed earth disintegrates rapidly, the smaller particle size silt and clay particles block the capillary pores, and the physical crusts are formed on the surface, resulting in the decrease of the infiltration capacity of the soil and the formation of obvious runoff on the surface. It has been found that the V-shaped or herringbone-shaped root system of the moss crust can increase the migration resistance of the soil body, improve the resistance of the soil layer to slippage, and effectively enhance the resistance of the soil layer to rainfall scouring [27]. In this study, it was also found in the field investigation that grooves left by rainfall scouring were prevalent on the surface of the physical crust compared to the biocrusts covered area, and that shrinkage and cracking formed on the surface of the bare soil during continuous drying, and further accelerated the deterioration of the site soils.

Biocrusts have a significant effect on water infiltration, which in turn drives ecosystem function and the hydrological cycle [50]. Mosses, algae, and microorganisms form a cover that reduces direct erosion from the surface of compacted soils, reduces runoff formation, and prevents soil erosion [51,52]. Usually, the biocrusts develops on the soil surface and destroys the original structure of the soil, in which the mycelium of fungi occupies the soil matrix pores and fissures, while organic matter such as extracellular polysaccharides secreted by algae clogs the pores, which in turn hinders the transport of water through the soil pores, and it has also been demonstrated that, with the succession of biocrusts, the content of organic matter, microorganisms and secondary metabolites increases in the soil, which in turn increases the biological abundance of soil in the crust layer, impeding precipitation infiltration [53]. Water infiltration in moss crust with large total porosity is significantly greater than other soil types, which have a high capacity to store precipitation and contain water, the biocrusts become active after wet-ting, the organisms can swell and also can uptake water (by capiliary on rhizoid for moss for instance), while compacted soils with small total porosity are not favorable to precipitation infiltration, which in turn enhances surface runoff and leads to increased soil erosion. The amount of water held in the field not only represents the effective water available in the farmland, but also is the main index to characterise the water retention performance of the soil. In the case of loess and sands, biocrusts significantly reduced precipitation infiltration [28], which is contrary to the results of the present study, because the soil selected in this study was artificially compacted, and the precipitation infiltration of the compacted soil was very small, whereas the growth of biocrusts was able to loosen the soil texture and increase the infiltration. However, unlike ordinary soil, compacted rammed soil has a very small number of pores, and in contrast, the development of biocrusts is more likely to lead to pore loosening, which reveals the microstructural aspects of the mechanism by which biocrusts promote water infiltration in rammed soil.

With the downward growth of moss rhizines and the burial of some stems and leaves by soil, the mechanical action of the roots and stems led to changes in soil particle displacement, which in turn drove changes in air and nutrients in the soil pore spaces and altered the macroscopic mechanical and hydraulic properties of the soil. In this study, it was found that the relative infiltration capacity of the biocrusts soil layer was high, but the infiltration capacity rapidly diminished as water entered the subsoil interface of the artificially compacted soil. The initial infiltration rate, stable infiltration rate, and average infiltration rate of the moss crust soil were significantly greater than those of the rest two types ($p < 0.05$). The microstructure indicates that compared to the rammed earth layer (parent material layer), the porosity around moss artificial roots is higher, and the soil particle arrangement is loose. It is easier to form precipitation pathways within the range of 5–8 mm depth of moss crusts roots, storing more water in the pores and increasing water infiltration. The soil macroporosity in the moss crust was closely related to the growth of its rhizines, and the soil structure was loose with a high stable infiltration rate. With the growth of moss rhizines, the friction between roots and soil particles gradually increased, and the gap between particles decreased, leading to a decrease in soil porosity, while the soil water-holding capacity was slightly increased, which slowed down the decrease in water content and increased soil infiltration. Therefore, the infiltration capacity of moss crust-covered soil in the surface layer of rammed soil is significantly better than that of algae-lichen crust and bare soil, and it is more suitable for improving the erosion resistance of soil on the surface of the Great Wall and preventing soil erosion.

Soil physicochemical properties are important parameters in determining the rate of water infiltration, and soil mechanical composition, bulk density, organic matter content, saturated water content and total porosity all play an important role in water infiltration [54]. In this study, soil physical and chemical properties are one of the main factors affecting soil porosity and water infiltration. Structural equation modelling and analysis of the main influencing factors revealed that CT porosity and clay content significantly contributed to the initial infiltration rate of the soil ($p < 0.001$), while organic matter content (most likely also contains EPS and other gels) may uptake water and also swell, and not conducive to the initial infiltration rate of the soil, and bulk density was detrimental to the development of CT porosity. Soil bulk density can indirectly hinder precipitation infiltration by affecting porosity. It has been found that organic matter content effectively promotes the formation of soil aggregates, which in turn improves soil pore structure, reduces soil bulk density, and has less resistance to water flow within the soil body, increasing soil infiltration capacity [50]. Plant roots also have a significant effect on soil water transport [55], moss rhizines can loosen and compact the soil, leading to a reduction in soil bulk density and an increase in pore space, while moss rhizines or stems and leaves buried in the compacted soil layer decay, converting them into organic matter, which promotes the formation of soil aggregates, improves the structure of the soil and its physicochemical properties, and increases the infiltration capacity of the soil.

Small rainfall events occur frequently in Weiyuan area, and when the intensity of a single small rainfall is greater than the infiltration rate of the soil surface, it will form excessive seepage runoff on the slope of the Great Wall. When a single rainfall lasts for a long time, the surface moisture content of the biocrusts tends to be saturated, and the infiltration rate of the surface soil gradually decreases to that of the lower soil. Long duration and low rainfall intensity are conducive to the accumulation of water in the biocrusts layer, slow evaporation and loss of water, and increased weight of compacted soil, especially at the slope of the wall. However, whether the generated gravity is sufficient to cause patchy peeling on the surface of the wall remains to be further confirmed by introducing mechanical methods. The higher water holding capacity of biocrusts is beneficial for the growth of vascular plants in the later stage [56], causing the shallow surface of compacted soil to become more loose, providing favorable conditions for plant growth, especially at the junction of compacted soil, which is more susceptible to erosion by rainwater and shallow root vascular plants, further increasing the risk of biological weathering. Therefore, the biological weathering caused by the growth of biocrusts on the surface of cultural relics cannot be ignored [57].

From this, it can be seen that biocrust increases the porosity of rammed earth and leads to a more complex pore structure, promoting soil moisture infiltration. With the development of biocrust, the changes in pore structure and physicochemical properties of rammed earth become more pronounced. From the perspective of biological degradation, the rhizoid of moss or fungal hyphae of lichens may penetrate the surface layer of rammed earth, forming biological channels, increasing local porosity, and potentially damaging the integrity of the rammed earth layer (parent material layer) in the long run. From the perspective of ecological function, biocrusts can stabilize soil through the binding of extracellular polysaccharides and root entanglement, and can also inhibit the development of surface diseases on soil sites by increasing rainfall infiltration and reducing runoff. However, the definition of biocrust protection and destruction still needs to be further validated through mechanics, simulation experiments, and other methods.

5. Conclusions

Compared to bare soil, biocrusts improved the porosity of the Great Wall rammed soil, increased precipitation infiltration rate of the rammed soil, trapped more water in the shallow surface layer of the site wall, and formed a microhabitat conducive to the growth of vascular plants. biocrusts significantly increased surface area density, mean pore volume, node density, length density and network density, and cyanobacterial-lichen crusts had a greater effect on the internal structure of the pores; Cyanobacterial-lichen crusts increased the degree of anisotropy of porosity, mean tortuosity, resulting in pores with stronger irregularity, and moss crusts decreased the degree of anisotropy of porosity and the mean tortuosity. Compared with bare soil, mean shape factor of cyanobacterial-lichen crust and moss crust decreased by 67.54% and 74.19%, flatness by 126.67% and 120.00%, and elongation by 103.23% and 96.77%, respectively. Biocrusts development also decreased soil pore sphericity and compactness, and increased pore gap; Biocrusts increased the number of connected pores, isolated pores and the number of channels, but decreased the average pore throat area and the average channel length; Biocrusts significantly increased the initial infiltration rate, the stable infiltration rate, the average infiltration rate, and the total infiltration rate, and changed the process of water infiltration in the surface compacted soil of the soil site; CT porosity and clay content accelerated the initial infiltration rate of the soil, the soil type, and the organic matter content slowed down the initial infiltration rate of the soil. This study provides new insights into the bio-degradation and conservation of open-air earthen sites in arid and semi-arid regions of China. Based on the above conclusion, the study suggests that biocrusts play an important role in regulating the soil pore structure of rammed earth sites in natural-artificial composite ecosystems, which in turn influences the role of biocrusts in precipitation infiltration and drives the erosion resistance of earth sites. Improved understanding of the physical, biological and chemical processes of biocrusts in earth sites can help to grasp their important role in the biological degradation of artefacts. Future research should aim to the balance between conservation and destruction of biocrusts in cultural heritage in arid and semi-arid region in order to more accurately predict the ecological function of biocrusts on artefact surfaces under global climate change.

Author Contributions: Writing-original draft, X.Y.; formal analysis, F.W.; investigation L.L.; software, R.S.; software, D.L.; investigation, L.X.; data curation, J.C.; writing-review & editing, X.Z. All authors have read and agreed to the published version of the manuscript.

Funding: This study was supported by the National Natural Science Foundation of China (32260292), The Science and Technology Plan of Gansu Province, China (24 ZDFF001), Key Laboratory of Urban and Architectural Heritage Conservation, Ministry of Education, Southeast University (KLUAHC2404).

Institutional Review Board Statement: Not applicable.

Informed Consent Statement: Not applicable.

Data Availability Statement: The datasets used and analyzed during the current study are available from Xiaoju Yang (yangxiaoju@dha.ac.cn) on reasonable request.

Acknowledgments: We sincerely thank the staff who provided assistance during the on-site investigation. Thank you to student Zheng Xiaotong for providing assistance during the sample collection period.

Conflicts of Interest: The authors declare no conflict of interest. Author Long Li was employed by Cultural Heritage Conservation and Design Consulting Co., Ltd. of Mogao Grottoes. The remaining authors declare that the research was conducted in the absence of any commercial or financial relationships that could be construed as a potential conflict of interest.

References

1. Li, Z.; Wang, X.; Sun, M.; Chen, W.; Guo, Q.; Zhang, H. Conservation of Jiaohe ancient earthen site in China. *J. Rock Mech. Geotech. Eng.* **2011**, *3*, 270–281. [CrossRef]
2. Holtorf, C. Can less be more? Heritage in the age of terrorism. *Public Archaeol.* **2006**, *5*, 101–109. [CrossRef]
3. Cao, Y.; Bowker, M.A.; Delgado-Baquerizo, M.; Xiao, B. Biocrusts protect the Great Wall of China from erosion. *Sci. Adv.* **2023**, *9*, eadk5892. [CrossRef]
4. Chen, H.; Chen, W.; Li, X.; Jia, B.; Zhang, S. The effect of SH-lime composite material on capillary water rise resistance in the Great Wall. *J. Build. Eng.* **2024**, *85*, 108512. [CrossRef]
5. Richards, J.; Guo, Q.; Viles, H.; Wang, Y.; Zhang, B.; Zhang, H. Moisture content and material density affects severity of frost damage in earthen heritage. *Sci. Total Environ.* **2022**, *819*, 153047. [CrossRef]
6. Guo, Z.; Wu, C.; Zhang, S.; Chen, W.; Qi, Q.; Wu, H. Characterization on Flaking of Rammed Earthen Sites Using SMO Algorithm and Surface Topography Analysis: A Case Study of Jiaohe Ruins. *ACM J. Comput. Cult. Herit.* **2024**, *17*, 1–20. [CrossRef]
7. Quagliarini, E.; Lenci, S.; Iorio, M. Mechanical properties of adobe walls in a Roman Republican domus at Suasa. *J. Cult. Herit.* **2010**, *11*, 130–137. [CrossRef]
8. Shao, M.; Li, L.; Wang, S.; Wang, E.; Li, Z. Deterioration mechanisms of building materials of Jiaohe ruins in China. *J. Cult. Herit.* **2013**, *14*, 38–44. [CrossRef]
9. Du, Y.; Chen, W.; Cui, K.; Gong, S.; Pu, T.; Fu, X. A model characterizing deterioration at earthen sites of the Ming Great Wall in Qinghai Province, China. *Soil Mech. Found. Eng.* **2017**, *53*, 426–434. [CrossRef]
10. Parracha, J.L.; Silva, A.S.; Cotrim, M.; Faria, P. Mineralogical and microstructural characterisation of rammed earth and earthen mortars from 12 th century Paderne Castle. *J. Cult. Herit.* **2020**, *42*, 226–239. [CrossRef]
11. Li, J.; He, Y.; He, C.; Xiao, L.; Wang, N.; Jiang, L.; Chen, J.; Liu, K.; Chen, Q.; Gu, Y. Diversity and composition of microbial communities in Jinsha earthen site under different degree of deterioration. *Environ. Res.* **2024**, *242*, 117675. [CrossRef]
12. Hueck, H. The biodeterioration of materials as part of hylobiology. *Mater. Und Org.* **1965**, *1*, 5–34.
13. Lim, A.B.; Matero, F.G.; Henry, M.C. Greening deterioration: Soft capping as preventive conservation for masonry ruin sites. *APT Bull. J. Preserv. Technol.* **2013**, *44*, 53–61.
14. Hosseini, Z.; Caneva, G. Evaluating hazard conditions of plant colonization in Pasargadae World Heritage Site (Iran) as a tool of biodeterioration assessment. *Int. Biodeterior. Biodegrad.* **2021**, *160*, 105216. [CrossRef]
15. Lee, Z.; Viles, H.A.; Wood, C.H. *Soft Capping Historic Walls: A Better Way of Conserving Ruins?* English Heritage: London, UK, 2009.
16. Liu, X.; Koestler, R.J.; Warscheid, T.; Katayama, Y.; Gu, J.-D. Microbial deterioration and sustainable conservation of stone monuments and buildings. *Nat. Sustain.* **2020**, *3*, 991–1004. [CrossRef]
17. Bettina Weber, B.; Belnap, J.; Büdel, B. What is a biocrust? A refined, contemporary definition for a broadening research community. *Biol. Rev.* **2022**, *97*, 1768–1785. [CrossRef]

18. Eldridge, D.J.; Reed, S.; Travers, S.K.; Bowker, M.A.; Maestre, F.T.; Ding, J.; Havrilla, C.; Rodriguez-Caballero, E.; Barger, N.; Weber, B. The pervasive and multifaceted influence of biocrusts on water in the world's drylands. *Glob. Change Biol.* **2020**, *26*, 6003–6014. [CrossRef]
19. Li, X.; Hui, R.; Tan, H.; Zhao, Y.; Liu, R.; Song, N. Biocrust research in China: Recent progress and application in land degradation control. *Front. Plant Sci.* **2021**, *12*, 751521. [CrossRef]
20. Zhang, J.; Wu, B.; Li, Y.; Yang, W.; Lei, Y.; Han, H.; He, J. Biological soil crust distribution in Artemisia ordosica communities along a grazing pressure gradient in Mu Us Sandy Land, Northern China. *J. Arid Land* **2013**, *5*, 172–179. [CrossRef]
21. Gao, L.; Bowker, M.A.; Xu, M.; Sun, H.; Tuo, D.; Zhao, Y. Biological soil crusts decrease erodibility by modifying inherent soil properties on the Loess Plateau, China. *Soil Biol. Biochem.* **2017**, *105*, 49–58. [CrossRef]
22. Xiao, B.; Hu, K. Moss-dominated biocrusts decrease soil moisture and result in the degradation of artificially planted shrubs under semiarid climate. *Geoderma* **2017**, *291*, 47–54. [CrossRef]
23. Liu, L.; Liu, Y.; Hui, R.; Xie, M. Recovery of microbial community structure of biological soil crusts in successional stages of Shapotou desert revegetation, northwest China. *Soil Biol. Biochem.* **2017**, *107*, 125–128. [CrossRef]
24. Richards, J.; Mayaud, J.; Zhan, H.; Wu, F.; Bailey, R.; Viles, H. Modelling the risk of deterioration at earthen heritage sites in drylands. *Earth Surf. Process. Landf.* **2020**, *45*, 2401–2416. [CrossRef]
25. Wu, F.; Ding, X.; Zhang, Y.; Gu, J.-D.; Liu, X.; Guo, Q.; Li, J.; Feng, H. Metagenomic and metaproteomic insights into the microbiome and the key geobiochemical potentials on the sandstone of rock-hewn Beishiku Temple in Northwest China. *Sci. Total Environ.* **2023**, *893*, 164616. [CrossRef] [PubMed]
26. Liu, Y.; Ren, J.; Wang, W.; Shi, Y.; Gao, Y.; Zhan, H.; Luo, Y.; Jia, R. Vascular plants and biocrusts ameliorate soil properties serving to increase the stability of the Great Wall of China. *Sci. Total Environ.* **2024**, *951*, 175506. [CrossRef]
27. Wang, T.; Guo, Q.; Pei, Q.; Chen, W.; Wang, Y.; Zhang, B.; Yu, J. Destruction or protection? Experimental studies on the mechanism of biological soil crusts on the surfaces of earthen sites. *Catena* **2023**, *227*, 107096. [CrossRef]
28. Sun, F.; Xiao, B.; Li, S.; Yu, X.; Kidron, G.J.; Heitman, J. Direct evidence and mechanism for biocrusts-induced improvements in pore structure of dryland soil and the hydrological implications. *J. Hydrol.* **2023**, *623*, 129846. [CrossRef]
29. Grose, M.J.; Gilligan, C.A.; Spencer, D.; Goddard, B. Spatial heterogeneity of soil water around single roots: Use of CT-scanning to predict fungal growth in the rhizosphere. *New Phytol.* **1996**, *133*, 261–272. [CrossRef] [PubMed]
30. Anderson, S.; Peyton, R.; Gantzer, C. Evaluation of constructed and natural soil macropores using X-ray computed tomography. *Geoderma* **1990**, *46*, 13–29. [CrossRef]
31. Lipiec, J.; Hatano, R. Quantification of compaction effects on soil physical properties and crop growth. *Geoderma* **2003**, *116*, 107–136. [CrossRef]
32. Perret, J.; Prasher, S.; Kantzas, A.; Langford, C. Three-dimensional quantification of macropore networks in undisturbed soil cores. *Soil Sci. Soc. Am. J.* **1999**, *63*, 1530–1543. [CrossRef]
33. Zhou, H.; Fang, H.; Mooney, S.J.; Peng, X. Effects of long-term inorganic and organic fertilizations on the soil micro and macro structures of rice paddies. *Geoderma* **2016**, *266*, 66–74. [CrossRef]
34. Carter, M.R.; Gregorich, E.G. *Soil Sampling and Methods of Analysis*; CRC Press: Boca Raton, FL, USA, 2007.
35. Xiao, B.; Bowker, M.A. Moss-biocrusts strongly decrease soil surface albedo, altering land-surface energy balance in a dryland ecosystem. *Sci. Total Environ.* **2020**, *741*, 140425. [CrossRef]
36. Jester, W.; Klik, A. Soil surface roughness measurement—Methods, applicability, and surface representation. *Catena* **2005**, *64*, 174–192. [CrossRef]
37. Zhao, H.; Li, Z.; Han, W.; Sun, M.; Wang, X. Research on the major diseases and protection of the Great Wall Site in Gansu Province. *Cult. Relics Prot. Archaeol. Sci.* **2007**, *19*, 28–32. (In Chinese) [CrossRef]
38. Komaei, A.; Soroush, A.; Fattahi, S.M.; Ghanbari, H. Influence of environmental stresses on the durability of slag-based alkali-activated cement crusts for wind erosion control. *Sci. Total Environ.* **2023**, *902*, 166576. [CrossRef] [PubMed]
39. Miralles, I.; Domingo, F.; Cantón, Y.; Trasar-Cepeda, C.; Leirós, M.C.; Gil-Sotres, F. Hydrolase enzyme activities in a successional gradient of biological soil crusts in arid and semi-arid zones. *Soil Biol. Biochem.* **2012**, *53*, 124–132. [CrossRef]
40. Aldughaishi, U.; Grattan, S.R.; Nicolas, F.; Peddinti, S.R.; Bonfil, C.; Ogunmokun, F.; Abou Najm, M.; Nocco, M.; Kisekka, I. Assessing the impact of recycled water reuse on infiltration and soil structure. *Geoderma* **2024**, *452*, 117103. [CrossRef]
41. Pan, J.; Xu, N.; Tang, Y.; Cheng, M.; Zhang, L.; Wang, B.; Lan, J. Quantitative evaluation of plants on top surface of the Great Wall in Dazhuangke using the analytical hierarchy process. *Herit. Sci.* **2023**, *11*, 191. [CrossRef]
42. Gregory, P.J. RUSSELL REVIEW Are plant roots only "in" soil or are they "of" it? Roots, soil formation and function. *Eur. J. Soil Sci.* **2022**, *73*, e13219. [CrossRef]
43. Wang, X.; Zhang, B.; Guo, Q.; Pei, Q. Discussion on the environmental adaptability of weather-resistant measures for earthen sites in China. *Built Herit.* **2022**, *6*, 19. [CrossRef]
44. Dhaliwal, J.K.; Kumar, S. 3 D-visualization and quantification of soil porous structure using X-ray micro-tomography scanning under native pasture and crop-livestock systems. *Soil Tillage Res.* **2022**, *218*, 105305. [CrossRef]

45. Dorau, K.; Uteau, D.; Hövels, M.P.; Peth, S.; Mansfeldt, T. Soil aeration and redox potential as function of pore connectivity unravelled by X-ray microtomography imaging. *Eur. J. Soil Sci.* **2022**, *73*, e13165. [CrossRef]
46. Wildenschild, D.; Vaz, C.; Rivers, M.; Rikard, D.; Christensen, B. Using X-ray computed tomography in hydrology: Systems, resolutions, and limitations. *J. Hydrol.* **2002**, *267*, 285–297. [CrossRef]
47. Dal Ferro, N.; Charrier, P.; Morari, F. Dual-scale micro-CT assessment of soil structure in a long-term fertilization experiment. *Geoderma* **2013**, *204*, 84–93. [CrossRef]
48. Katuwal, S.; Norgaard, T.; Moldrup, P.; Lamandé, M.; Wildenschild, D.; de Jonge, L.W. Linking air and water transport in intact soils to macropore characteristics inferred from X-ray computed tomography. *Geoderma* **2015**, *237*, 9–20. [CrossRef]
49. Du, Y.; Dong, W.; Cui, K.; Chen, W.; Yang, W. Research progress on the development mechanism and exploratory protection of the scaling off on earthen sites in NW China. *Sci. China Technol. Sci.* **2023**, *66*, 2183–2196. [CrossRef]
50. Chen, W.; Chen, H.; Jia, B.; Bi, J.; Li, X. Study of salt migration on the upper part of the Great Wall under the rainfall-radiation cycle. *Bull. Eng. Geol. Environ.* **2022**, *81*, 419. [CrossRef]
51. Belnap, J. The potential roles of biological soil crusts in dryland hydrologic cycles. *Hydrol. Process. Int. J.* **2006**, *20*, 3159–3178. [CrossRef]
52. Li, S.; Bowker, M.A.; Xiao, B. Biocrust impacts on dryland soil water balance: A path toward the whole picture. *Glob. Change Biol.* **2022**, *28*, 6462–6481. [CrossRef]
53. Pande, S.; Kost, C. Bacterial unculturability and the formation of intercellular metabolic networks. *Trends Microbiol.* **2017**, *25*, 349–361. [CrossRef] [PubMed]
54. Shao, Q.; Baumgartl, T. Field evaluation of three modified infiltration models for the simulation of rainfall sequences. *Soil Sci.* **2016**, *181*, 45–56. [CrossRef]
55. Bodner, G.; Leitner, D.; Kaul, H.-P. Soil, Coarse and fine root plants affect pore size distributions differently. *Plant* **2014**, *380*, 133–151. [CrossRef]
56. Tripathi, I.M.; Mahto, S.S.; Kushwaha, A.P.; Kumar, R.; Tiwari, A.D.; Sahu, B.K.; Jain, V.; Mohapatra, P.K. Dominance of soil moisture over aridity in explaining vegetation greenness across global drylands. *Sci. Total Environ.* **2024**, *917*, 170482. [CrossRef]
57. Liu, Y.; Wang, W.; Ren, J.; Gao, Y.; Wu, F.; Shi, Y.; Zhan, H.; Chi, C.; Jia, R. Microtopography-induced hydrological heterogeneity promotes the co-assembly of vascular plant and biocrust communities, providing synergistic protective functions for the Great Wall. *Sci. Total Environ.* **2025**, *963*, 178513. [CrossRef] [PubMed]

coatings

MDPI

Article

Durability Analysis of Brick-Faced Clay-Core Walls in Traditional Residential Architecture in Quanzhou, China

Yuhong Ding [1,2], Ruiming Guan [3,4,*], Li Chen [3,4], Jinxuan Wang [2], Yangming Zhang [5], Yili Fu [6] and Canjin Zhang [7]

1 Faculty of Humanities and Arts, Macau University of Science and Technology, Avenida Wailong, Taipa, Macau 999078, China; 3240006977@student.must.edu.mo
2 College of Civil Engineering, Liming Vocational University, No. 298 Tonggang West Street, Quanzhou 362000, China; 19970701@lmu.edu.cn
3 School of Architecture and Urban-Rural Planning, Fuzhou University, No. 2 Xueyuan Road, Minhou University City New District, Fuzhou 350108, China; t10002@fzu.edu.cn
4 College of Human Settlements and Environment, Quanzhou Vocational and Technical University, Daxue Road, Neikeng Town, Jinjiang 362268, China
5 Institute for Research on Portuguese-Speaking Countries, City University of Macau, Xu Risheng Yin Gong Road, Taipa, Macau 999078, China; a24092100223@cityu.edu.mo
6 Quanzhou Artisan Architectural Design Co., Ltd., No. 16, Fuchuo Lane, Licheng District, Quanzhou 362000, China; fu@artisanjz.cn
7 School of Management, Xiamen University, No. 422 Siming South Road, Siming District, Xiamen 361005, China; zhangcanjin@stu.xmu.edu.cn
* Correspondence: t08127@fzu.edu.cn

Abstract: This study analyzes the durability of brick-faced clay-core walls (BCWs) in the traditional residential architecture of Quanzhou—a UNESCO World Heritage City. Taking the northern gable of Ding Gongchen's former residence as an example, the mechanical properties, microscopic structure, and changes in chemical symbol, oxides and minerals of the red bricks and clay-cores were analyzed using finite element mechanics analysis (FEM), scanning electron microscopy (SEM), X-ray fluorescence (XRF), and X-ray diffraction (XRD). The results indicate a triple mechanism: (1) The collaborative protection and reinforcement mechanism of "brick-wrapped-clay". (2) The infiltration and destruction mechanism of external pollutants. (3) The material stability mechanism of silicate minerals. Therefore, the key to maintaining the durability of BCWs lies in the synergistic effect of brick and clay materials and the stability of silicate mineral materials, providing theoretical and methodological support for sustainable research into brick and clay constructions.

Keywords: brick-faced clay-core wall; brick–clay construction; durability; mortise and tenon; FEM analysis; weathering mechanism; world heritage site; traditional residential architecture of Quanzhou

1. Introduction

1.1. Background and Significance

The brick-faced clay-core wall (BCW, also known as a "cavity wall" or "sealed brickwork") is a common brick–clay construction form of traditional residential architecture in China that is widely used in Fujian and Taiwan, as well as surrounding areas in Putian, Quanzhou, Xiamen, Zhangzhou, Chaoshan, and Taiwan [1]. The load-bearing strength of a BCW is poorer than a solid brick wall [2]. Therefore, the Code for Design of Masonry Structures, GB50003, has prohibited the use of BCWs since 2001 [3]. However, as a traditional building material, BCWs have profound historical and cultural value. They are widely present in Chinese traditional architecture and difficult to eliminate on a large scale in the short term [4]. Compared with other regions in China, the BCWs in Fujian and Taiwan bear not only the red brick cultural characteristics of Southern Fujian but also the collective wisdom of the people of Quanzhou—a UNESCO World Heritage City. Therefore, studying the internal factors of the BCW's durability (the ability to maintain one's own carrying capacity) in Quanzhou, in order to explain its widespread vitality in folk use, is of great significance for the safety of "brick and clay heritage" in subtropical regions.

1.2. Review of Durability Research

Studies on BCWs' durability have focused mainly on compressive and seismic resistance and reinforcement. Thenmozhi, R.; Saranya, K. et al. compared the compressive strength of 18 manually cast BCW samples, which included block materials such as red bricks, fly ash bricks, and solid bricks in different proportions [5]. Zhu, B.; Wu, M. et al. conducted compression tests on BCWs at a load of 27 kg/cm^2 and reported that the ultimate load of different masonry methods was approximately 60% to 85% of that of actual masonry walls in the 1980s [6]. Meanwhile, Romina, S. et al. studied ways to improve durability through the addition of accessories for BCWs during the period from 2023 to 2024 [7].

In addition to mechanical studies, researchers have also proposed various filler improvement methods, yet the intrinsic material mechanisms remain insufficiently explored. For example, Zhou, T.; Tian, P.; Deng, M.; Zhao, X.; Liang, Z. proposed the use of engineered cementitious composite (ECC) reinforcement technology in 2018 [8]; Argentina, O. et al. proposed filling the cavity of an empty bucket wall with PCM to enhance its thermal insulation performance in 2021 [9]; and Deng, J. studied the role of microorganisms in filling clays in 2023 [10]. Meanwhile, a small number of studies also began to pay attention to the waterproof performance of BCWs' skin brick material in 2024, for example, D'Ayala, D. et al. [11].

1.3. Research Objective

Current research mainly focuses on using external intervention to enhance the compressive, seismic, or thermal insulation performance of BCWs by using reinforcement fittings or materials. However, the BCW has been able to be used by the public for a long time without technological improvements. The inherent characteristics of its structure and materials inevitably have advantages in terms of durability. Therefore, it is feasible and necessary to thoroughly and comprehensively analyze the internal factors of BCWs' durability from the perspectives of construction and material properties.

2. Materials and Methods

2.1. BCW Construction and Sampling

A BCW is a wall structure constructed of overlapping "mortise and tenon" joints made of a mixture of red bricks and clay cores [12]. Among them, the "cavity" built with red bricks is filled with clay cores (rammed earth, pebbles, or brick fragments) inside, providing effective insulation and sound insulation functions for residential spaces [13,14]. BCWs often use "Xiaoxiu bricks (150 mm × 100 mm × 20 mm)" or "salt bricks (210 mm × 150 mm × 20 mm)" as horizontal ties (Mian bricks) and "Shixing bricks (210 mm × 170 mm × 20 mm)" as vertical ties (Li bricks), forming a wall with a width of 210 mm using "Mian-and-Li" mortise and tenon overlap. The masonry methods can be divided into two main types: horizontal bonding and vertical bonding. The horizontal bonding is composed of a "shun" (stretcher) and a "ding" (header), including patterns of ding-and-shun, ding-and-double-shun, or ding-and-triple-shun (Figure 1a–c); the vertical bonding method is composed of a "mian" (solid layer) and a "dou" (hollow layer), i.e., including patterns of mian-and-dou, mian-and-double-dou, or mian-and-triple-dou (Figure 1d–f) [15].

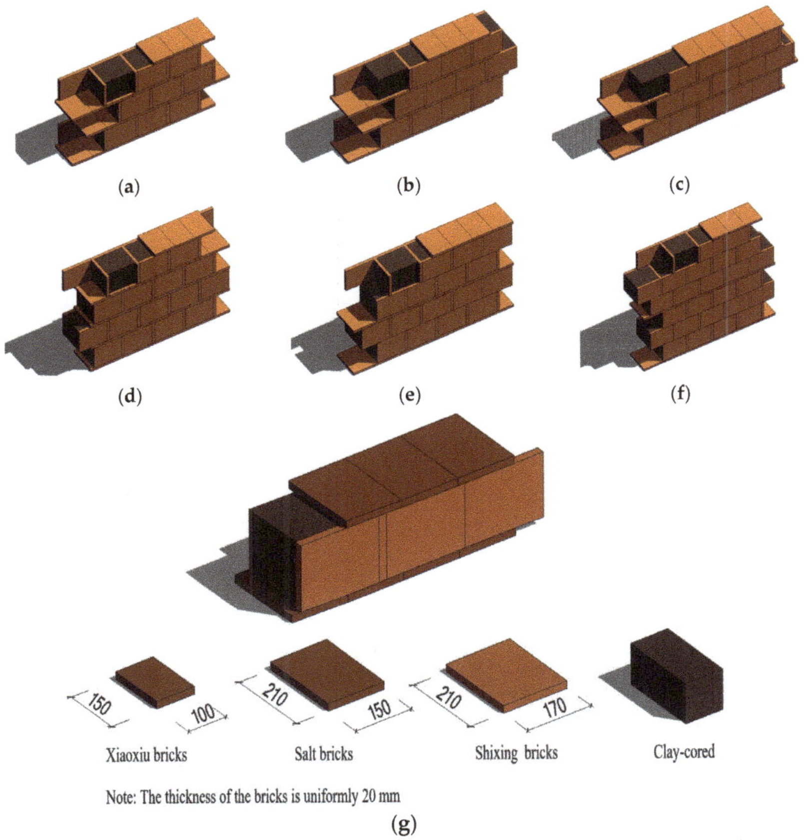

Figure 1. The masonry method and size of BCWs. (**a**) Ding-and-shun. (**b**) Ding-and-double-shun. (**c**) Ding-and-triple-shun. (**d**) Mian-one-dou. (**e**) Mian-and-double-dou. (**f**) Mian-and-triple-dou. (**g**) Size specifications of red bricks (chart source: Revit 2022. Produced by Autodesk, San Francisco, CA, USA).

Based on the above construction features, the following text evaluates its durability performance in real environments through on-site sampling and experimental analysis. The research team collected and extracted four types of red brick and clay-core samples, categorized by a lower and higher degree of weathering, from sampling points A and B on the northern gable of Ding Gongchen's former residence.

After field sampling, the samples were promptly transferred to the laboratory, where they were packaged, initially ground, and finely sieved (with a powder size less than 0.25 mm [16]). The samples were then compressed into tablets using a Ruishenbao tablet press set at 220 kN to provide experimental samples. They were coded as red brick (Z), clay-core (T), with a low degree of weathering (01), and with a high degree of weathering (02). Thus, Z-01 (red brick with a low degree of weathering), Z-02 (red brick with a high degree of weathering), T-01 (clay-core with a low degree of weathering), and T-02 (clay-core with a high degree of weathering) were used to ensure the systematic and traceable analysis of the subsequent experimental results (Table A1).

2.2. Basis for Sample Selection

The former residence of Ding Gongchen (No. 107 Guanyin Lane, Anduo Village, Chendai Town), located near the mouth of Quanzhou Bay in the Jinjiang River Basin, was selected as the research sample (Figure 2) to reveal the internal factors affecting the durability of BCWs in terms of the structure and materials of red bricks and clay cores, based on the following:

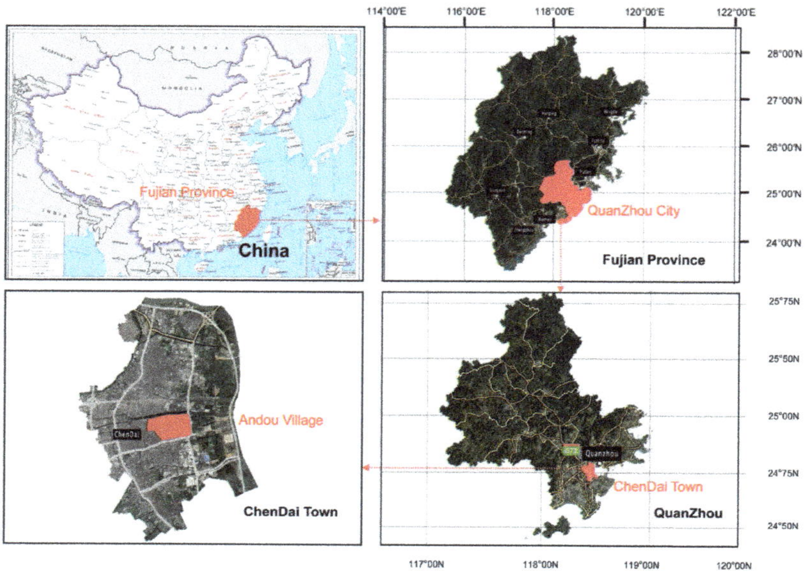

Figure 2. Location of study area (chart source: http://www.bigemap.com/source/tree/satel-1.html (accessed on 8 June 2025) and http://bzdt.ch.mnr.gov.cn/ (accessed on 8 June 2025)).

(1) Quanzhou—as a world marine commerce center and a World Heritage City in China during the Song and Yuan dynasties—can be considered a model for the construction of traditional residential architecture in Southern China [17]. The research object of this study is the historic "three-bay double-courtyard" residence in Quanzhou built by Ding Gongchen, a scientist in the Qing dynasty, during the Jiaqing era (1800–1875). This is a typical case for the study of BCWs (Figure 3).

(2) The sample collection area was located on the south bank of Jinjiang River in Quanzhou City, belonging to a subtropical humid climate zone. Due to its proximity to the sea, the salinity in both the air and water is high. In addition, the region has experienced severe environmental pollution in the past 30 years [18]. Therefore, the weathering of BCW materials in the region makes it an ideal case for durability research.

(3) The north mountain wall of the sample building shows three different states of BCW: (A) fully preserved wall, (B) wall damaged due to years of disrepair, and (C) a temporary reinforced wall completed in 1991. According to the principle of "recognizability" in the protection and restoration of ancient architecture, contemporary crafts and materials can be compared, distinguished, and excluded based on the current situation of point (C), in order to accurately identify the crafts and materials of points (A) and (B), which is a traditional practice in the Qing Dynasty of China [19] (Figures 3–5).

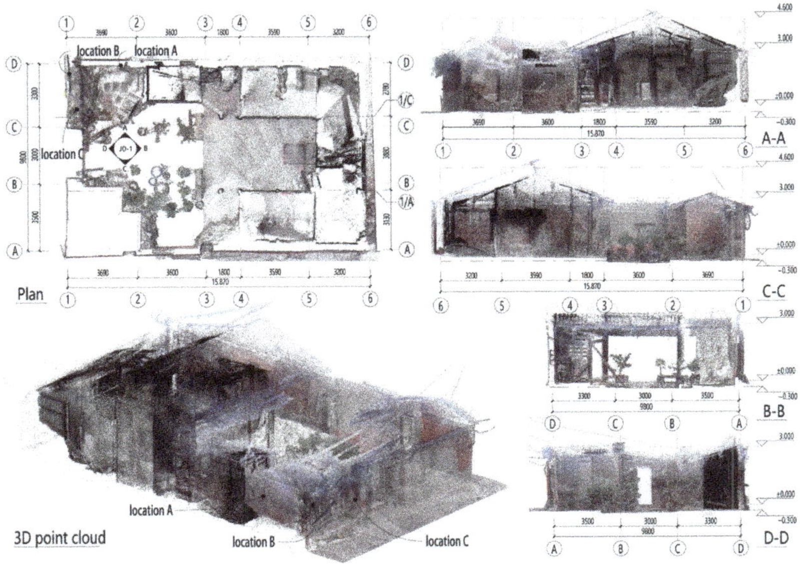

Figure 3. Sampling point identification diagram of Ding Gongchen's former residence. Note that, due to the fact that the area is a no-fly zone, it was not possible to use drones for aerial photography (chart source: we obtained point cloud data by the XGRIDS Lixel L 3D laser scanner (Produced by Shenzhen Qiyu Innovation Technology Co., Ltd., Shenzhen City, China) and then imported the point cloud data into Revit 2022 for layout and annotation).

(**a**) (**b**) (**c**)

Figure 4. Current condition of sampling points: (**a**) current condition of sampling point A; (**b**) current condition of sampling point B; and (**c**) current condition of sampling point C (chart source: photography during on-site inspections).

(**a**) (**b**) (**c**)

Figure 5. Field survey and sampling site: (**a**) on-site sample collection; (**b**) on-site sample classification; and (**c**) sealing samples in transparent plastic containers sterilized with anhydrous ethanol (chart source: photography during on-site inspections).

2.3. Experimental Methods

Finite element mechanical analysis (FEM) was used to analyze the structural relationship between red bricks and clay cores in the BCW. Then, scanning electron microscopy (SEM), X-ray fluorescence (XRF), and X-ray diffraction (XRD) were used to determine the microstructure, chemical symbols, and changes in oxide and mineral composition of the two materials from the perspective of material chemistry, in order to demonstrate the durability advantage and internal factors of BCWs' "brick-wrapped soil" structure. The specific method is as follows:

2.3.1. FEM Analysis

Using SOLIDWORKS 2024 (SW. software is produced by Dassault Systemes company, Concord, MA, USA) software, finite element mechanics (FEM) simulation analysis was conducted on three structural samples of the BCW (Figures 6–8). The analysis utilized triangular meshes with an element size of 35 mm, as set within the software. The sample models were created using Revit 2022, with boundary conditions ranging from approximately 1 m^3 to 1.5 m^3 (710 mm × 970 mm × 210 mm), and the Revit model data were then converted to the IFC format for importation into SW.

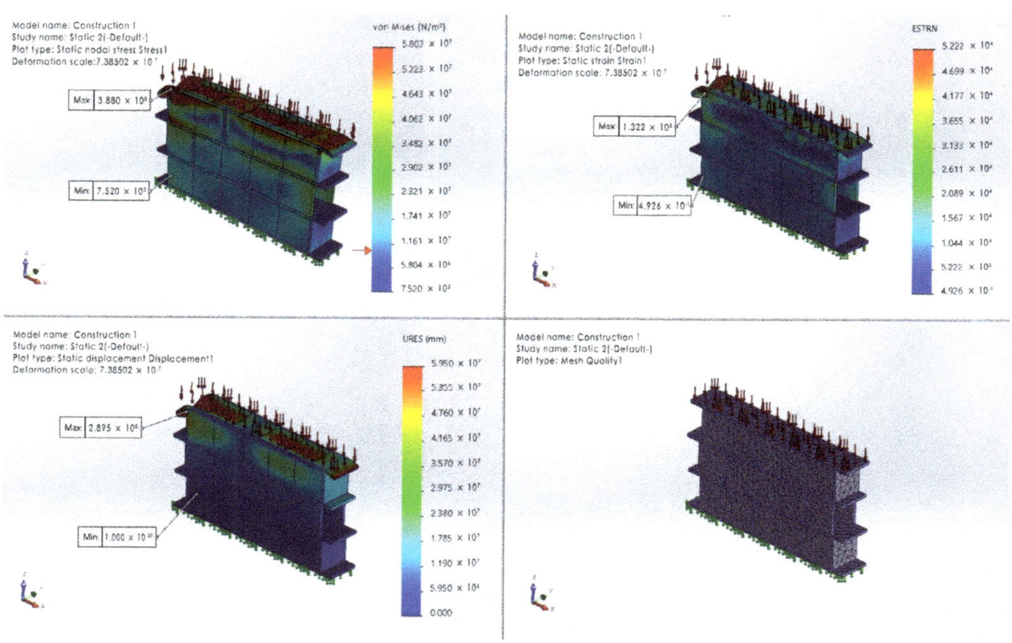

Figure 6. Mechanical FEM analysis of Construction I. (Chart source: Revit 2022 modeling, exporting IFC format model, analyzed using SOLIDWORKS software.)

Considering the nonlinear relationship between the brick and clay materials, special attention was given to the interfaces between these different materials during the modeling process. The material properties for both the brick and clay were defined according to the software's material library and relevant building standards (Table 1). Fixed clamps were placed at the bottom of each experimental sample, and a uniform pressure of 27 kg/cm^2 was applied at the top [10]. The analysis focused on the von Mises stress, static strain, and static displacement of the samples in order to evaluate the mechanical interactions between the red brick and clay materials and their impact on the durability of the BCW.

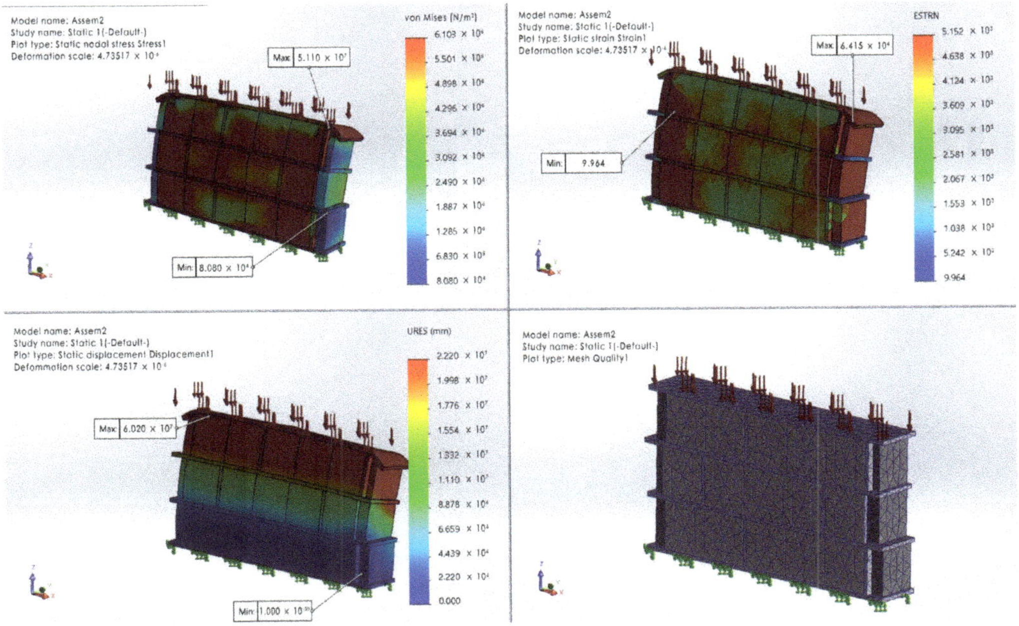

Figure 7. Mechanical FEM analysis of Construction II. (Chart source: Revit 2022 modeling, exporting IFC format model, analyzed using SOLIDWORKS software.)

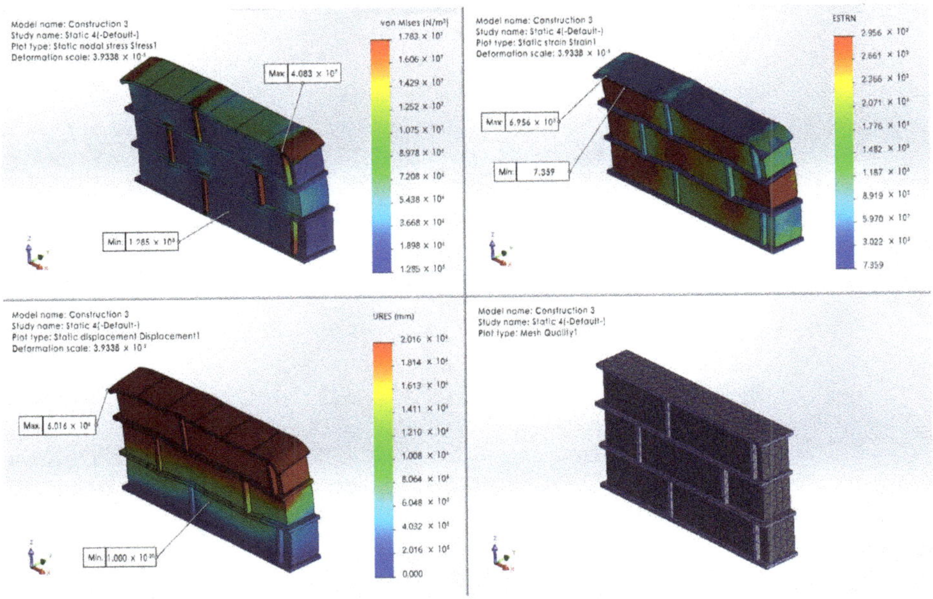

Figure 8. Mechanical FEM analysis of Construction III. (Chart source: Revit 2022 modeling, exporting IFC format model, analyzed using SOLIDWORKS software.)

Table 1. Material properties settings for SW. Experimental samples.

Material	Elastic Modulus	Poisson's Ratio	Shear Modulus	Mass Density	Yield Strength
Unit	MPa	(μ)	N/m^2	kg/m^3	N/m^2
Red brick	1300	0.2	5.15×10^8	2100	10^7
Clay core	100	0.3	3.81×10^7	1600	10^5

Chart source: the property settings of the material library in SOLIDWORKS 2024 and References [3,5].

2.3.2. SEM Analysis

A trace amount of powder was affixed on the conductive adhesive, and the powder was plated using a Quorum SC7620 sputtering coater (Produced by Quorum Technologies, Laughton, East Sussex, UK) or 45 s with the current set to 10 mA. SEM was performed using a ZEISS Sigma 360 (Produced by ZEISS, Jena City, Germany) The magnification was set from 5000× to 500×, the WD was 6.7 mm to 9.4 mm, and the microstructure topography of the sample was determined. During imaging, the accelerating voltage was set to 5 kV, and the detector was an SE2 secondary electron detector (Produced by ZEISS, Jena City, Germany) to observe the microstructural changes during weathering [16]. Then, based on the principle that the more pores, cracks, and crystal expansion in the SEM images, the higher the average gray level of the images, the density changes in the microstructure of the four samples were analyzed using ImageJ software (version: 1.50d, Produced by National Institutes of Health, Bethesda, MD, USA) combined with Origin 2022 (Produced by Electronic Arts Inc., EA, Redwood City, CA, USA) [20].

2.3.3. XRF Analysis

A ZSX Primus III+ X-ray fluorescence spectrometer (Produced by Rigaku, Akishima City, Kyoto, Japan) was employed, with a total scan time of 10 min, to semi-quantitatively detect the changes in the content of various chemical symbols in the samples, and the migration and enrichment patterns of the elements during the weathering process were analyzed by Rigaku (Produced by Rigaku, Akishima City, Kyoto, Japan) to reveal the main internal factors influencing their durability [21]. Meanwhile, the C content of the experimental samples did not exceed 10% and the carbon element content was relatively low, which did not affect the accuracy of the test; therefore, the LOI treatment was not carried out.

2.3.4. XRD Analysis

A D/MAX-2600 X-ray diffractometer (Produced by Rigaku, Akishima City, Kyoto, Japan) was used to identify the mineral composition and changes in the amorphous phase of the samples and to semi-quantitatively analyze their structural deterioration and weathering mechanisms. The experimental parameters were as follows: scan mode, 2 theta; scan type, continuous scan; X-ray voltage, 40 kV; current, 150 mA; DivSlit, 1°; DivH.L Slit, 10 mm; SctSlit, 1°; and RecSlit, 0.3 mm. The total scan time was 21 min, the testing range was 3–64 degrees, and the testing rate was 3 degrees per minute.

JADE 9 (Produced by Materials Data Inc., Livermore, CA, USA) software was used for mineral composition analysis, while Origin software was used to generate diffraction peak patterns. For qualitative phase analysis, a powder diffraction database was utilized for both manual and computer retrieval, enabling the identification of matching phases from the database (PDF cards compiled and published by the International Centre for Diffraction Data, which contain standard powder diffraction spectra for various phases) [22].

During the analysis, it is important to consider factors such as overlapping peaks and preferred orientation, which can lead to spectral anomalies. Additionally, the presence of amorphous phases should be assessed. For phases that are difficult to identify, a comprehensive evaluation can be made using sample information or alternative methods.

3. Results

3.1. FEM Simulation Analysis

FEM analysis experiments were conducted for clay-filled masonry (Construction I), cavity-filled earth masonry (Construction II), and masonry with weathered or peeled surface bricks (Construction III). The results are as follows (Figures 6–8).

3.1.1. Construction I

The maximum value of stress is 3.88×10^8 N/m^2, which is far greater than the design strength of conventional building materials, indicating that the material in some areas has already undergone plastic deformation, and that the structure is at risk of local yielding under this load. Moreover, the stress distribution is uneven; the stress is concentrated in some locations in the upper and middle parts of the wall, whereas the lower part is relatively small, and the minimum is 7.52×10^2 N/m^2.

The maximum value of static strain reaches 1.32×10^5, indicating that under the action of load, the deformation of the wall is relatively large, the microstructure inside the material underwent prominent changes, and there has been a tendency for cracking or local failure [10]. The minimum value is 4.92×10^{-1}. The strain distribution echoes the stress distribution, which further highlights the nonuniform deformation characteristics of the unfilled soil masonry under loading.

The maximum static displacement reaches 2.89×10^8 mm, which means that the overall displacement of the wall is relatively large, and that the structural stability is poor. Under this load, the wall can tilt significantly or displacement can accumulate, and the long-term stability and safety of the structure cannot be guaranteed. The minimum displacement is 1.00×10^{-30} mm, which can be disregarded.

This uneven stress distribution is due to the lack of clay filling in the masonry, which leads to significant stress concentrations and exacerbates the nonlinear behavior of material strain and displacement [10].

3.1.2. Construction II

The maximum value of stress is 5.11×10^7 N/m^2, which is within a reasonable range, indicating that the overall stress level of the wall is reasonable; the fill soil effectively shares part of the load, making the stress distribution more uniform; the stress coordination of each part of the wall is good, and there is no obvious stress concentration area; and the minimum value is 8.08×10^4 N/m^2. This shows that the fill soil can significantly improve the load-bearing performance of the wall and enhance its bearing capacity [12].

The maximum static strain is 6.41×10^4, which is within a reasonable range and far lower than the maximum strain value of masonry without filled soil. The above results show that with the filled soil, the internal damage of the wall material is light, the deformation is controllable, and the structural integrity is good. The minimum value is 9.96. The strain distribution is uniform, indicating that the filling soil could coordinate the deformation of the wall under loading and reduce the risk of local damage [12].

The maximum value of the static displacement is 6.02×10^7 mm, which is significantly lower than that of masonry without filled soil, indicating that the filled soil has a positive effect on the overall stability of the wall. The wall has a lower displacement under load and can maintain a better geometric shape and position. The minimum value is 1.00×10^{-30} mm.

There is a certain frictional force between red bricks and clay, which helps to disperse stress and reduce the nonlinear response of material strain and displacement.

3.1.3. Construction III

The maximum stress value is 4.08×10^7 N/m^2; Although it does not exceed the ultimate strength of traditional building structural materials, it is close to the upper limit [23]. The results indicate that after the surface bricks fall off, the local stress of the wall increases, and the stress distribution becomes uneven to a certain extent. The peripheral stress of the separation position is relatively concentrated, while the internal stress is relatively small. The minimum value is 1.28×10^5 N/m^2. Its load-bearing capacity is between Construction I and Construction II.

The maximum value of static strain is 6.95×10^3, which is less than the maximum strain value of the masonry without filled soil, indicating that although the detachment of the surface bricks caused some damage to the wall, the global deformation is still acceptable under the action of the filled soil, the wall still has a certain bearing capacity [12], and the minimum value is 7.35. The strain distribution is uneven, reflecting that the wall is affected by the shedding of surface bricks, and that the local deformation capacity of the wall is reduced.

The maximum value of static displacement is 6.01×10^6 mm, which is between Constructions I and II, indicating that the detachment of surface bricks has a certain impact on the wall displacement, but the fill soil can still have a partial confinement effect, ensuring that the wall displacement within a certain range is a minimum of 1.00×10^{-30} mm.

The detachment of surface bricks weakens the friction and interaction between red bricks and fill soil, resulting in poor stress transmission in certain areas, increased local stress, an uneven stress distribution, and an appropriate increase in strain and displacement, reflecting the nonlinear characteristics at the material interface.

3.1.4. Summary: The Mechanical Relationship Between Bricks and Clay

Experiments have shown that Construction II (filled soil masonry) has the best mechanical properties; The maximum von Mises stress is 5.11×10^7 N/m^2, the maximum static strain is 6.41×10^4, and the maximum static displacement is 6.02×10^7 mm, which is lower than the values of structures I and III. This indicates that filling the cavity with soil can make the deformation of the wall more uniform and controllable and improve the stability of the wall.

In addition, the mechanical properties of structure III (masonry with stripped-skin bricks) are between structures I and II, with a maximum von Mises stress of 4.08×10^7 N/m^2, a maximum static strain of 6.95×10^3, and a maximum static displacement of 6.01×10^6 mm. Both are superior to structure I but not as good as structure II. The detachment of surface bricks has a certain negative impact on the performance of the wall. But the fill in the cavity can still barely support the force distribution and overall performance of the wall [8].

In summary, BCWs' "brick-wrapped soil" mortise and tenon structure improves the overall durability of the wall. At the same time, even if the surface bricks become hollow and peel off, the clays in the inner core can still bear weight before high weathering.

3.2. SEM Microstructure Analysis

3.2.1. SEM-Red Brick

A comparison of the SEM images of the Z-01 and Z-02 samples (Tables A2 and 2, Figure 9) reveals that the grain boundaries of the Z-01 sample are clear, the grains are small and dense, and the average gray value of the SEM image is low, which is attributed to the manual brick-making process involving clay kneading, low heat dehydration, and high-temperature firing [23]. However, there are local areas with marked gray value fluctuations. For example, when the Mag (KX) value is 0.2, the gray value of Z-01 drops from 127,889 to 96,865, revealing local loose structures that are due to over-firing or under-firing. However, these issues generally do not impact normal use [24]. In contrast, the Z-02 sample has a loose structure with blurred grain boundaries and a higher (average value: 143.459) overall average gray value, reflecting severe chemical element loss and mineral transformation from weathering, as well as crystal expansion and physical damage caused by external material absorption [25].

Table 2. SEM image average mean grayscale value.

No. of Observations	Mag (KX)	Z-01	Z-02	T-01	T-02
1	5	109,124	159,747	149,542	127,539
2	1	131,942	158,001	137,266	148,312
3	0.5	114,391	154,923	122,237	136,525
4	0.2	96,865	119,649	109,963	121,846
5	5	129,237	152,001	148,924	155,796
6	1	115,287	145,134	149,112	142,807
7	0.5	123,102	145,394	140,929	130,344
8	0.2	127,889	112,821	111,434	108,938
Average value		118,480	143,459	133,676	134,013

Chart source: statistics based on Table A2 data. Grayscale value unit: dimensionless.

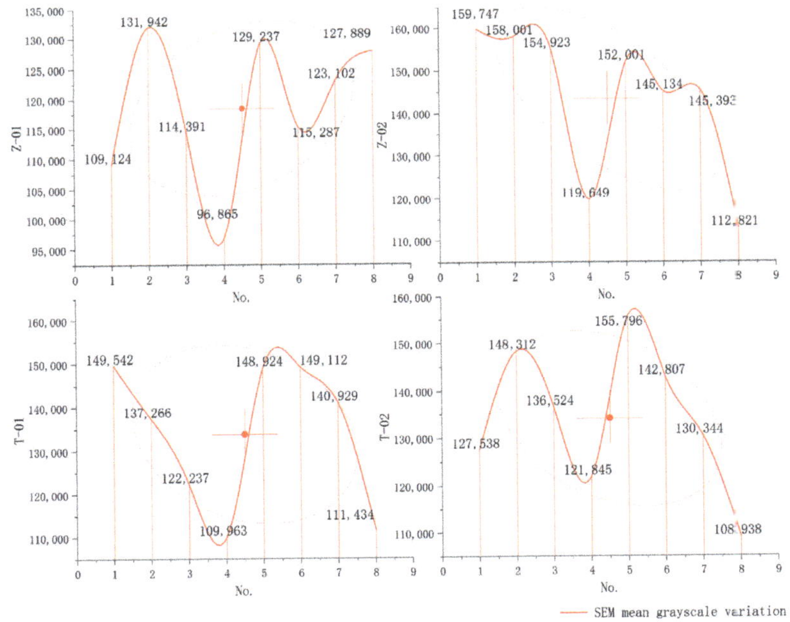

Figure 9. Curve graph variation in SEM image average mean grayscale value (chart source: drawing by Origin 2022).

3.2.2. SEM-Clay Core

A comparison of the SEM images and gray values of the T-01 and T-02 samples (Tables 2 and A2, Figure 9) shows that the T-01 sample has clearer grain boundaries, a smoother surface, and a more uniform particle distribution than the T-02 sample, but it also has local crack areas. This is due to material mixing, the introduction of masonry cavities, and uneven compaction during the artificial clay-core manufacturing process [26]. The average grayscale value of the T-02 sample image is almost the same as that of T-01 but slightly higher. For instance, at a Mag (KX) value of 5, the gray value of the T-01 sample is 148,924, while that of the T-02 sample is 155,796. This indicates that clay, protected within the masonry cavity, experiences less weathering but still suffers from natural aging and environmental impacts, leading to crystal expansion, structural damage, and crack growth [27].

3.2.3. Summary: Microscopic Images of Materials

Both the red bricks and clay cores are handmade, making regions with large pores and cracks inevitable. Natural weathering and the ingress of external pollutants cause the loss of chemical symbols, mineral transformation, and crystal expansion in both materials (including a small amount of dark mineral enrichment). These factors result in structural damage and a negative impact on their durability.

Meanwhile, comparing the four samples, it was found that the average gray value of the bricks has increased significantly (118,480 to 143,459), while the parameters of clay have not changed significantly under the protection of the bricks (Table 2). Therefore, we can conclude that clay cores generally suffer less damage than red bricks.

3.3. XRF Chemical Composition Analysis

3.3.1. XRF-Red Brick

X-ray fluorescence (XRF) spectroscopy analysis revealed significant changes in the major chemical composition of the red bricks due to the combined effects of long-term disrepair and external environmental erosion (Figure 10, Tables A3 and A4). From low weathering to high weathering, the SiO_2 content decreased from 60.906% to 54.211%, a reduction of 10.99%; Al_2O_3 decreased from 20.672% to 19.212%, a reduction of 7.06%; and Fe_2O_3 dropped from 8.398% to 7.816%, a decrease of 6.93%. MnO, K_2O, and P_2O_5 showed slight losses. The Na_2O content surged from 1.21% to 5.88%, demonstrating an enrichment of 386.35%; the Cl content also increased from 0.32% to 3.84%, indicating an enrichment of 1096%. Trace elements such as Co and Pb also showed some degree of enrichment. These changes suggest that the increase in microcracks within the red bricks made it easier for external salts to penetrate, accelerating the hydrolysis and dissolution of silicate minerals. At the same time, the dry and wet deposition of sea salt aerosols further accelerated the weathering of the red bricks, and secondary oxides such as NaCl's formation weakened the structural stability of the red bricks further [26].

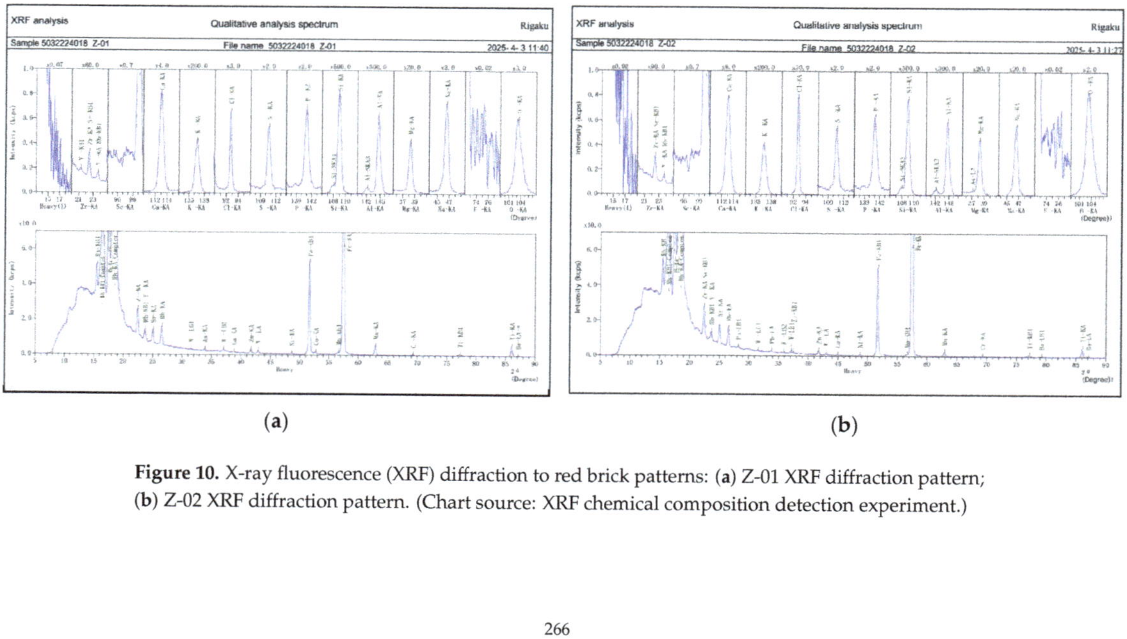

(a) (b)

Figure 10. X-ray fluorescence (XRF) diffraction to red brick patterns: (**a**) Z-01 XRF diffraction pattern; (**b**) Z-02 XRF diffraction pattern. (Chart source: XRF chemical composition detection experiment.)

3.3.2. XRF-Clay Core

Similarly, the clay core was also influenced by external environmental factors, but the changes in the major elemental oxides were relatively smaller (Figure 11, Tables A5 and A6). The SiO_2 content decreased from 60.080% to 58.06%, a reduction of 3.36%; Al_2O_3 decreased from 22.46% to 21.98%, a reduction of 2.14%; and Fe_2O_3 dropped from 5.53% to 4.10%, a decrease of 25.86%. The CaO content increased from 1.14% to 1.39%, indicating an enrichment of 21.90%; the Na_2O content increased from 2.34% to 3.47%, demonstrating an enrichment of 48.44%. Trace elements such as Zn, Ba, and Pb also showed some degree of loss. Although the clay core exhibited a relatively lower weathering, secondary Ca oxides and trace elements resulting from rainwater or industrial pollution negatively impacted its structural stability [27]. The SO_3 content increased from 0.55% to 0.92%, a 65.50% increase, reflecting the infiltration of external pollutants from the pores of red bricks into clay and their impact on its durability.

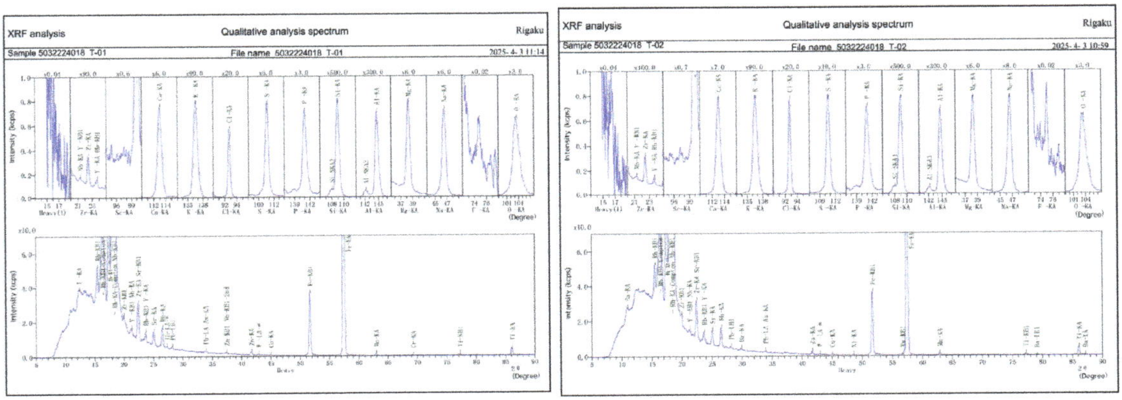

(a) (b)

Figure 11. X-ray fluorescence (XRF) diffraction to clay-core patterns: (**a**) T-01 XRF diffraction pattern; (**b**) T-02 XRF diffraction pattern. (Chart source: XRF chemical composition detection experiment.)

3.3.3. Summary: Changes in Chemical Composition

In summary, both the red bricks and the clay cores exhibited increased weathering, with the red bricks showing a more significant degree of weathering. This was primarily reflected in the loss of SiO_2 and Al_2O_3, which have a "skeleton effect," making it easier for external NaCl and SO_3 to penetrate and exacerbate the hydrolysis and dissolution of silicate minerals [26]. In contrast, due to the protective effect of the outer layer of red bricks, the clay core has undergone a relatively low weathering, resulting in less damage to its "skeleton effect". However, external pollution leads to the infiltration of secondary calcium oxides and trace elements, which also has a negative impact on the durability of clay.

3.4. XRD Mineral Analysis

3.4.1. XRD-Red Brick

The mullite content in the red bricks decreased from 6.5% to 0%. This is due to the breakage of the Al–O bond of mullite and the limited migration ability of Al^{3+}, which almost completely eliminates the "skeleton protection effect" of red bricks and significantly reduces the structural stability of the material, which gradually becomes powdery [27]. The hematite content decreased from 3.1% to 2.9%, a loss of 0.2%, reflecting the dissolution of hematite during the weathering process [28]. The content of the amorphous phase decreased significantly from 51.6% to 45.5%, a loss of 6.1%, which was mainly due to mineral substance dissolution and transformation and physical damage during the weathering process [27,28].

Moreover, the illite content was enriched by 1.3%, increasing from 1.7% to 3.0%; the quartz content was enriched by 5.2%, increasing from 27.6% to 32.8%; and the potassium feldspar content was enriched by 5.5%, increasing from 9.0% to 14.5%. These findings indicate that minerals such as illite and quartz are relatively enriched in the weathering process because of their stable structure [27]. In addition, the halite content was enriched by 0.8%, increasing from 0.5% to 1.3%, indicating that external salt affects red brick through rainwater or air and accelerates damage to the material [26] (Figure 12, Table 3).

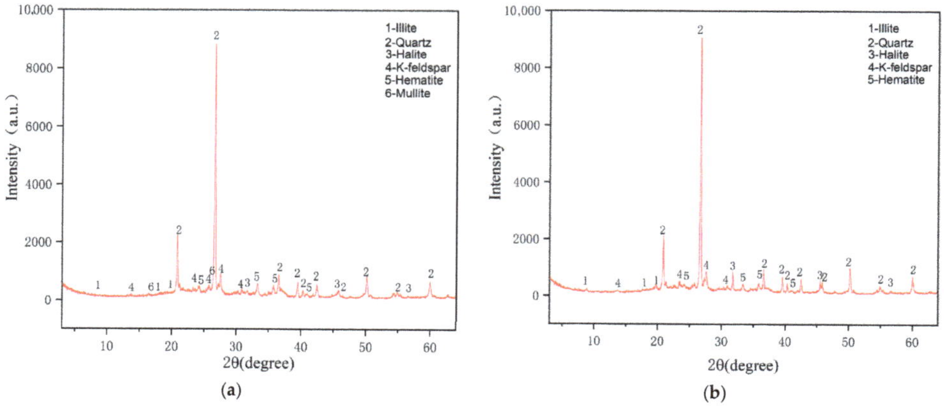

Figure 12. X-ray diffraction (XRD) to red brick; (**a**) Z-01 XRD; (**b**) Z-02 XRD. (Chart source: XRD mineral analysis detection experiment).

Table 3. Statistical analysis of the mineral content of the samples.

No.	Kaolinite	Chlorite	Illite	Amphibole	Gypsum	Quartz	K-Feldspar	Plagioclase	Mullite	Halite	Hematite	A.ph.
Z-01	/	/	1.7	/	/	27.6	9.0	/	6.5	0.5	3.1	51.6
Z-02	/	/	3.0	/	/	32.8	14.5	/	/	1.3	2.9	45.5
T-01	32.8	3.7	11.2	/	/	46.5	2.2	3.6	/	/	/	/
T-02	32.0	3.6	10.9	0.5	0.5	44.9	2.2	5.4	/	/	/	/

Unit: mass%; To ensure a reliable and comparable analysis results, the K value method was used, adding 10% corundum (α-Al_2O_3) as an internal standard to correct diffraction intensity and standardize the results. Although XRD was used for semi-quantitative analysis in this study, JADE software was used to effectively control the relative deviation for different mineral content ranges. Specifically, when the mineral content is >40% (20%–40%, 5%–20%, ≤5%), the relative deviation is <10% (<20%,<30%,<40%) [29]. Chart source: Author 6 used instrument detection, and team members jointly analyzed and verified it.

3.4.2. XRD-Clay Core

The chemically stable quartz content in the clay cores decreased from 46.5% to 44.9%. The illite content decreased from 11.2% to 10.9%, representing a loss of 0.3%; the kaolinite content decreased from 32.8% to 32.0%, indicating a loss of 0.8%; and the chlorite content decreased from 3.7% to 3.6%, demonstrating a loss of 0.1%. Although the content of all of the above minerals decreased, the loss of minerals that affected the compression resistance and viscosity of the soil was lower due to the protection of the external red bricks, and only a small part was converted and lost in the weathering process. Moreover, the augite content ranged from 0 to 0.5, which was the result of the weathering transformation of chlorite [30]. The plagioclase content increased from 3.6% to 5.4%, with an enrichment of 1.8%. This is because plagioclase is relatively stable during the weathering process, and its proportion is relatively enriched due to the decrease in other minerals [31]. In addition, the gypsum content ranged from 0 to 0.5%, which was due to the fusion of Ca dust and clay core to form gypsum during environmental pollution [32] (Figure 13, Table 3).

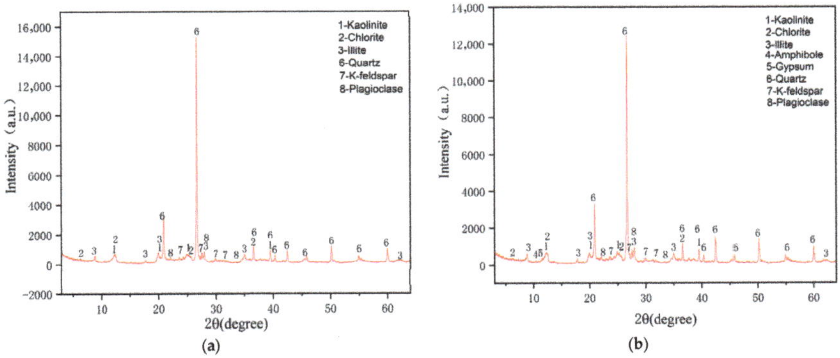

Figure 13. X-ray diffraction (XRD) to clay core: (**a**) T-01 XRD; (**b**) T-02 XRD. (Chart source: XRD mineral analysis detection experiment).

3.4.3. Summary: Changes in Minerals

Under the influence of weathering mechanisms, the mineral composition of the red bricks and clay cores has undergone significant changes. In the red brick samples, the content of mullite, hematite, and amorphous materials decreased, whereas illite and quartz were relatively enriched, and the halite content increased under the influence of external salinity. Meanwhile, the clay core retained a good compression resistance and bond strength under the protection of red bricks. However, there were still some losses due to the influence of illite, kaolinite, etc., forming new minerals such as ordinary pyroxene and gypsum, which reduces the viscosity of the clay ores. As a result, the viscosity between the red brick and Clay decreases and it falls off.

4. Discussion

Based on the above findings, under the influence of natural weathering and external environmental factors, three mechanisms can be summarized that affect the durability of BCWs: (1) The protection and reinforcement mechanism of "brick-wrapped clay" joint and tenons' coordination. (2) The infiltration and destruction mechanism of external pollutants. (3) The material stability mechanism of silicate minerals. The details are as follows:

4.1. The Protection and Reinforcement Mechanism

The mortise and tenon overlap between the red bricks and the clay cores significantly enhances the stability of the structure through the complementarity of the materials and the reasonable distribution of the load. FEM and material analysis jointly verified its advantages: brick–clay masonry (Construction II) exhibited superior integrity and load-sharing capacity. At the same time, the soil fill provides "internal support," and its plastic deformation ability can absorb local stress and avoid stress concentration in the red brick. In addition, the following points can also illustrate the improvement effect of red bricks on the durability of clay.

The SEM results revealed that the changes in the mean grayscale values of both materials indicate that the brick has undergone significant damage, while the change in clay is not significant. Clay is not easily affected by external environmental factors and did not weather significantly under the protection of bricks.

XRF also revealed that the loss rate of elements in soil is generally lower than that in red bricks (e.g., the loss rate of Al is 6.98%, but the clay core performs better, because it is protected by red bricks), which confirms that structural design has an important role in delaying materials' weathering. Meanwhile, it is evident from Figure 14 that Z-02 is closer to "K_2O-Al_2O_3-CaO-Na_2O" than T-02, indicating that bricks are more prone to weathering than clays.

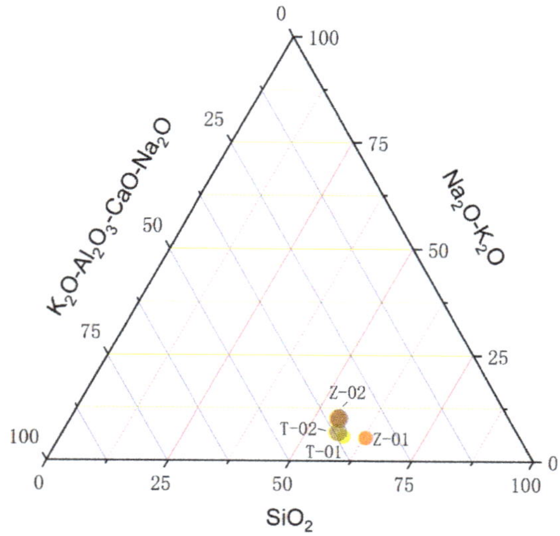

Figure 14. SiO_2 and Na_2O-Ka_2O on a K_2O-Al_2O_3-CaO-Na_2O ternary diagram. (Chart source: Drawing based on XRD mineral analysis results.)

Moreover, XRD analysis confirmed that mineral loss in the clay core was relatively slight (kaolinite only decreased by 0.8%), which was mainly ascribed to the effective blocking of external salt by the red bricks (in XRF, the Na enrichment in the clay veneer was 149%, which was far lower than the 387% of red brick) and acid precipitation.

Combining the mechanical FEM analysis, the soil filled in the cavity exhibited a uniform stress distribution in Construction II, whereas the loss of amorphous red brick material (XRD showed a decrease of 6.1%) led to embrittlement. These two structures form a sharp contrast and highlight the significant effect of the "combination of rigidity and flexibility" in improving durability.

Therefore, through red bricks for weathering resistance, clay core for load bearing, and the mechanical synergy of the two types materials, the design of "brick-encased soil" can extend the life of the clay core far beyond its independent use limit, reflecting the outstanding wisdom of traditional construction technology.

4.2. The Infiltration and Destruction Mechanism

External environmental factors such as sea wind erosion and rising groundwater lead to mineral decomposition, increasing halite; improper industrial waste treatment raises gypsum content; and heavy metal pollutants in industrial waste cause dual damage to BCWs' materials, including physical expansion and chemical degradation. The specific analysis is as follows:

SEM analysis revealed the consequences of pollution: the crystal expansion of red brick (Z-02) was caused by the infiltration of rock salt, resulting in an increase in the mean gray value. Meanwhile, compared with T-01, the average grayscale value of clay (T-02) remained relatively unchanged. However, crack propagation and pore enlargement were clearly observed in the microstructure, fully confirming the physical damage of pollutants.

XRF element migration was used to quantify the pollution paths. In areas with a frequent dry–wet alternation and severe industrial pollution, chlorine was enriched by 1096% (Na↑387%), and the capillary adsorption of salt exacerbated the propagation of microcracks in red bricks; sulfur was enriched by 68.38% (Ca↑23.75%), which provided the material base for the formation of gypsum in the clay core.

XRD analysis of the mineral transformation confirmed the chemical metamorphic mechanism. The halite content in the red bricks increased by 0.8%, which accelerated the destruction of the dissolution–crystallization cycle. The newly formed gypsum (0.5%) in the clay core squeezed the silicate mineral structure, resulting in the transformation of chlorite into gypsum. Mineral modification directly reduces the bond strength [27].

FEM demonstrated that the performance of masonry with peeling bricks (Construction III) is more degraded than Construction II, and the stress is reduced by 1.03×10^7 N/m^2, reflecting the gradual loss of material viscosity and structural bearing capacity after the accumulation of pollution.

In summary, the intervention of pollutants formed a chain reaction of "salt infiltration → element migration → mineral deterioration → structural loosening," eventually leading to a decrease in compressive strength and durability.

4.3. The Material Stability Mechanism

Silicate minerals (such as mullite, illite, quartz, and chlorite) play critical supporting roles in the compressive strength of red brick and clay core.

From the perspective of the SEM microstructure, compared with Z-02, due to the uniform distribution of silicate minerals in the raw materials, Z-01 has the characteristic of small particles and pores (low mean grayscale value); however, Z-02 has a loose structure with expanded pores (high mean grayscale value) due to mineral transformation, which fully confirms the important influence of mineral stability on microscopic integrity.

XRF elemental analysis showed that Al_2O_3 in Z-01 decreased from 20.67% to 19.21% in Z-02, and SiO_2 decreased from 60.91% to 54.21%. The increase of $Na_2O + K_2O$ from 5.37% to 9.95% indicates that with the increase in weathering, the loss of Al_2O_3 and SiO_2 and mineral degradation weaken the strength and stability of the material. The weathering trend from 01 to 02 indicates (Figure 14) that the mass fraction of K_2O-Al_2O_3-$CaO + Na_2O$ decreased for brick and clay materials, while $Na_2O + K_2O$ and SiO_2 increased, once again indicating accelerated salt acid synergistic destruction, the decomposition of silicate minerals, and the release of alkaline earth minerals, which weaken the durability of the material (Table 4). Therefore, the degradation of silicate minerals during weathering is the main cause of reduced strength in brick and clay.

Table 4. The ternary analysis diagram of the main oxide conversion relationship.

No.	Oxidate	Z-01	Z-02	T-01	T-02
1	SiO_2	60.91	54.21	60.08	58.06
2	$Na_2O + K_2O$	5.37	9.95	5.99	7.14
3	K_2O-Al_2O_3-$CaO + Na_2O$	30.42	3430	36.69	37.70
Note	Al_2O_3	20.67	19.21	22.46	21.98

The changes in the main oxide content in the brick and clay samples were determined according to Tables A4 and A6.

XRD analysis further confirms the importance of silicate minerals in BCW materials. Firstly, the complete disappearance of mullite resulted in the complete pulverization of red bricks. At the same time, quartz (enriched 5.2%) and illite (1.3%) in red bricks were relatively enriched during weathering, indicating that silicate minerals also play a role in slowing down weathering efficiency. At the same time, there was slight loss of chlorite and kaolinite in the soil, but due to the protection of the outer layer of red bricks, the degree of mineral change only slightly decreased [27].

Therefore, the stability of materials formed by silicate minerals through micro compactness, crystal stability, and chemical inertness is the core element for maintaining the durability of BCW materials.

5. Conclusions and Expectation

5.1. Conclusions

Focusing on building construction and material analysis, the internal factors of BCWs' durability were analyzed, and the effects of minerals, external pollutants, and "brick-wrapped clay" structures in raw materials on overall building durability were elucidated. The results show that

(1) The synergy of "brick-encased soil" mortise and tenon construction greatly enhances structural stability. The protective function of the red bricks and the internal support function of the soil wall complement each other, optimize the load dispersion effect, and delay the process of clay cores' weathering and deterioration, thereby extending their service life.
(2) External pollutants (such as moisture, halite, gypsum, and acidic substances) destroy the material structure through physical expansion and chemical deterioration, resulting in a reduction in compressive strength and durability.
(3) Silicate minerals (such as illite, quartz, mullite, et al.) improve the compressive strength and durability of the red brick and clay core through their compact microstructure, crystal stability, and chemical inertness.

Therefore, the key to maintaining the durability of BCWs lies in the reinforcement and protection provided by the "brick-wrapped soil" structure. Although a BCW is often subjected to infiltration and erosion by external pollutants, the stability of silicate minerals slows down the weathering rate of the BCW and enhances the durability of the structure.

5.2. Expectation

BCWs are also distributed in Zhejiang, Anhui, Guangdong, and other places in China, as well as in countries such as the United Kingdom, Italy, and Canada. The chemical composition and minerals of BCW brick soil materials in these regions vary, but their mortise and tenon structures are highly similar [7,11,25]. However, existing research often overlooks material properties, and there are significant differences in climate and weathering mechanisms between different regions. In modern engineering, similar structures are only used as non-load bearing infill walls, and their performance urgently needs systematic evaluation.

This study only focuses on the aspects of structure and material properties, and four subsequent works can be carried out:

(1) Multiple factors can be considered to affect the durability of BCWs and expand research, such as "tie brick reinforcement", "wood column wall interlocking", etc.
(2) Cross-regional comparisons should be made to elucidate the adaptability of BCWs to different climates.
(3) By combining wind load testing, fatigue resistance testing, or interface detachment nonlinear contact simulation, the durability mechanism under multiple degradation conditions will be revealed.
(4) Based on the results of this study and the emphasis on the principle of "minimal intervention" in the ICOMOS Authenticity Guidelines, further research will be conducted on protection and restoration strategies for BCWs [33].

In summary, the durability of a BCW refers to its ability not only to withstand compressive and seismic forces but also to maintain its own load and extend its service life. From ancient times to the present, the BCW has been mostly used as a maintenance structure and less as a load-bearing structure. In most cases, it only needs to maintain its own stability and durability. Therefore, the GB50003 standard in China prohibits it without conducting comprehensive and in-depth research on it, and the comparison with the bearing capacity of solid load-bearing walls may not be appropriate [3,34]. Therefore, the established "FEM-SEM-XRF-XRD" quadruple analysis method, which is a comprehensive evaluation method composed of mechanical properties, microstructure, chemical composition, and mineral changes, provides theoretical and methodological support for the sustainable development of BCWs and brick and soil buildings.

Author Contributions: Conceptualization, Y.D., R.G. and L.C.; methodology, Y.D., R.G. and J.W.; software, Y.D. and Y.F.; validation, Y.D., C.Z. and Y.F.; formal analysis, Y.D. and C.Z.; investigation, Y.D., C.Z., Y.Z. and Y.F.; resources, Y.D., C.Z., R.G. and L.C.; data curation, Y.D. and Y.F.; writing—original draft preparation, Y.D., Y.Z. and C.Z.; writing—review and editing, R.G., L.C. and J.W.; visualization, Y.D. and Y.F.; supervision, R.G., L.C. and J.W.; project administration, Y.D. and R.G.; funding acquisition, Y.D. All authors have read and agreed to the published version of the manuscript.

Funding: Quanzhou Science and Technology Plan Project: Research on Gao Chi of Traditional Minnan Architecture Based on Digital Technology (2024NS007).

Institutional Review Board Statement: Not applicable.

Informed Consent Statement: Not applicable.

Data Availability Statement: The datasets used and analyzed during the current study are available from Y.D. (3240006977@student.must.edu.mo) upon reasonable request.

Conflicts of Interest: Author Yili Fu were employed by the company Quanzhou Artisan Architectural Design Co., Ltd. and field investigations, data collection, and analysis were conducted for this article. The authors declare that the research was conducted in the absence of any commercial or financial relationships that could be construed as a potential conflict of interest.

Abbreviations

The following abbreviations are used in this manuscript:

BCW	Brick-faced clay-core wall
ESL.	Elemental spectral lines
Ele.	Element
w/o N.	w/o Normal
D. limit	Detection limit
SW FEM	SOLIDWORKS FEM
A.ph.	Amorphous phase

Appendix A

Table A1. Sample production process table.

No.	Z-01	Z-02	T-01	T-02
Entity				

Table A1. *Cont.*

No.	Z-01	Z-02	T-01	T-02
Preliminary grinding	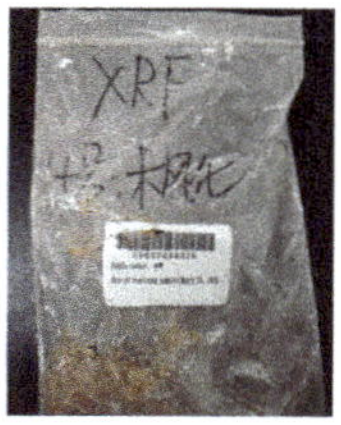	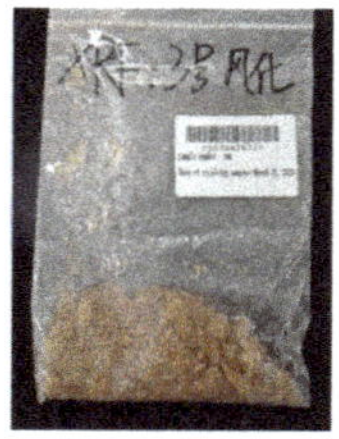	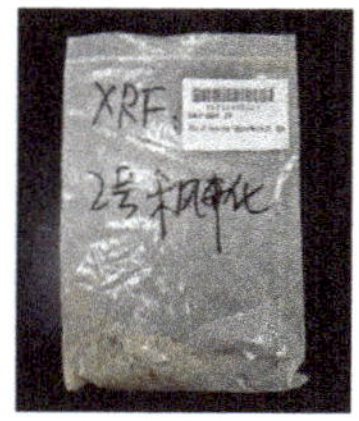	
Fine sieving	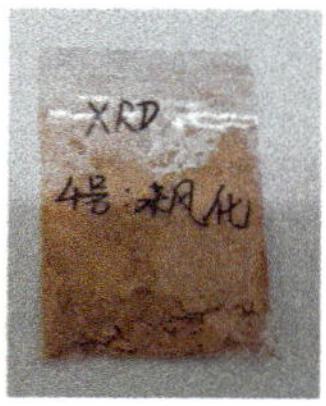	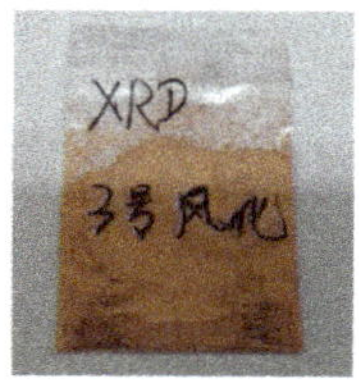		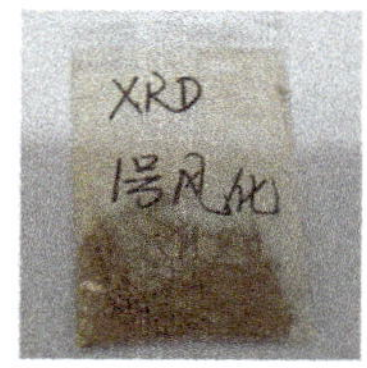
Tablet compression				

The sample code is not related to the marker number in the image. Source of chart: Sampling, production, and photography by Authors 1 and 6.

Table A2. A statistical table of the microstructure analysis.

Microstructure Analysis of SEM Images				
Mag	5.00 K X	1.00 K X	500 X	200 X
Z-01 (WD = 9.8 mm)	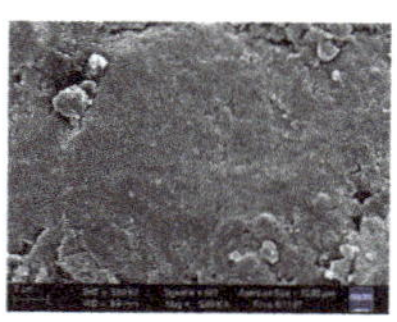			
Mean grayscale fluctuation	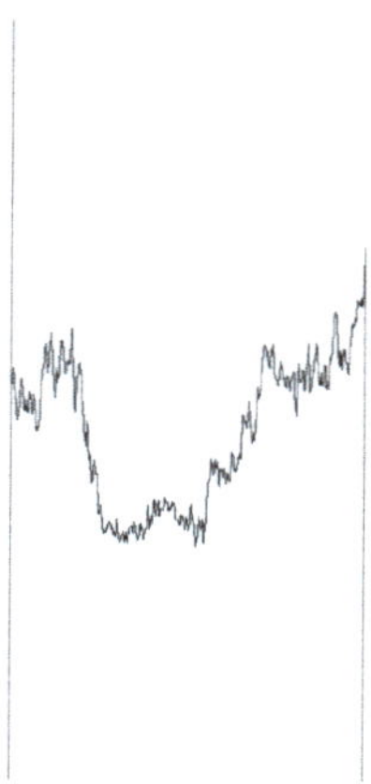	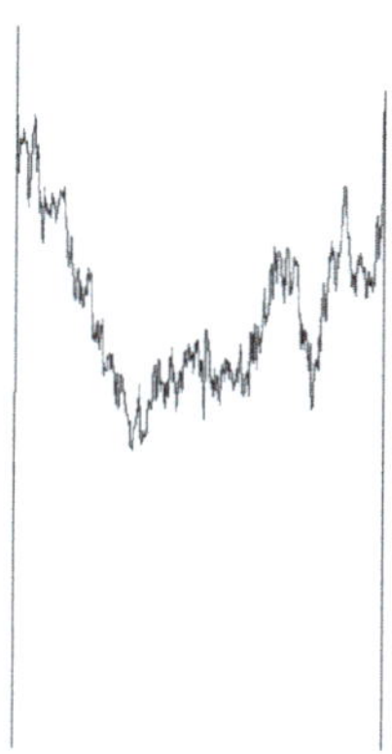	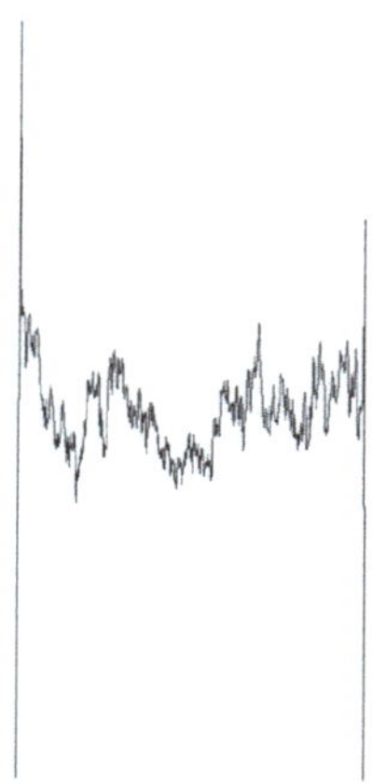	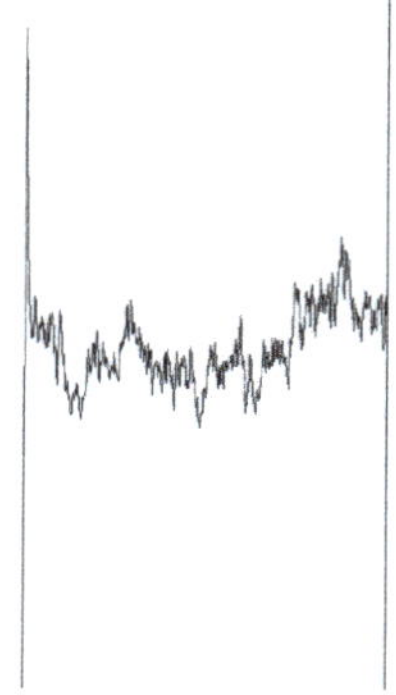

Table A2. *Cont.*

Microstructure Analysis of SEM Images				
Mean grayscale	109,124	131,942	114,391	96,866
Z-01 (WD = 9.2 mm)				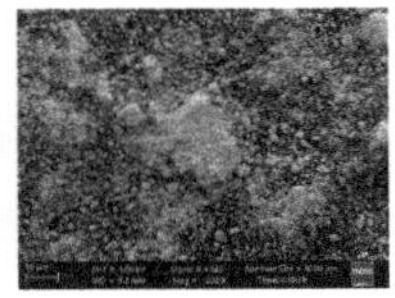
Mean grayscale fluctuation	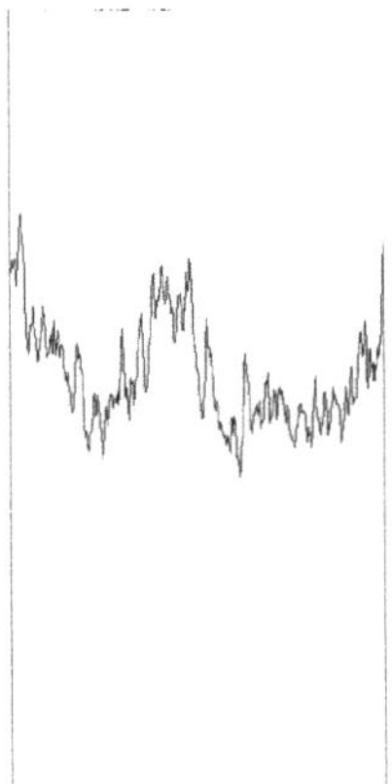	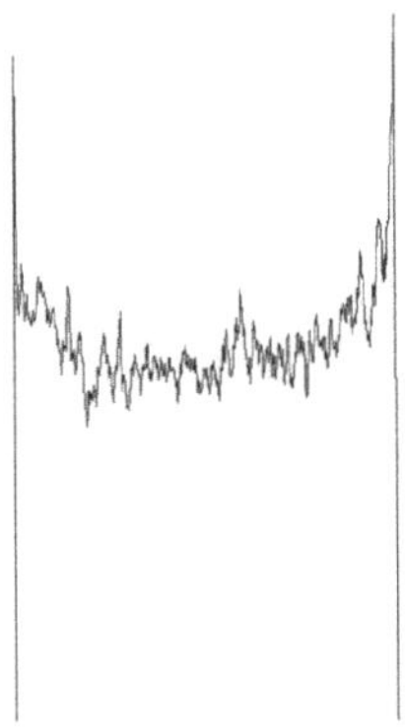	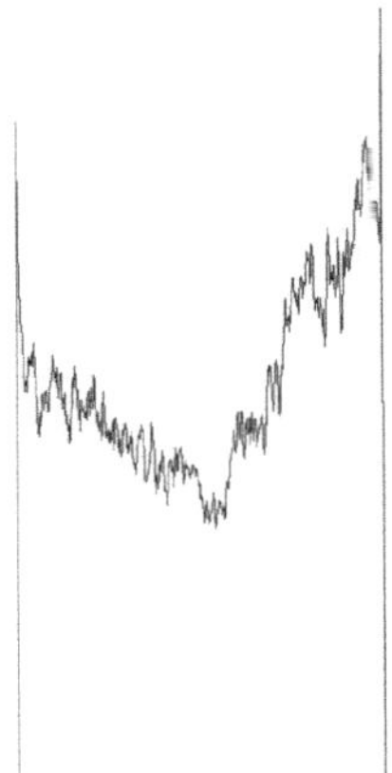	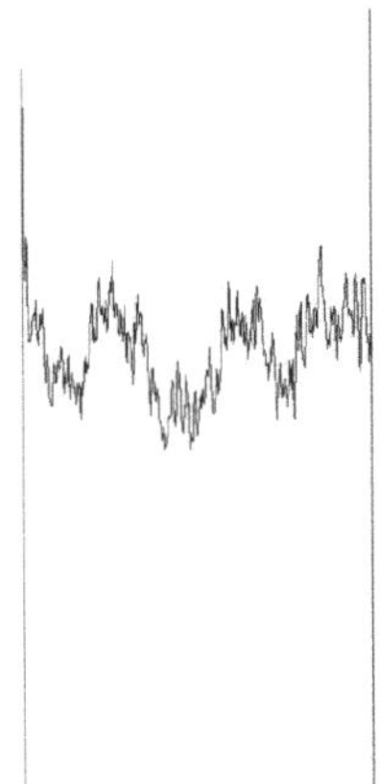
Mean grayscale	129,237	115,287	123,102	127,890
Z-02 (WD = 7.6 mm)				
Mean grayscale fluctuation	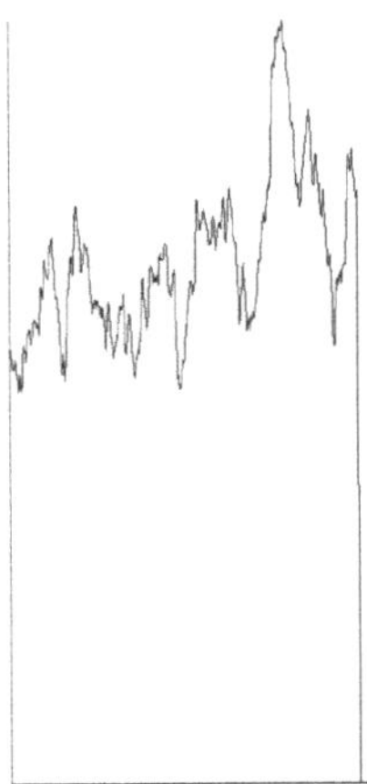	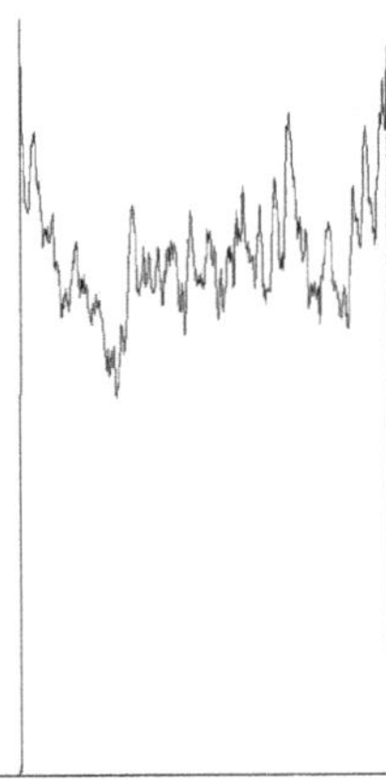	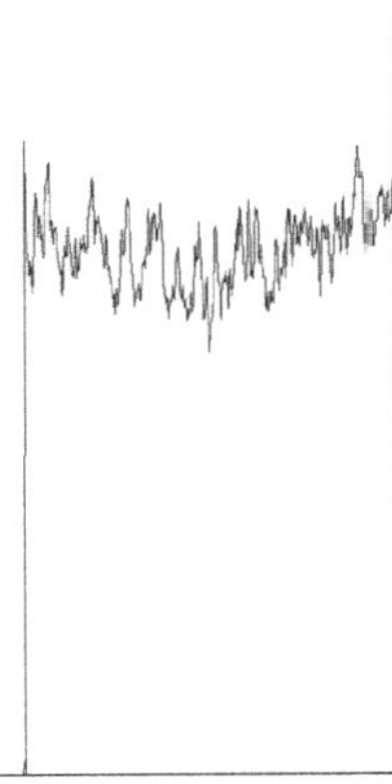	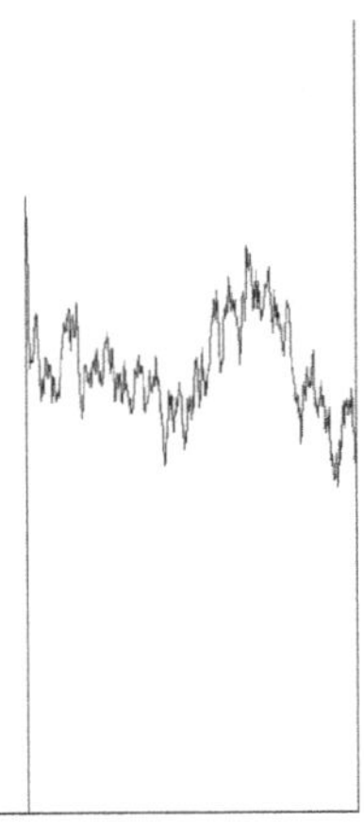
Mean grayscale	159,747	158,001	154,923	119,649

Table A2. *Cont.*

Microstructure Analysis of SEM Images				
Z-02 (WD = 6.9 mm)		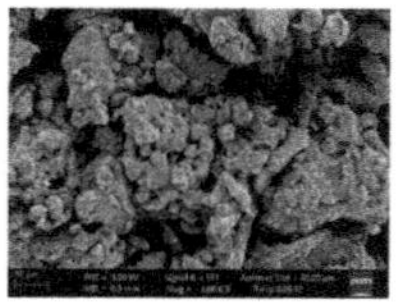		
Mean grayscale fluctuation	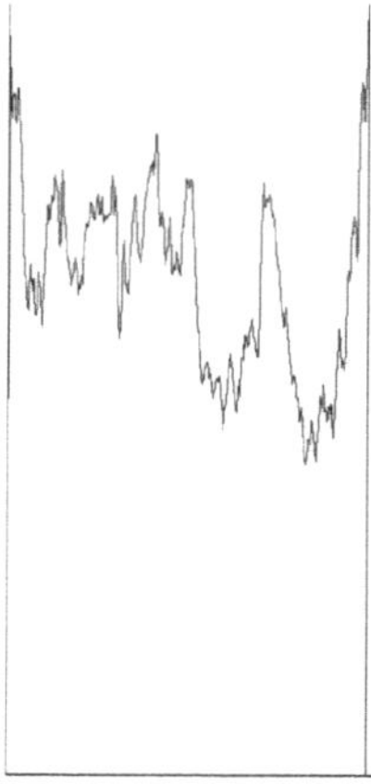	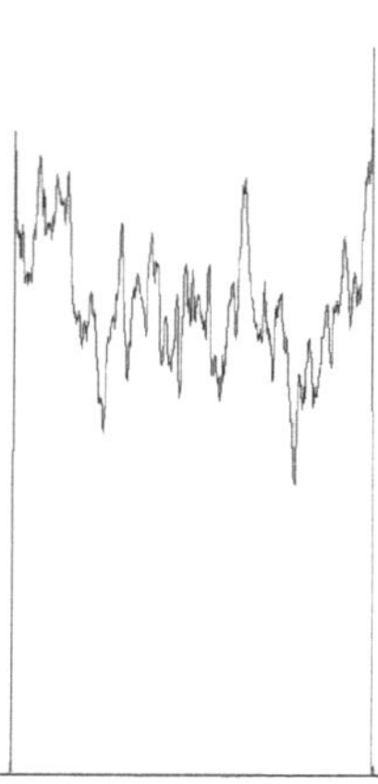	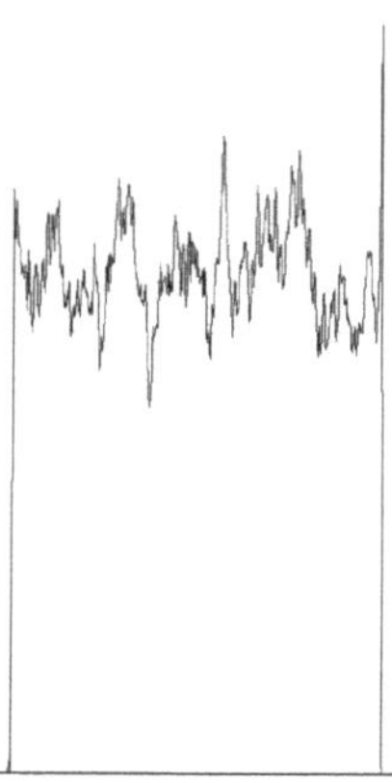	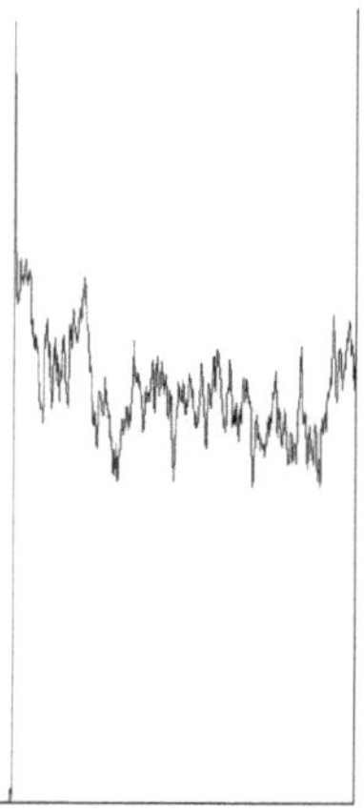
Mean grayscale	152,001	145,134	145,394	112,821
T-01 (WD = 6.8 mm)				
Mean grayscale fluctuation	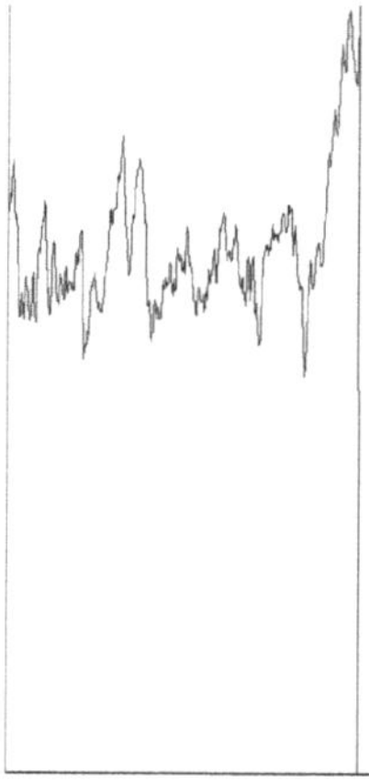	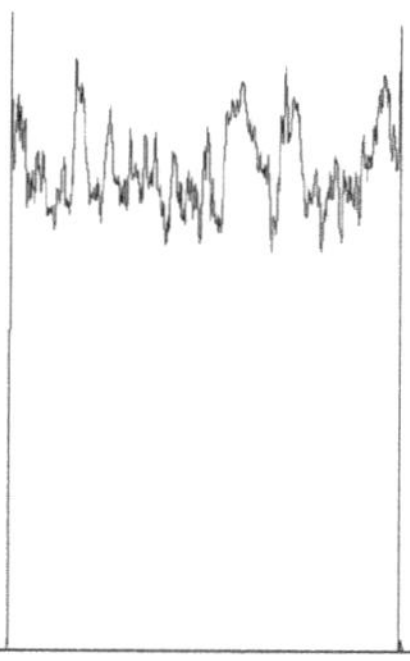	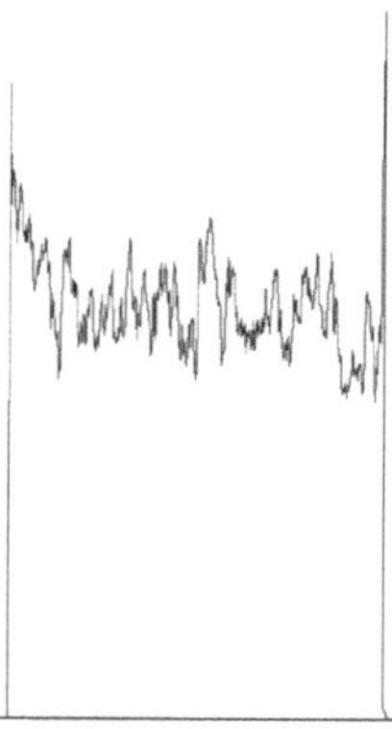	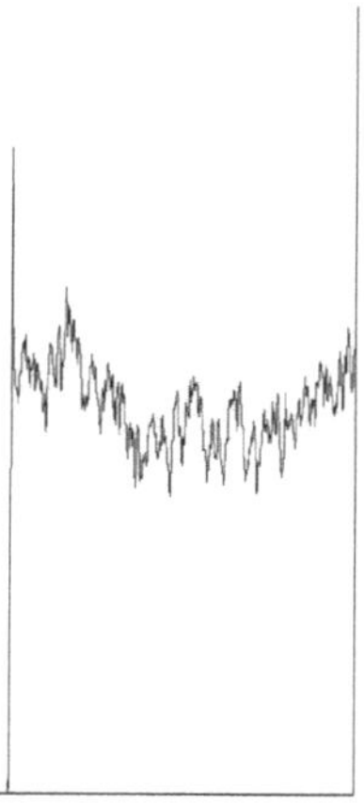
Mean grayscale	159,542	137,266	122,237	109,963
T-01 (WD = 6.7 mm)				

Table A2. *Cont.*

Microstructure Analysis of SEM Images				
Mean grayscale fluctuation	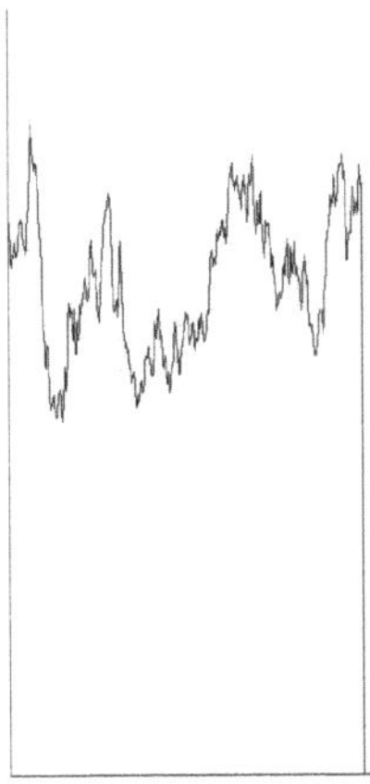	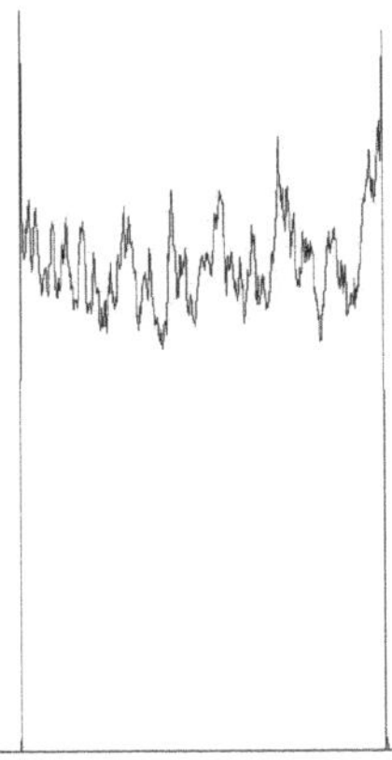	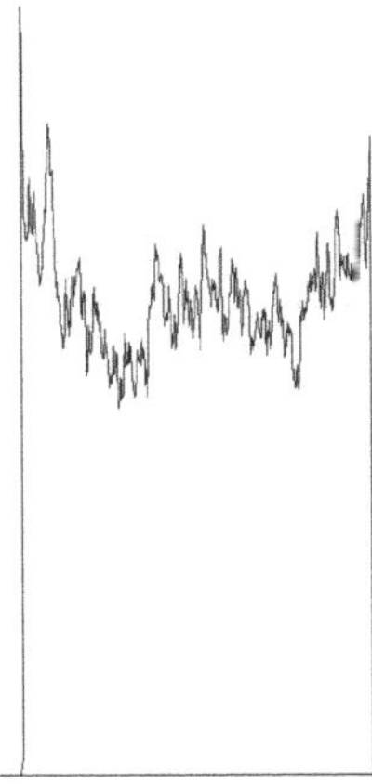	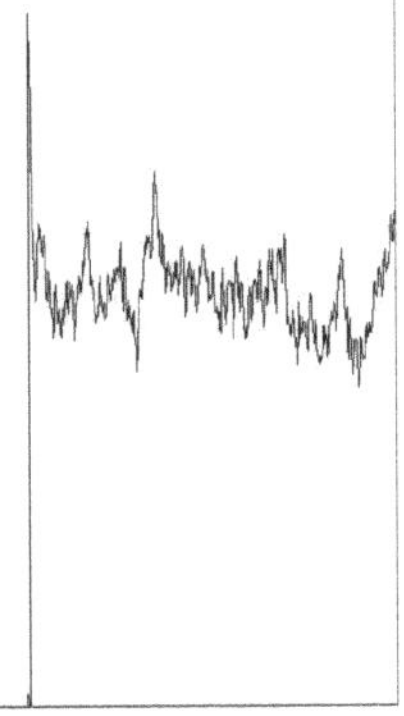
Mean grayscale	148,924	149,112	140,929	111,434
T-02 (WD = 7.4 mm)	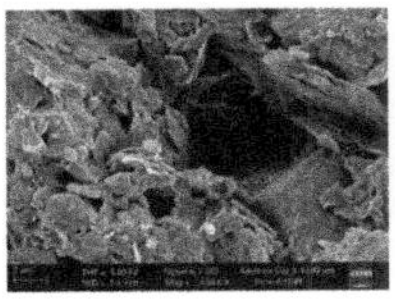			
Mean grayscale fluctuation	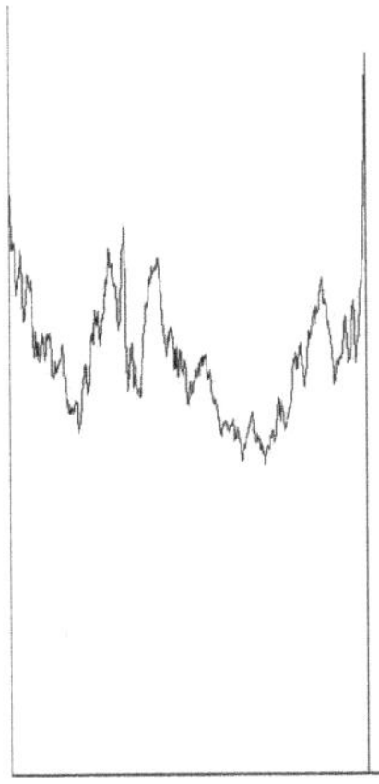	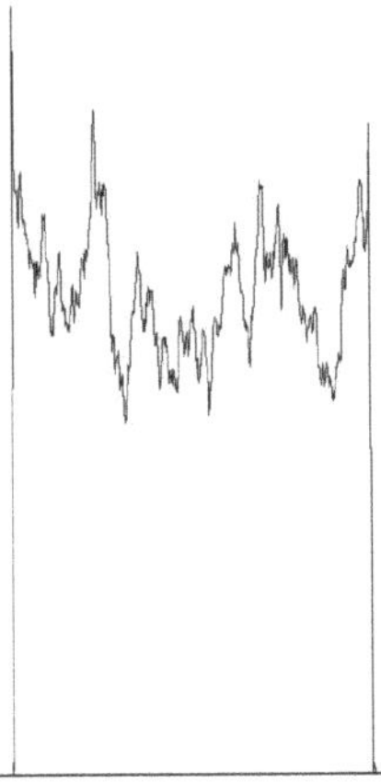	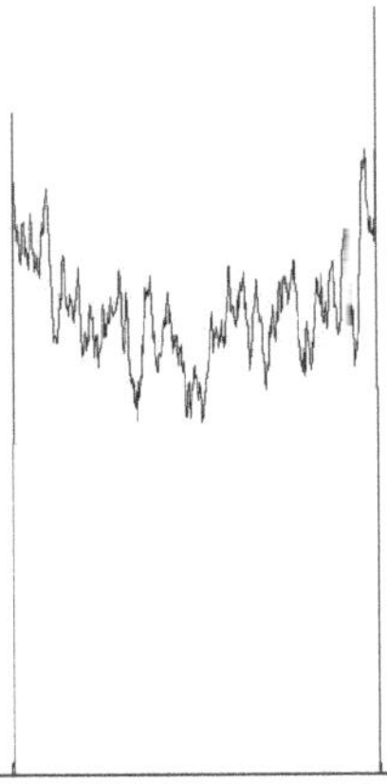	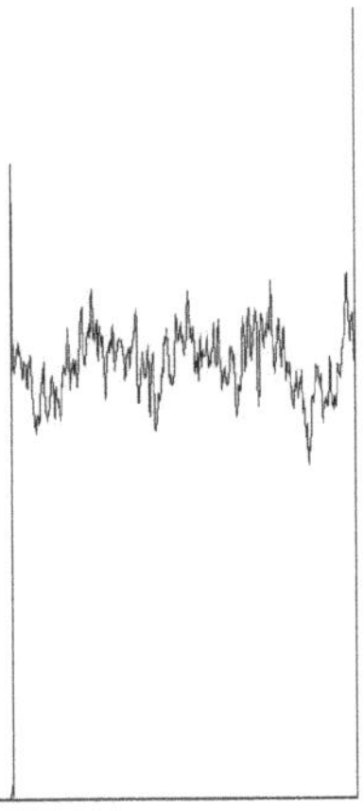
Mean grayscale	127,539	148,312	136,525	121,846
T-02 (WD = 7.1 mm)				

Table A2. *Cont.*

Microstructure Analysis of SEM Images				
Mean grayscale fluctuation	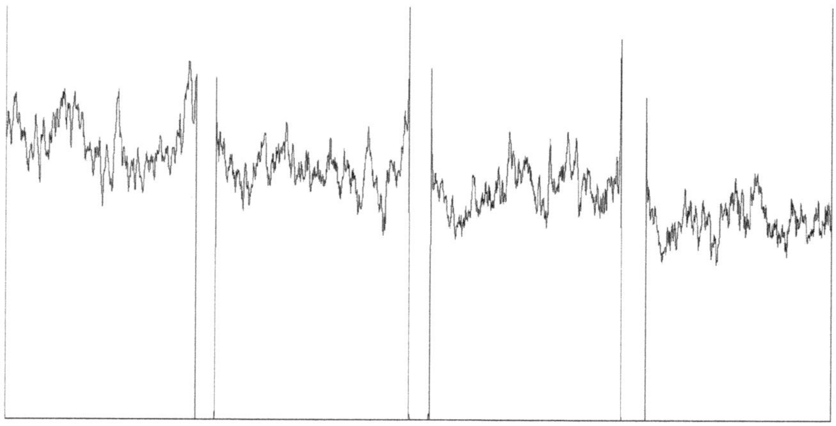			
Mean grayscale	155,796	142,807	130,344	108,938

The unit of mean grayscale is dimensionless. Chart source: Author 6 used an electron microscope to capture samples; Author 1 used ImageJ software for grayscale recognition.

Table A3. Z-01 and Z-02 XRF element detection result.

Analytical: Element Analysis Result by Rigaku												
Sample designation: 25032224018 Sample Z-01 Red Brick (low degree of weathering)							Sample designation: 25032224018 Sample Z-02 Red Brick (high degree of weathering)					
No.	Ele.	Result	D. limit	ESL.	Intensity	w/o N.	Ele.	Result	D. limit	ESL.	Intensity	w/o N.
					The main element (mass%)							
1	O	47.09	0.633	O-KA	1.773	35.318	O	43.95	0.569	O-KA	1.614	34.544
2	Na	0.90	0.010	Na-KA	2.131	0.674	Na	4.38	0.013	Na-KA	11.071	3.439
3	Mg	1.07	0.007	Mg-KA	8.321	0.805	Mg	1.11	0.007	Mg-KA	8.648	0.876
4	Al	10.97	0.006	Al-KA	190.77	8.232	Al	10.21	0.006	Al-KA	178.697	8.024
5	Si	28.59	0.007	Si-KA	417.964	21.446	Si	25.47	0.007	Si-KA	385.808	20.018
6	P	0.06	0.002	P-KA	1.261	0.042	P	0.05	0.002	P-KA	1.127	0.036
7	S	0.05	0.001	S-KA	1.005	0.035	S	0.04	0.001	S-KA	0.993	0.034
8	Cl	0.40	0.007	Cl-KA	1.894	0.241	Cl	3.84	0.015	Cl-KA	24.037	3.022
9	K	3.48	0.005	K-KA	85.59	2.607	K	3.40	0.005	K-KA	83.387	2.672
10	Ca	0.58	0.005	Ca-KA	3.249	0.431	Ca	1.11	0.006	Ca-KA	6.258	0.872
11	Ti	0.65	0.003	Ti-KA	5.404	0.49	Ti	0.62	0.005	Ti-KA	4.973	0.481
12	Cr	0.01	0.004	Cr-KA	0.31	0.011	Cr	0.01	0.003	Cr-KA	0.286	0.011
13	Mn	0.17	0.004	Mn-KA	5.189	0.126	Mn	0.08	0.003	Mn-KA	2.508	0.065
14	Fe	5.92	0.004	Fe-KA	278.659	4.434	Fe	5.51	0.004	Fe-KA	256.718	4.333
					The trace elements (mg/kg)							
15 *	Co	80	30.2	Co-KA	5427	60	Pb	86	36.3	Pb-LBI	9995	68
16 *	Ni	87	18.7	Ni-KA	6372	65	Ni	67	18.7	Ni-KA	4987	53
17 *	Zn	177	15.8	Zn-KA	22,166	133	Cu	41	16.5	Cu-KA	3905	33
18 *	Ga	39	15.4	Ga-KA	5675	29	Zn	178	15.6	Zn-KA	22,683	140
19 *	As	44	12.9	As-KA	9435	33	Rb	259	12	Rb-KA	10,149	203
20 *	Rb	281	11.6	Rb-KA	10,812	211	Sr	248	12	Sr-KA	11,272	195
21 *	Sr	169	11	Sr-KA	75,414	127	Y	31	12	Y-KA	5168	24
22 *	Y	36	12	Y-KA	57,706	27	Zr	241	11	Zr-KA	17,526	189
23 *	Zr	229	10.5	Zr-KA	158,257	172	Ba	855	197.6	Ba-KA	2071	672
24 *	W	439	50.9	W-LA	15,936	329	W	281	52.4	W-LA	10,383	221

* Chart source: Author 6 used instrument detection, and team members jointly analyzed and verified this.

Table A4. Z-01 and Z-02 XRF oxidate detection result.

Analytical: Oxidate Analysis Result by Rigaku												
Sample designation: 25032224018 Sample Z-01 Red Brick (low degree of weathering)							Sample designation: 25032224018 Sample Z-02 Red Brick (high degree of weathering)					
No.	Ele.	Result	D. limit	ESL.	Intensity	w/o N.	Ele.	Result	D. limit	ESL.	Intensity	w/o N.
					The main element (mass%)							
1	SiO_2	60.91	0.015	Si-KA	417.964	45.822	SiO_2	54.21	0.015	Si-KA	385.808	42.771
2	TiO_2	1.08	0.005	Ti-KA	5.404	0.815	TiO_2	1.01	0.008	Ti-KA	4.973	0.799
3	Al_2O_3	20.67	0.012	Al-KA	190.77	15.553	Al_2O_3	19.21	0.012	Mg-KA	178.697	15.158
4	MgO	1.78	0.012	Mg-KA	8.321	1.335	MgO	1.84	0.012	Mg-KA	8.648	1.452
5	CaO	0.80	0.007	Ca-KA	3.249	0.601	CaO	1.54	0.009	Ca-KA	6.258	1.215
6	MnO	0.22	0.004	Mn-KA	5.189	0.162	MnO	0.11	0.004	Mn-KA	2.508	0.083
7	Fe_2O_3	8.40	0.005	Fe-KA	278.659	6.318	Fe_2O_3	7.82	0.006	Fe-KA	256.718	6.166
8	Na_2O	1.21	0.014	Na-KA	2.131	0.909	Na_2O	5.88	0.018	Na-KA	11.071	4.64
9	K_2O	4.16	0.005	K-KA	85.59	3.132	K_2O	4.07	0.006	K-KA	83.387	3.208
10	P_2O_5	0.13	0.004	P-KA	1.261	0.095	P_2O_5	0.10	0.003	P-KA	1.127	0.082
11	Cl	0.32	0.007	Cl-KA	1.894	0.241	Cl	3.82	0.015	Cl-KA	24.037	3.013
					The trace elements (mg/kg)							
12 *	SO_3	17	33.2	S-KA	10,049	885	SO_3	1067	32.7	P-KA	9933	842
13 *	Cr_2O_3	209	58.9	Cr-KA	3099	157	Cr_2O_3	196	49.4	Cr-KA	2859	155
14 *	Co_2O_3	112	42.4	Vo-KA	5427	84	NiO	85	23.7	Ni-KA	4987	67
15 *	NiO	110	23.7	Ni-KA	6372	83	CuO	51	20.6	Cu-KA	3905	41
16 *	ZnO	219	19.6	Zn-KA	22,166	165	ZnO	220	19.3	Zn-KA	22,683	173
17 *	Ga_2O_3	52	20.7	Ga-KA	5675	39	Rb_2O	280	13	Rb-KA	10,149.5	221
18 *	As_2O_3	57	17	As-KA	9435	43	SrO	290	13.6	Sr-KA	11,272.9	229
19 *	Rb_2O	306	12.6	Rb-KA	10,812.8	230	Y_2O_3	39	15.5	Y-KA	51,682	31
20 *	SrO	199	12.9	Sr-KA	75,414	150	ZrO_2	322	14.9	Zr-KA	175,264	254
21 *	Y_2O_3	45	15.2	Y-KA	57,706	34	BaO	947	219.7	Br-LA	2071	747
22 *	ZrO_2	307	14.2	Zr-KA	158,257	231	WO_3	351	65.8	W-KA	10,383	277
23 *	WO_3	549	64	W-LA	15,936	413	Pb	92	38.9	Pb-LBI	9995	73

* Chart source: Author 6 used instrument detection, and team members jointly analyzed and verified this.

Table A5. T-01 and T-02 XRF element detection result.

Analytical: Element Analysis Result by Rigaku												
Sample designation: 25032224018 Sample T-01 Clay Core (low degree of weathering)							Sample designation: 25032224018 Sample T-02 Clay Core (high degree of weathering)					
No.	Ele.	Result	D. limit	ESL.	Intensity	w/o N.	Ele.	Result	D. limit	ESL.	Intensity	w/o N.
					The main element (mass%)							
1	O	49.10	0.599	O-KA	1.896	37.551	O	48.10	0.647	O-KA	1.834	37.093
2	Na	1.70	0.011	Na-KA	4.225	1.297	Na	2.53	0.012	Na-KA	6.358	1.95
3	Mg	0.56	0.007	Mg-KA	4.498	0.428	Mg	0.54	0.007	Mg-KA	4.349	0.42
4	Al	11.54	0.007	Al-KA	210.583	8.825	Al	11.35	0.006	Al-KA	206.087	8.751
5	Si	27.00	0.007	Si-KA	408.626	20.651	Si	26.27	0.007	Si-KA	398.829	20.261
6	P	0.08	0.002	P-KA	1.986	0.064	P	0.82	0.002	P-KA	1.996	0.064
7	S	0.21	0.002	S-KA	4.667	0.159	S	0.35	0.002	S-KA	7.968	0.271
8	Cl	1.82	0.01	Cl-KA	11.132	1.391	Cl	2.44	0.011	Cl-KA	15.071	1.884
9	K	2.87	0.004	K-KA	71.501	2.196	K	2.91	0.004	K-KA	72.123	2.242
10	Ca	0.77	0.005	Ca-KA	4.446	0.586	Ca	0.95	0.005	Ca-KA	5.478	0.732
11	Ti	0.54	0.007	Ti-KA	4.573	0.414	Cr	0.04	0.006	Cr-KA	1.597	0.425
12	Cr	0.01	0.004	Cr-KA	0.308	0.011	Mn	3.73	0.004	Mn-KA	1.171	0.028
13	Mn	0.03	0.003	Mn-KA	1.096	0.026	Fe	0.01	0.003	Fe-KA	183.21	2.875
14	Fe	3.63	0.003	Fe-KA	180.518	2.774	Ni	0.01	0.002	Ni-KA	0.496	0.005
					The trace elements (mg/kg)							
15 *	Cu	36	15.5	Cu-KA	3854	27	Cu	36	16.1	Cu-KA	3845	28
16 *	Zn	109	14.1	Zn-KA	15,921	84	Zn	131	14.3	Zn-KA	18,772	101
17 *	As	0	18	As-KA	10,477	0	As	0	18.3	As-KA	10,740	0
18 *	Rb	243	10.3	Rb-KA	11,040.4	186	Br	27	11.2	Br-KA	9232	21
19 *	Sr	147	9.9	Sr-KA	77,575	112	Rb	231	10.7	Rb-KA	104,127	178
20 *	Y	26	10.5	Y-KA	55,914	20	Sr	149	10.1	Sr-KA	77,018	115
21 *	Zr	265	9	Zr-KA	211,458	203	Y	42	10.4	Y-KA	61,741	33
22 *	Nb	44	8.7	Nb-KA	35,156	33	Zr	282	9.1	Zr-KA	219,914	217
23 *	I	358	70.4	I-KA	56,843	274	Nb	27	9	Nb-KA	24,010	21
24 *	Pb	99	31.4	Pb-LBI	13,282	76	Ba	646	234.3	Ba-KA	1605	498
25 *	--	--	--	--	--	--	Pb	106	31.6	Pb-LBI	13,952	82

* Chart source: Author 6 used instrument detection, and team members jointly analyzed and verified this.

Table A6. T-01 and T-02 XRF oxidate detection result.

Analytical: Oxidate Analysis Result by Rigaku												
Sample designation: 25032224018 Sample T-01 Clay Core (low degree of weathering)							Sample designation: 25032224018 Sample T-02 Clay Core (high degree of weathering)					
No.	Ele.	Result	D. limit	ESL.	Intensity	w/o N.	Ele.	Result	D. limit	ESL.	Intensity	w/o N.
The main element (mass%)												
1	SiO_2	60.08	0.015	Si-KA	408.626	44.56	SiO_2	58.06	0.015	Al-KA	398.829	43.661
2	TiO_2	0.96	0.012	Ti-KA	4.573	0.71	TiO_2	0.97	0.011	Ti-KA	4.597	0.727
3	Al_2O_3	22.46	0.012	Al-KA	210.583	16.658	Al_2O_3	21.98	0.012	Al-KA	206.087	16.526
4	MgO	0.95	0.011	Mg-KA	4.498	0.706	MgO	0.92	0.012	Mg-KA	4.349	0.693
5	CaO	1.14	0.007	Ca-KA	4.446	0.843	CaO	1.40	0.007	Ca-KA	5.478	1.048
6	MnO	0.05	0.004	Mn-KA	1.096	0.035	Fe_2O_3	5.62	0.005	Fe-KA	183.21	4.225
7	Fe_2O_3	5.52	0.005	Fe-KA	180.518	4.095	NiO	0.01	0.002	Ni-KA	0.496	0.006
8	Na_2O	2.34	0.014	Na-KA	4.225	1.735	Na_2O	3.47	0.016	Na-KA	6.358	2.611
9	K_2O	3.65	0.005	K-KA	71.501	2.71	K_2O	3.66	0.005	K-KA	72.123	2.756
10	P_2O_5	0.20	0.004	P-KA	1.986	0.149	P_2O_5	0.20	0.004	P-KA	1.996	0.148
11	SO_3	0.55	0.004	S-KA	4.667	0.406	SO_3	0.91	0.004	S-KA	7.968	0.688
12	CI	1.92	0.01	Cl-KA	11.132	1.422	CI	2.55	0.012	Cl-KA	15.071	1.918
The trace elements (mg/kg)												
13 *	Cr_2O_3	218	56.3	Cr-KA	3083	161	Cr_2O_3	500	46.7	Mn-KA	11,706	376
14 *	CuO	48	20.2	Cu-KA	3854	36	CuO	48	20.9	Cu-KA	3845	36
15 *	ZnO	146	18.2	Zn-KA	15,921	108	ZnO	173	18.4	Zn-KA	18,772	130
16 *	As_2O_3	0	25.5	As-KA	10,477	0	As_2O_3	0	25.6	As-KA	10,740	0
17 *	Rb_2O	285	11.7	Rb-KA	11,040.4	211	Br	29	11.6	Br-KA	9232	22
18 *	SrO	186	12.2	Sr-KA	77,575	138	Rb_2O	268	12.1	Rb-KA	104,127	201
19 *	Y_2O_3	33	13.9	Y-KA	55,914	24	SrO	186	12.3	Sr-KA	77,018	140
20 *	ZrO_2	383	12.6	Zr-KA	211,458	284	Y_2O_3	55	13.7	Y-KA	61,741	41
21 *	Nb_2O_5	67	12.9	Nb-KA	35,156	50	ZrO_2	402	12.8	Zr-KA	219,914	303
22 *	I	382	72.9	I-KA	56,843	283	Nb_2O_5	41	13.3	Nb-KA	24,010	31
23 *	PbO	114	35.1	Pb-LBI	13,282	85	BaO	758	268	Ba-LA	1605	570
24 *	--	--	--	--	--	--	PbO	121	35.1	Pb-LBI	13,952	91

* Chart source: Author 6 used instrument detection, and team members jointly analyzed and verified this.

References

1. Huang, X. *Cleverly Cut and Fine Sticky: Appreciation of Wang Baoyuan's Sticky Sculpture Works*; Cultural Assets Bureau: Taiwan, China, 2010.
2. Lu, E.; Lu, H. Preliminary report on the model test of hollow brick wall. *J. South China Inst. Technol.* **1957**, *1*, 63–72.
3. Li, K.; Sun, J.K.; Wang, S.P.; Wang, C.; Pan, H.Q. Experimental study on compressive strength of high titanium slag concrete hollow block walls. *Sichuan Archit.* **2019**, *39*, 261–263.
4. Nan, J. Study on Reinforcement and Seismic Performance of Cavity Walls' Housing. *J. Archit. Res. Dev.* **2023**, *1*, 7–8. [CrossRef]
5. Thenmozhi, R.; Saranya, K. Study on Compressive Strength of Reinforced and Cavity Wall Brickwork. *IOP Conf. Ser. Earth Environ. Sci.* **2022**, *1086*, 12048. [CrossRef]
6. Zhu, B.; Wu, M.; Jiang, Z. Experimental Study on the Basic Behavior of Brick Masonry Under Cyclic Loading. *J. Tongji Univ.* **1980**, *2*, 1–14.
7. Romina, S.; Marco, C.; Alberto, B.; Di Ludovico, M.; Prota, A.; Speranzini, E.; Molinari, A. Analysis of the seismic vulnerability and innovative retrofit solutions of cavity brickwork walls. *Constr. Build. Mater.* **2023**, *409*, 134018. [CrossRef]
8. Zhou, T.; Tian, P.; Deng, M.; Zhao, X.; Liang, Z. Research on shaking table test of cavity-wall building strengthened with engineered cementitious composite. *J. Build. Struct.* **2018**, *39*, 147–152.
9. Argentina, O.; Jesus, H.; Kathia, A. The Effect of a Segmented Wall Filled with PCM on Heat Transfer and Airflow in a Closed Cavity. *J. Heat Transf.* **2021**, *143*, 93001.
10. Deng, J.R. Study on the Mechanical Behavior and Microscopic Characteristics of Microbial Remediation of Rock Materials. Ph.D. Thesis, Central South University, Changsha, China, 2023. Available online: https://oversea.cnki.net/KCMS/detail/detail.aspx?dbcode=CDFD&dbname=CDFDLAST2024&filename=1024367415.nh&uniplatform=OVERSEA&v=xxILHkb5uLlZV2YR9B-IF5dosD72YAit7cAJ8PR2psEtkFzsrN543eujKbG6l7wR (accessed on 8 June 2025).
11. D'Ayala, D.; Zhu, H.; Aktas, Y. The Impact of Wind-Driven Rain on Surface Waterproofed Brick Cavity Walls. *Buildings* **2024**, *14*, 447. [CrossRef]
12. Mehmood, T.; Abid, M.A.; Chatveera, B.; Sua-Iam, G.; Saingam, P.; Ejaz, A.; Hussain, Q.; Joyklad, P.; Suparp, S. Sustainable Reinforcement Methods for Brick Masonry Walls: An Experimental and Finite Element Analysis Approach. *Buildings* **2025**, *15*, 2180. [CrossRef]

13. Wang, T.; Ma, J.; Zhang, W.; Wei, S.W. Feasibility of a Local Enclosure Structure Approach, Adaptability, and Promotion: Empty Bucket Walls in Jinhua and Wenzhou, Zhejiang Province. *Small Town Constr.* **2005**, *4*, 56–57.
14. Hou, L.; Liu, Y.; Zhu, Y.; Zhao, X.; Yang, L. Thermal performance of cavity masonry walls and thermal design investigation in Western China. *Energy* **2024**, *292*, 130298. [CrossRef]
15. Lai, S.; Zheng, Z. Traditional masonry techniques of Minnan red bricks and their inspiration. *Huazhong Archit.* **2007**, *2*, 154–157.
16. Wang, Y.J.; Song, J.X.; Zhang, J.H.; Huang, Y.; Yang, S. Analysis of Rammed Earth Wall Erosion in Traditional Village Dwellings in Zhuhai City. *Coatings* **2025**, *15*, 526. [CrossRef]
17. Wang, X.P.; Fang, K.; Chen, L.; Furuya, N. The influence of atrium types on the consciousness of shared space in amalgamated traditional dwellings—A case study on traditional dwellings in Quanzhou City, Fujian Province, China. *J. Asian Archit. Build. Eng.* **2019**, *18*, 335–350. [CrossRef]
18. Yang, S.; Chen, Y.; Lin, X. Analysis of characteristics of climate change and agricultural disastrous weatherin Jinjiang City over the past 50 years. *Fujian Agric. Sci. Technol.* **2017**, *1*, 9–14. [CrossRef]
19. Liu, R.; Gao, W.; Yang, F. Authenticity, Integrity, and Cultural–Ecological Adaptability in Heritage Conservation: A Practical Framework for Historic Urban Areas—A Case Study of Yicheng Ancient City, China. *Buildings* **2025**, *15*, 1304. [CrossRef]
20. Schroeder, A.B.; Dobson, E.T.A.; Rueden, C.T.; Tomancak, P.; Jug, F.; Eliceiri, K.W. The ImageJ Ecosystem: Open-source Software for Image Visualization, Processing, and Analysis. *Protein Sci.* **2021**, *30*, 234–249. [CrossRef]
21. Zhang, L.; Chen, Y.; Zheng, L.; Zheng, R. Research on the Composition and Casting Technology of Bronze Arrowheads Unearthed from the Ruins of the Imperial City of the Minyue Kingdom. *Materials* **2025**, *18*, 402. [CrossRef]
22. Kabekkodu, S.; Blanton, T. Importance of powder diffraction raw data archival in a curated database for materials science applications. *Acta Crystallogr. Sect. B Struct. Sci. Cryst. Eng. Mater.* **2024**, *80*, 364–369. [CrossRef]
23. Shen, J.; Zhang, Z. Influence of firing temperatures on the physical and mechanical properties of red bricks. *Sci. Conserv. Archaeol.* **2019**, *31*, 105–111.
24. Zhen, L. The Traditional Firing Craft of Red Brick Kiln in South Fujian. *Chin. J. Hist. Sci. Technol.* **2005**, *3*, 249–256.
25. Toufigh, V.; Ghasemalizadeh, S.; Mhkaramian, M.K. Experimental Investigation of Mixture Design and Durability Performance of Alkali-Activated Rammed Earth. *Int. J. Geomech.* **2022**, *22*, 4022029. [CrossRef]
26. Han, W.; Gong, R.; Liu, Y.; Gao, Y. Influence mechanism of salt erosion on the earthen heritage site wall in Pianguan Bastion. *Case Stud. Constr. Mater.* **2022**, *17*, e01388. [CrossRef]
27. Yi, Z.; Fu, L.J.; Zhan, S.; Yang, H.; Tang, A.; Jiang, T. Substitutional Doping for Aluminosilicate Mineral and Superior Water Splitting Performance. *Nanoscale Res. Lett.* **2017**, *12*, 456. [CrossRef]
28. Hua, J.; Jia, Y.G.; Lv, Y.; Yang, X.; Gao, Y.; Sun, Y.C.; Shang, J.Y.; Yan, X.L. Co-transport and influence mechanism of hematite colloids and Cr(VI)in saturated porous media under acidic conditions. *Environ. Chem.* **2024**, *43*, 3845–3853. [CrossRef]
29. *SY/T 5163-2010*; Method for X-Ray Diffraction Analysis of Clay Minerals and Common Non Clay Minerals in Sedimentary Rocks. National Energy Administration: Beijing, China, 2010. Available online: https://oversea.cnki.net/KCMS/detail/detail.aspx?dbcode=SCSD&dbname=SCSD&filename=SCSD000006151917&uniplatform=OVERSEA&v=Oph4xQC-wrVy3DBnmOhA9_Ce5kn6OqFNkd2dHkKx8Wfjq7awTg7D22yIkuAgASqOZiY3riBzqXo%3d (accessed on 8 June 2025).
30. He, H.; Guo, L.; Xie, X.; Peng, J. Experimental studies on the selective adsorption of cu2+,pb2+,zn2+,cd2+,cr3+ ions on montmorillonite, illite and kaolinite and the influence of medium conditions. *Acta Mineral. Sin.* **1999**, *2*, 231–235.
31. Wang, J. Chemical weathering mechanism of Albite-rich rocks in Grottoes under an acidic environment: An atomistic view from ReaxFF simulation. *Comput. Mater. Sci.* **2022**, *211*, 111498. [CrossRef]
32. Jin, Z.L.; Liu, D.D.; Zhang, Y.K.; Chen, G.Q.; Xia, Y.; Su, B.; Zhou, T.; Lv, G.X.; Luo, H.J. Salt migrations and damage mechanism in cultural heritage objects. *Sci. Conserv. Archaeol.* **2017**, *29*, 102–116.
33. Lin, M.; Roders, A.P.; Nevzgodin, I.; de Jonge, W. Values and interventions: Dynamic relationships in international doctrines. *J. Cult. Herit. Manag. Sustain. Dev.* **2025**, *15*, 728–749. [CrossRef]
34. *GB 50003-2011*; Code for Design of Masonry Structures. General Administration of Quality Supervision, Inspection and Quarantine: Beijing, China, 2011.

coatings

MDPI

Editorial

New Insights into Earthen Site Conservation: Multidisciplinary Collaboration and Development

Yile Chen * and Liang Zheng

Faculty of Humanities and Arts, Macau University of Science and Technology, Avenida Wai Long, Tapai, Macau 999078, China; zhengliang@must.edu.mo
* Correspondence: chenyile@must.edu.mo

Earthen sites are ancient sites that hold significant historical, cultural, and scientific value, with earth serving as the primary building material [1–3]. As places of great importance for civilization, earthen sites not only account for a large proportion of cultural heritage resources but also contain extremely rich historical information. They form an important basis for studying the origin and development of civilization and have great historical, scientific, and artistic value. Taking China as an example, earthen sites are distributed throughout the country [4]; as of the confirmation of the sixth batch of national key cultural relic protection sites, there are about 400 earthen sites on this list, distributed over 30 provinces (municipalities and autonomous regions) across the country. Their construction technology includes adobe, rammed earth, cast earth, and other methods, and their construction methods mainly include raw earth excavation, silt ramming, adobe masonry, wet earth, and mud pile construction [5–7]. These sites include ancient cities (such as the rammed-earth ancient city wall of Pingyao, Shanxi), the Great Wall, passes, beacons, earth towers (such as the Yumen Pass earthen site), tombs (such as the Western Xia royal tombs), earth pits, earth caves, earth kilns, earth cellars (such as the Qin Shihuang Terracotta Warriors and Horses Earth Pit), and rammed-earth buildings (such as the Hakka earth buildings in Fujian).

However, due to the multiple influences of environmental factors, microbial activity, and physical wear, the surface corrosion problem of earthen sites is becoming increasingly serious [8]. Irreversible damage constantly erodes their historical and archeological value. The protection of earthen sites urgently requires interdisciplinary innovative ideas and technological breakthroughs. The protection of earthen sites has developed into a multidisciplinary field combining physics, chemistry, archeology, geology, humanities, and art [9–12]. Relevant research fields encompass protection concepts, value interpretation, disease mechanisms, protection technologies, preventive protection, and the digitization of cultural relics. As a result, numerous excellent scientific research results have been obtained.

Against this backdrop, we organized and launched a Special Issue titled New Insights into Earthen Site Conservation: Methods, Techniques, Management, and Key Case Studies (https://www.mdpi.com/journal/coatings/special_issues/IK22749J11, accessed on 24 October 2025). In early research progress, Xiao and other scholars reviewed the technological evolution of earthen site surface protection from 2000 to 2025 [13]. They believed that with the deep integration of digital twin technology, spatial information technology, intelligent systems, and sustainable concepts, earthen site surface protection technology is transforming from single-point application to multi-dimensional integration. They also pointed out that the development of surface protection technology for earthen sites has gone through three stages: from early point-based monitoring technologies such as remote sensing (RS) and the Global Positioning System (GPS) to spatial modeling technologies such as Laser Detection and Ranging (LiDAR) and Geographic Information Systems (GIS), and finally to today's integrated intelligent monitoring systems based on multi-source fusion. Key surface technology systems include GIS-based spatial data management, high-precision modeling based on LiDAR, 3D reconstruction based on oblique photography, and Building Information Modeling (BIM) for structural protection. Cutting-edge areas focus on digital twins (DT) and the Internet of Things (IoT) for intelligent monitoring, augmented reality (AR) for immersive visualization, and blockchain technology for digital authentication. Future research on surface protection technologies for earthen sites is expected to integrate big data and cloud computing to achieve multidimensional predictions of surface degradation. Virtual reality (VR) will overcome spatial and temporal constraints and advance the conservation paradigm towards automation, intelligence, and sustainability.

These are all areas that deserve our in-depth attention and future research The implementation of multidisciplinary collaborative innovation and practical application in earthen site conservation requires mechanisms as a foundation, technology as a core, practice as a guide, and international cooperation as a catalyst. Only by breaking down disciplinary barriers, overcoming technical bottlenecks, adapting to regional needs, and sharing global experience can we achieve the long-term protection of earthen sites. These heritage sites, carrying millennia of civilization, can be revitalized through scientific conservation and become a cultural link connecting the past and the future.

Author Contributions: Conceptualization, Y.C.; methodology, Y.C.; software, Y.C.; validation, Y.C.; formal analysis, Y.C.; investigation, Y.C.; resources, Y.C.; data curation, Y.C.; writing—original draft preparation, Y.C.; writing—review and editing, Y.C. and L.Z.; visualization, Y.C.; supervision, Y.C.; project administration, Y.C. and L.Z.; funding acquisition, Y.C. and L.Z. All authors have read and agreed to the published version of the manuscript.

Funding: This research was funded by the (1) Faculty Research Grants funded by Macau University of Science and Technology (FRG-MUST grant number: FRG-25-041-FA; FRG-25-067-FA); (2) Guangdong Provincial Department of Education's key scientific research platforms and projects for general universities in 2023: Guangdong, Hong Kong, and Macau Cultural Heritage Protection and Innovation Design Team (grant number: 2023WCXTD042); (3) Guangdong Provincial Philosophy and Social Sciences Planning 2025 Lingnan Cultural Project (grant number: GD25LN30). The funders had no role in study conceptualization, data curation, formal analysis, methodology, software, decision to publish, or preparation of the manuscript. There was no additional external funding received for this study.

Institutional Review Board Statement: Not applicable.

Informed Consent Statement: Not applicable.

Data Availability Statement: Not applicable.

Acknowledgments: We sincerely thank the authors who contributed to this Special Issue and submitted high-quality manuscripts. We also thank the reviewers for their responsible and rigorous feedback, which has greatly improved each manuscript. Finally, we sincerely thank the editorial team of *Coatings* for their professional support, which ensured the success of this Special Issue.

Conflicts of Interest: The Guest Editors declare that they have no known competing financial interests or personal relationships that could have appeared to influence the work reported in this paper.

References

1. Avrami, E.; Guillaud, H.; Hardy, M. *Terra Literature Review: An Overview of Research in Earthen Architecture Conservation*; The Getty Conservation Institute: Los Angeles, CA, USA, 2008; Available online: http://hdl.handle.net/10020/gci_pubs/terra_literature_review (accessed on 24 October 2025).
2. Li, Z.; Wang, X.; Sun, M.; Chen, W.; Guo, Q.; Zhang, H. Conservation of Jiaohe ancient earthen site in China. *J. Rock Mech. Geotech. Eng.* **2011**, *3*, 270–281. [CrossRef]

3. Mousourakis, A.; Arakadaki, M.; Kotsopoulos, S.; Sinamidis, I.; Mikrou, T.; Frangedaki, E.; Lagaros, N.D. Earthen Architecture in Greece: Traditional Techniques and Revaluation. *Heritage* **2020**, *3*, 1237–1268. [CrossRef]
4. Guo, Z.; Qi, Q.; Zhang, S.; Chen, W.; Yang, L.; Zhang, Y. Characterization of spatial and chrono-cultural evolution of earthen sites construction technology in China based on GIS. *Archaeol. Anthr. Sci.* **2025**, *17*, 142. [CrossRef]
5. Huang, H. Ancient and Unique Construction Techniques. In *Fujian's Tulou: A Treasure of Chinese Traditional Civilian Residence*; Springer: Singapore, 2019; pp. 197–212. [CrossRef]
6. Gallipoli, D.; Bruno, A.W.; Perlot, C.; Mendes, J. A geotechnical perspective of raw earth building. *Acta Geotech.* **2017**, *12*, 463–478. [CrossRef]
7. Kong, D.; Wan, R.; Chen, J.; Jing, Y.; Huang, W.; Wang, Y. The study on engineering characteristics and compression mechanisms of typical historical earthen site soil. *Constr. Build. Mater.* **2019**, *213*, 386–403. [CrossRef]
8. Guo, Q.; Wang, Y.; Chen, W.; Pei, Q.; Sun, M.; Yang, S.; Zhang, J.; Du, Y. Key Issues and Research Progress on the Deterioration Processes and Protection Technology of Earthen Sites under Multi-Field Coupling. *Coatings* **2022**, *12*, 1677. [CrossRef]
9. Pan, C.; Chen, K.; Chen, D.; Xi, S.; Geng, J.; Zhu, D. Research progress on in-situ protection status and technology of earthen sites in moisty environment. *Constr. Build. Mater.* **2020**, *253*, 119219. [CrossRef]
10. Puertas, E.; Ávila, F.; García-Macías, E.; Gallego, R. Preventive Preservation of Rammed Earth Historical Heritage Through Continuous Monitoring, Architectural Inspections, and Data Fusion. *Buildings* **2024**, *14*, 3294. [CrossRef]
11. Luo, X.; Gu, Z.; Yu, C. Desiccation cracking of earthen sites in archaeology museum–a viewpoint of chemical potential difference of water content. *Indoor Built Environ.* **2015**, *24*, 147–152. [CrossRef]
12. Pollard, A.M. *Analytical Chemistry in Archaeology*; Cambridge University Press: Cambridge, UK, 2007.
13. Xiao, Y.; Chen, Y.; Huang, Y.; Yan, Y. Surface Protection Technologies for Earthen Sites in the 21st Century: Hotspots, Evolution, and Future Trends in Digitalization, Intelligence, and Sustainability. *Coatings* **2025**, *15*, 855. [CrossRef]

MDPI AG
Grosspeteranlage 5
4052 Basel
Switzerland
Tel.: +41 61 683 77 34

Coatings Editorial Office
E-mail: coatings@mdpi.com
www.mdpi.com/journal/coatings

www.ingramcontent.com/pod-product-compliance
Lightning Source LLC
Chambersburg PA
CBHW061816210926
51850CB00006B/725
* 9 7 8 3 7 2 5 8 6 4 7 2 0 *